Materials for Energy Production, Conversion, and Storage

This volume provides a comprehensive review of energy production, management, and its challenges pertaining to various materials. It covers different material fabrication strategies involved in the processes such as laser-assisted fabrication, electrospinning strategy, and so forth, including a review of the different nanostructured materials and challenges in energy management. Factors affecting energy storage and conversion focussing on high entropy and phase change-based materials are covered. The concepts in the book are supported by illustrations and case studies.

Features:

- Covers different fabrication strategies for various energy materials.
- Focusses on emerging materials such as MXenes, aerogels, and so forth.
- Provides a detailed study of laser-assisted fabrication, electrospinning strategy, and 3D-printed materials.
- Includes a comprehensive study of energy management from biomass.
- Reviews current strategies for electronic waste management.

This book is aimed at researchers and graduate students in chemical engineering, electrochemistry, and materials science.

Materials for Energy Production, Conversion, and Storage

Edited by
Jenitta Johnson M, Nisa Salim, and Sabu Thomas

CRC Press
Taylor & Francis Group
Boca Raton London New York

CRC Press is an imprint of the
Taylor & Francis Group, an **informa** business

Designed cover image: © Jenitta Johnson Mapranathukaran

First edition published 2024
by CRC Press
2385 NW Executive Center Drive, Suite 320, Boca Raton FL 33431

and by CRC Press
4 Park Square, Milton Park, Abingdon, Oxon, OX14 4RN

CRC Press is an imprint of Taylor & Francis Group, LLC

© 2024 selection and editorial matter, Jenitta Johnson M, Nisa Salim, and Sabu Thomas; individual chapters, the contributors

ISBN: 9781032313047 (hbk)
ISBN: 9781032332574 (pbk)
ISBN: 9781003318859 (ebk)

DOI: 10.1201/9781003318859

Typeset in Times
by Newgen Publishing UK

Dedication

For JO

Contents

Acknowledgements

I owe my deepest gratitude to several individuals who have played vital roles in the creation of this book. First and foremost, I am indebted to Prof. Dr. Sabu Thomas, the Vice Chancellor of MG University, India. His unwavering commitment to providing systematic guidance and support has been valuable to me throughout this journey. I must also express my heartfelt appreciation to Dr. Nisa Salim, Senior Lecturer at Swinburne University of Technology, Australia. Her constant motivation and encouragement have been a driving force behind the completion of this project. I am profoundly grateful to Dr. Jose Varghese R, my inspiring senior and beloved husband, for his selfless assistance and unwavering support, which have been instrumental in bringing this book to fruition. Additionally, I extend my sincere thanks to all my collaborators, chapter authors, and contributors, whose invaluable contributions have enriched the content of this book. Your dedication and expertise have greatly enhanced its quality. Thank you very much

About the Editors

Jenitta Johnson Mapranathukaran is working as a doctoral researcher at Center of Advanced Photonics & Process Analysis (CAPPA) in Munster Technological University (MTU), Ireland. She worked as research assistant in the research group of Prof. Sabu Thomas at Mahatma Gandi University (MG), India. She is a postgraduate of the 2018–2020 batch of postgraduates in Optoelectronics and Laser Technology from the International School of Photonics, Cochin University of Science and Technology, Kochi. She has more than a year of experience in nanophotonics, nanomaterials, and material science as a master thesis research fellow in the Laser Material Processing Division, Raja Ramanna Center for Advanced Technology, Indore, and a Summer research fellow of the Indian Academy of Sciences at Saha Institute of Nuclear Physics, Calcutta.

Nisa Salim is a Vice Chancellors Initiative Research Fellow at the Swinburne University of Technology. She received her PhD from Deakin University in 2013 on nanostructured polymer materials and joined Carbon Nexus as a Research Fellow in 2014. Her research has been focused on next-generation carbon fibres; porous carbon materials; and functional fibres. Nisa has won many awards as she advances her research career, including the AINSE Gold Medal, Royal Society of Victoria – Philp Law Award, Smart Geelong Early Researcher award and a SPE-ANZ Award. She has been awarded several prestigious fellowships including a Victoria Fellowship and Endeavour Fellowship. Nisa's vision is to develop smart, engineered materials that are enablers for digitalization and the Internet of Things – living materials that sense, actuate and harvest energy.

Sabu Thomas is currently Vice-Chancellor of Mahatma Gandhi University, Kerala, India, and the Founder Director and Professor of the International and Interuniversity Centre for Nanoscience and Nanotechnology, Mahatma Gandhi University, Kerala, India. He is also a full professor of Polymer Science and Engineering at the School of Chemical Sciences of Mahatma Gandhi University, Kottayam, Kerala, India. Professor Thomas is an outstanding leader with sustained international acclaim for nanoscience, polymer Science and engineering, polymer nanocomposites, elastomers, polymer blends, and interpenetrating polymer networks, polymer membranes, green composites, nanomedicine, and green nanotechnology. Professor Thomas's ground-breaking inventions in polymer nanocomposites, polymer blends, bionanotechnological and nano-biomedical sciences, have made transformative differences in the development of new materials for automotive, space, housing and biomedical fields. In collaboration with India's premier tyre company, Apollo Tyres, Professor Thomas's group invented new high-performance barrier rubber nanocomposite membranes for inner tubes and inner liners for tyres. Professor Thomas has received a number of national and international awards including the Fellowship of the Royal Society of Chemistry,

London FRSC, Distinguished Professorship from Josef Stefan Institute, Slovenia, MRSI medal, Nano Tech Medal, CRSI medal, Distinguished Faculty Award, Dr. APJ Abdul Kalam Award for Scientific Excellence – 2016, Mahatma Gandhi University – Award for Outstanding Contribution – Nov. 2016, Lifetime Achievement Award of the Malaysian Polymer Group, Indian Nano Biologists Award, 2017 and the Sukumar Maithy Award for the best polymer researcher in the country.

Contributors

Abdul Mateen (Department of Physics and Beijing Key Laboratory of Energy Conversion and Storage Materials, Beijing Normal University, Beijing, 100084, China)

Ahmad Allahbakhsh (Department of Materials and Polymer Engineering, Faculty of Engineering, Hakim Sabzevari University, Sabzevar, Iran)

Ailing Song (Hebei Key Laboratory of Applied Chemistry, College of Environmental and Chemical Engineering, Yanshan University)

Akito Takasaki (Department of Engineering Science and Mechanics, Shibaura Institute of Technology, 3-7-5 Toyosu, Koto-ku, Tokyo, 135-8548, Japan)

Alexandr Dubina (Department of Industrial Ecology, Belarusian State Technological University, Belarus)

Ali Akbari Sehat (Department of Materials Science and Engineering, University of Virginia, USA)

Alice Alex (Department of Chemistry, CMS College (Autonomous), Kottayam, Kerala 686 001, India)

Alicia Gomis-Berenguer (Institute of Electrochemistry, University of Alicante, 03080 Alicante, Spain)

Amitava Mandal (Department of Mechanical Engineering, Indian Institute of Technology (Indian School of Mines), Dhanbad, 826004, Jharkhand, India)

Amrita Vishwa Vidyapeetham (Amrita University, Amritapuri, Kollam-690525, Kerala, India)

Ana Casanova (Department of Nanoengineering, University of California San Diego, La Jolla, California 92093, United States)

Athira A R (Centre for Advanced Materials Research, Department of Physics, Government College for Women, University of Kerala, Thiruvananthapuram, Kerala 695014, India)

Conchi O. Ania (CEMHTI, CNRS (UPR 3079), University of Orléans, 45071 Orléans, France)

Deepti Ranjan Sahu (Department of Mechanical Engineering, Indian Institute of Technology (Indian School of Mines), Dhanbad, 826004, Jharkhand, India)

Dipanwita Majumdar (Department of Chemistry, Chandernagore College, Chandannagar, Hooghly, Pin-712136, West Bengal, India)

Dmitry Moskovskikh (Center of Functional Nano-Ceramics, National University of Science and Technology "MISIS", Russia)

Elsayed Tag Eldin (Faculty of Engineering and Technology, Future University in Egypt, New Cairo 11835, Egypt)

Encarnación Raymundo-Piñero (CEMHTI, CNRS (UPR 3079), University of Orléans, 45071 Orléans, France)

Fathollah Pourfayaz (Department of Renewable Energies and Environment, Faculty of New Sciences and Technologies, University of Tehran, Tehran, Iran)

Francisco Trivinho-Strixino (Federal University of São Carlos (UFSCar) – Brazil)

Gigi George (Department of Chemistry, CMS College (Autonomous), Kottayam, Kerala 686 001, India)

Guoxiu Wang (Centre for Clean Energy Technology, School of Mathematical and Physical Sciences, Faculty of Science, University of Technology Sydney)

Hao Tian (Centre for Clean Energy Technology, School of Mathematical and Physical Sciences, Faculty of Science, University of Technology Sydney)

Janaina S. Santos (Chulalongkorn University – Thailand)

Jesús Iniesta (Institute of Electrochemistry and Department of physical chemistry, University of Alicante, 03080 Alicante, Spain)

Jinoop AN (School of Computing, Engineering & Digital Technologies, Teesside University, Middlesbrough TS1 3BX, United Kingdom)

Majed A. Bajaber (Chemistry Department, Faculty of Science, King Khalid University, P.O. Box 9004, Abha 61413)

Manchu Mohan Krishna Sai (Department of Mechanical Engineering, Indian Institute of Technology (Indian School of Mines), Dhanbad, 826004, Jharkhand, India)

Mariana Sikora (Federal University of Technology – Paraná (UTFPR) – Brazil)

Mohd Asyadi Azam (Fakulti Kejuruteraan Pembuatan, Universiti Teknikal Malaysia Melaka, Hang Tuah Jaya, 76100 Durian Tunggal, Melaka)

Muhammad Sufyan Javed (School of Physical Science and Technology, Lanzhou University, Lanzhou 730000, China)

Munmun Mondal (Department of Chemistry, Indian Institute of Technology, Kharagpur, Pin-721302, West Bengal, India)

Nur Ezyanie Safie (Fakulti Kejuruteraan Pembuatan, Universiti Teknikal Malaysia Melaka, Hang Tuah Jaya, 76100 Durian Tunggal, Melaka)

Raja Noor Amalina Raja Seman (Fakulti Kejuruteraan Pembuatan, Universiti Teknikal Malaysia Melaka)

Revathy K.P (Centre for Integrated Studies, Cochin University of Science and Technology (CUSAT), Kochi, Kerala, India)

Sajjad Moradpour (Department of Renewable Energies and Environment, Faculty of New Sciences and Technologies, University of Tehran, Tehran, Iran)

Sebastian Torres (Institute of Electrochemistry, University of Alicante, 03080 Alicante, Spain)

Soghra Ghorbanzadeh (Department of Materials and Polymer Engineering, Faculty of Engineering, Hakim Sabzevari University, Sabzevar, Iran)

S Alwin (Department of Chemistry the American College, Madurai-625002, Tamilnadu, India)

Sunish K Sugunan (Department of Chemistry, CMS College (Autonomous), Kottayam, Kerala 686 001, India)

Valentin Romanovski (Department of Materials Science and Engineering, University of Virginia, USA, Center of Functional Nano-Ceramics, National University of Science and Technology "MISIS", Russia)

Vinod V.T Padil (Amrita School for Sustainable Development (AST))

X Sahaya Shajan (Center for Scientific and Applied research, PSN College of Engineering and Technology, Melathediyoor, Tirunelveli-627152, Tamilnadu, India)

Xavier T S (Centre for Advanced Materials Research, Department of Physics, Government College for Women, University of Kerala, Thiruvananthapuram, Kerala 695014, India)

Xintai Su (School of Environment and Energy, Guangdong Provincial Key Laboratory of Solid Wastes Pollution Control and Recycling, South China University of Technology, China)

Zeinab Jarrahi (Department of Materials and Polymer Engineering, Faculty of Engineering, Hakim Sabzevari University, Sabzevar, Iran)

Introduction
Materials for Energy Production, Conversion, and Storage

Jenitta Johnson M, Nisa Salim, and Sabu Thomas

ENERGY

Energy is the capacity to do work. Energy is the fundamental property of the universe which is responsible for motion, heat and light. In physics, energy is explained as the capacity to do the work or cause a change in a system: kinetic energy, potential energy, thermal energy, electrical energy, etc. Energy types can be classified into primary and secondary. The energy types that are directly available from natural resources are known as primary energy. The energy type that is derived from the primary energy is known as secondary energy (Figure I.1).

   (i) *Classification of primary energy:*

      (a) Renewable energy: Renewable energy is derived from the resources from the environment that are replenished over time and will not be depleted. Examples of renewable energy include sunlight, wind, rain and tides.

      (b) Non-renewable energy: Non-renewable energy refers to sources of energy that are finite and can not be replenished as quickly as they are consumed. Examples of non-renewable energy are coal, crude oil, natural gas, nuclear fuel.

      (c) Waste: Waste energy refers to energy that is produced as a by-product of a process or activity. Examples of energy from waste are municipal waste, food waste, and animal manure.

  (ii) *Classification of secondary energy:*

      (a) Electrical Energy: This is the most commonly used form of secondary energy. It is produced by the conversion of primary energy sources such as coal, natural gas, and nuclear power, and renewable energy sources like wind, solar, and hydroelectric power.

      (b) Mechanical Energy: This type of energy is generated by the use of machines or engines that convert primary energy sources such as gasoline, diesel, or natural gas into motion or mechanical power. Examples

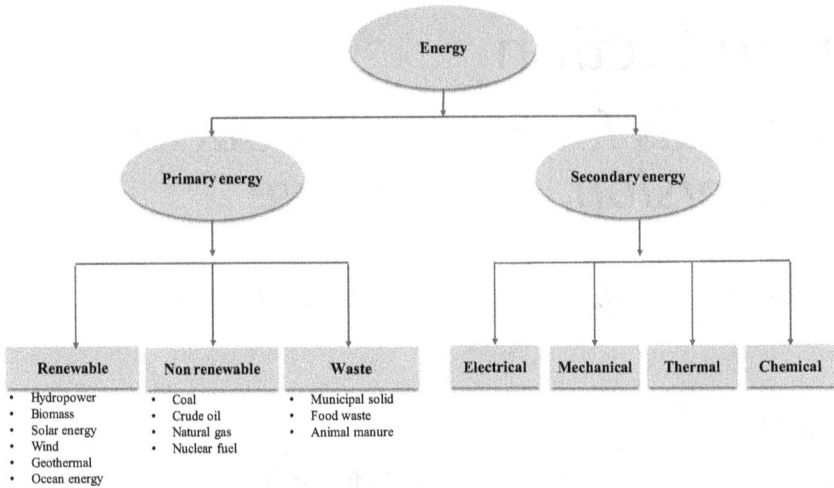

FIGURE I.1   The classification of energy.

of mechanical energy include the movement of vehicles, the rotation of turbines, and the operation of industrial machinery.

(c) Thermal energy: This type of energy is generated by the conversion of primary energy sources into heat. Examples include the heating of buildings and the generation of steam for power generation.

(d) Chemical Energy: This type of energy is stored in the bonds of molecules and can be released through chemical reactions. Examples include batteries, fuels such as gasoline, and explosives.

The consumption of these energy resources in the past centuries led to one of the main problems faced by the modern world, which is the energy crisis. Also, the demand for energy has increased due to population growth. The population touched the 7 billion mark in 2011 and is expected to increase by 2 billion in the next 25 years[1]. Recent investigations state that energy consumption will increase by 1.1% every year. With the energy consumption of 5.3×10^{20} J in 2006, it might reach to 7.5×10^{20} J by 2030[2]. So the answer for today's major concern 'energy crisis' can be obtained through highly efficient and advanced materials. The improvement of energy efficiency and reduction of energy waste is the best solution to fill the energy crisis since the traditional energy resources such as solar, wind, tide and fossil fuels. Advanced materials are engineered to have specific properties that make them suitable for a variety of applications including energy production, conversion and storage. Advanced materials like phase-change materials[3], fullerene[4], CNT[5], graphene[6], quantum dots[7], natural clay[8], anodic oxide-based materials[9], green and bio-waste-based materials[10], metal-organic materials[11], 3D-printed materials[12]–[14], inorganic and organic materials[15], [16], mesoporous metal-based materials[17], MXene based 2D materials[18], [19], aerogels-based materials[20], hydrogels-based materials[21] and E-waste nanomaterials[22] for different energy production, conversion and storage are discussed in this review book.

ENERGY PRODUCTION

In actuality, efficient technical approaches for energy creation are just as crucial as reducing energy use. The production of energy is crucial to the health of the modern economy and society. Unavoidable energy needs include those for homes, businesses, factories, and hospitals. Without energy generation, the world will face several challenges. The primary goal of energy production, in the modern world, can be said to be to balance elements like economic growth, increased quality of life, energy security, environmental advantages, and technological advancements. For basic human requirements including reliable and economical lighting, heating and cooling, and food preparation, energy production is necessary. Additionally, regional energy production can increase a nation's energy security by lowering its reliance on foreign nations. Modern technologies, including necessities of daily life like computers, mobile phones, and vehicles, are dependent on energy use. The reduction of greenhouse gases and other environmentally hazardous pollution, which eventually has an impact on climate change and global warming, is another crucial aspect. One of the main forces behind the world of today and the better world of tomorrow is global energy production.

ENERGY CONVERSION

The relevance of energy conversion is primarily due to the high demand for our energy requirements. Not only does energy need to be produced, but it also needs to be transformed into the proper form. We can meet the need for electricity to run homes, businesses, and industries by converting energy into numerous forms of energy. However, energy conversion methods have made it possible to employ less conventional energy sources, such as fossil fuels, which are limited and the main source of greenhouse gas emissions. We can lessen our reliance on non-renewable energy by converting renewable energy sources like wind, solar, and hydropower into useful energy. On the other side, with the aid of energy conversion, energy efficiency can also be increased. One example is the ability to produce useful electricity from waste heat from industrial processes. Therefore, the primary benefit of energy conversion is cogeneration. As a result, we can argue that energy conversion is essential for supplying our energy needs while effectively minimizing their negative effects on the environment.

ENERGY STORAGE

Energy storage is essential to maintaining a balance between supply and demand. Researchers became interested in this intriguing energy concept because it had the potential to store energy during times of abundance and release it when needed. We may avoid costly backup generators and power interruptions by storing energy. The utilization of renewable energy sources, which are often cleaner than fossil fuels, can also dramatically cut greenhouse gas emissions. Batteries, pumped hydro storage, compressed air energy storage, thermal energy storage, and other types of energy storage technologies are all in use today.

This review book covers the recent developments in the applications of energy production, conversion and storage-based advanced materials.

The synthesis of micro and nanostructured materials by laser-assisted methods for energy production, conversion, and storage applications is discussed in Chapter 1. This chapter discusses the significance of laser-assisted techniques, the factors that must be taken into account when applying lasers, various materials that can be utilized with these techniques, and the uses of these materials in many fields.

Chapter 2 focuses on the various phase-change materials for energy harvesting and management for energy production, storage and conversion. This chapter covers the various energy storage system types as well as the design criteria and materials employed. These materials' crucial characteristics, including their thermal, physical, kinetic, poor stability, chemical, and economical ones, are also included.

Porous carbon-based materials for energy production, storage and conversion have been discussed in Chapter 3. Based on the porous carbon-based materials described throughout the chapter, several modern energy storage devices, such as batteries and supercapacitors, and various energy conversion processes, such as electrocatalysis and photocatalysis, are now available.

In Chapter 4, the synthetic strategies for the preparation of nano-porous carbons are demonstrated. The standard and new synthesis techniques are explored and contrasted. The benefits of these techniques for creating porous carbon-based materials are discussed.

A summary of perspectives on natural clay composites for energy generation, storage, and conversion is provided in Chapter 5. The essential function of naturally occurring clay-based materials in the application of energy and the combination of natural clay with various polymers are explained.

Chapter 6 explores the use of graphene-based materials for energy production, conversion, and storage. The most recent research based on graphene-derived materials is explained, as well as its most recent uses. Another highlight of the chapter is how crucial graphene is because of its special features.

The most recent advancements in anodic oxide-based active layers for energy generation, conversion, and storage are covered in Chapter 7. Different anodic oxide-based devices and their use in various energy domains through the inclusion of nanostructure have been described. This chapter also examines the role that anodic oxide plays when combined with other materials.

The focus of Chapter 8 is on green and bio-waste-based materials for energy production, conversion, storage, and hybrid technologies. Wastes with both organic and inorganic bases, various green strategies, waste recycling, and waste sources have all been covered.

In Chapter 9, current advancements in materials based on metal-organic frameworks for energy generation, conversion, and storage are discussed. It has also been described how to synthesize metal-organic frameworks using various methods, how to use them in various energy applications, and what they have to offer in terms of essential qualities and advantages.

The focus of Chapter 10 is on electrospinning as an efficient strategy for the designing and fabrication of architectures for energy production, conversion, and storage structures. The process for setting up the electrospinning as well as the most recent uses are described, along with the designing and fabricating of architectures from electrospinning for diverse energy applications.

3D-printed materials for high energy production, conversion, and storage devices are discussed in Chapter 11. The chapter covers 3D printing technologies, the elements that must be taken into account when choosing materials, various types of energy devices, based 3D-printed materials, and 3D-printed materials-based major design architecture for energy devices.

Chapter 12 provides an explanation of two-dimensional molybdenum disulphide-based materials: Synthesis, modification, and applications in supercapacitor technology. This chapter covers the structure, numerous synthesis techniques, and several energy uses of molybdenum disulphide-based materials.

Recent advances in mesoporous metal-based materials for energy generation, storage, and conversion are discussed in Chapter 13. This chapter explains various synthesis techniques and the energy application of several derivatives of mesoporous metal-based materials. Mesoporous metal-based materials' influence has been noted because of their improved properties.

Chapter 14 focuses on current developments using MXene-based 2D materials for applications related to energy production, conversion, and storage. The significance of 2D materials in the energy application sectors is discussed, as well as several catalytic applications such ORR, OER, HER, CO2RR, and NRR. Recent developments in the various energy applications described using MXene-based 2D materials.

Aerogels-based nanostructured materials for energy generation, conversion, and storage applications are summarized in Chapter 15. We've looked over the characteristics of aerogel structures and several application possibilities, such as fuel cells, solar cells, batteries, and supercapacitors. This chapter provides examples of the intriguing aspects of aerogel in regard to electrochemical energy applications.

Nanostructured materials based on hydrogels are the focus of Chapter 16's discussion of energy production, conversion, and storage applications. This chapter focuses on the significant hydrogel properties, various production techniques, morphology, structure, and the impact on energy applications.

REFERENCES

[1] "iiasa-eprint-12557."

[2] J. Van Bavel, "The world population explosion: Causes, backgrounds and -projections for the future.," *Facts, Views Vis. ObGyn*, vol. 5, no. 4, pp. 281–291, 2013.

[3] Z. Hu *et al.*, "Stabilized multifunctional phase change materials based on carbonized Cu-coated melamine foam/reduced graphene oxide framework for multiple energy conversion and storage," *Carbon Energy*, vol. 4, no. 6, pp. 1214–1227, Nov. 2022. doi: https://doi.org/10.1002/cey2.218

[4] J. Joseph *et al.*, "Borophene and boron fullerene materials in hydrogen storage: Opportunities and challenges," *ChemSusChem*, vol. 13, no. 15, pp. 3754–3765, Aug. 2020. doi: https://doi.org/10.1002/cssc.202000782

[5] J. Zheng and X. Bai, "Preparation of Ni-Co PBA-derived beaded NiSe2/CoSe2/CNT for high-performance supercapacitors," *J. Alloys Compd.*, vol. 944, pp. 169110, 2023. doi: https://doi.org/10.1016/j.jallcom.2023.169110

[6] W. Cui *et al.*, "Thermal performance of modified melamine foam/graphene/paraffin wax composite phase change materials for solar-thermal energy conversion and storage," *J. Clean. Prod.*, vol. 367, pp. 133031, 2022. doi: https://doi.org/10.1016/j.jclepro.2022.133031

[7] Q. Xu *et al.*, "Recent progress of quantum dots for energy storage applications," *Carbon Neutrality*, vol. 1, no. 1, pp. 13, 2022. doi: 10.1007/s43979-022-00002-y

[8] R. Song, X. Meng, C. Yu, J. Bian, and J. Su, "Oil shale in-situ upgrading with natural clay-based catalysts: Enhancement of oil yield and quality," *Fuel*, vol. 314, pp. 123076, 2022. doi: https://doi.org/10.1016/j.fuel.2021.123076

[9] M. Kim, J. Ha, Y.-T. Kim, and J. Choi, "Stainless steel: A high potential material for green electrochemical energy storage and conversion," *Chem. Eng. J.*, vol. 440, pp. 135459, 2022. doi: https://doi.org/10.1016/j.cej.2022.135459

[10] M. Abd Elkodous *et al.*, "Cutting-edge development in waste-recycled nanomaterials for energy storage and conversion applications," vol. 11, no. 1, pp. 2215–2294, 2022. doi: doi:10.1515/ntrev-2022-0129

[11] J. Choi, T. Ingsel, D. Neupane, S. R. Mishra, A. Kumar, and R. K. Gupta, "Metal-organic framework-derived cobalt oxide and sulfide having nanoflowers architecture for efficient energy conversion and storage," *J. Energy Storage*, vol. 50, pp. 104145, 2022. doi: https://doi.org/10.1016/j.est.2022.104145

[12] J. Ma, T. Ma, J. Cheng, and J. Zhang, "Polymer encapsulation strategy toward 3D printable, sustainable, and reliable form-stable phase change materials for advanced thermal energy storage," *ACS Appl. Mater. Interfaces*, vol. 14, no. 3, pp. 4251–4264, Jan. 2022. doi: 10.1021/acsami.1c23972

[13] K. Ghosh, S. Ng, C. Iffelsberger, and M. Pumera, "2D MoS2/carbon/polylactic acid filament for 3D printing: Photo and electrochemical energy conversion and storage," *Appl. Mater. Today*, vol. 26, pp. 101301, 2022. doi: https://doi.org/10.1016/j.apmt.2021.101301

[14] V. Dogra, D. Verma, G. K. Dalapati, M. Sharma, and M. Okhawilai, "Chapter 28 – Special focus on 3D printing of sulfides/selenides for energy conversion and storage," G. Dalapati, T. Shun Wong, S. Kundu, A. Chakraborty, and S. B. T.-S. and S. B. M. for E. A. Zhuk, Eds., Elsevier, 2022, pp. 757–772. doi: https://doi.org/10.1016/B978-0-323-99860-4.00012-5

[15] H. Cui, L. Ma, Z. Huang, Z. Chen, and C. Zhi, "Organic materials-based cathode for zinc ion battery," *SmartMat*, vol. 3, no. 4, pp. 565–581, Dec. 2022. doi: https://doi.org/10.1002/smm2.1110

[16] J. Mei, T. Liao, H. Peng, and Z. Sun, "Bioinspired materials for energy storage," *Small Methods*, vol. 6, no. 2, pp. 2101076, Feb. 2022. doi: https://doi.org/10.1002/smtd.202101076

[17] Y. Feng, Y. Chen, Z. Wang, and J. Wei, "Synthesis of mesoporous carbon materials from renewable plant polyphenols for environmental and energy applications," *New Carbon Mater.*, vol. 37, no. 1, pp. 196–222, 2022. doi: https://doi.org/10.1016/S1872-5805(22)60577-8

[18] N. H. Solangi *et al.*, "MXene-based phase change materials for solar thermal energy storage," *Energy Convers. Manag.*, vol. 273, pp. 116432, 2022. doi: https://doi.org/10.1016/j.enconman.2022.116432

[19] T. Rasheed, "MXenes as an emerging class of two-dimensional materials for advanced energy storage devices," *J. Mater. Chem. A*, vol. 10, no. 9, pp. 4558–4584, 2022. doi: 10.1039/D1TA10083A

[20] Y. Cai, N. Zhang, X. Cao, Y. Yuan, Z. Zhang, and N. Yu, "Ultra-light and flexible graphene aerogel-based form-stable phase change materials for energy conversion and energy storage," *Sol. Energy Mater. Sol. Cells*, vol. 252, pp. 112176, 2023. doi: https://doi.org/10.1016/j.solmat.2022.112176

[21] S. Sardana, A. Gupta, K. Singh, A. S. Maan, and A. Ohlan, "Conducting polymer hydrogel based electrode materials for supercapacitor applications," *J. Energy Storage*, vol. 45, pp. 103510, 2022. doi: https://doi.org/10.1016/j.est.2021.103510

[22] R. Seif, F. Z. Salem, and N. K. Allam, "E-waste recycled materials as efficient catalysts for renewable energy technologies and better environmental sustainability," *Environ. Dev. Sustain.*, 2023. doi: 10.1007/s10668-023-02925-7

1 Synthesis of Micro and Nanostructured Materials by Laser-Assisted Methods Towards Energy Production, Conversion, and Storage Applications

Alice Alex, Sunish K Sugunan, and Gigi George

1.1 INTRODUCTION

The utilisation of lasers in the synthesis of materials has led to the discovery of novel nanostructures. Traditional methods of synthesis and microfabrication are not as effective as laser-assisted processing approaches. Traditional approaches to the production of nanomaterials (NMs) include wet chemical procedures as well as thermal treatment operations carried out in environments that conduct solutions and gases [1]. The conventional approaches were unable to stimulate the development of particles at precise locations [2]. Synthesis of nanomaterials on a large scale necessitates the development of new technology. Synthesising nanomaterials typically involves the utilisation of lasers. The laser is a preferable alternative because of its low impact on the surrounding ecosystem, improved reproducibility, scalability, position controllability, cost-effectiveness, high productivity, less energy waste, and direct contribution to material synthesis [3]. Advantages in synthesis can be gained by being able to treat heat-sensitive substrates [4–6].

Recent developments that enhanced the applicability of laser microfabrication in material synthesis involve mask unemployment which allows complex patterning on a wide range of nanomaterials. Traditional wet chemical procedures, in particular as a synthetic technique, have the potential to form distinct morphologies in NMs [7]; however, harmful chemicals were often used. In contrast, a significant peculiarity of aqueous laser synthesis methods is the use of a target starting material, thereby preventing the use of hazardous chemicals [8]. Additionally, the size of nanoparticles

DOI: 10.1201/9781003318859-1

can be reduced to the desired size by regulating different laser parameters such as laser frequency, laser intensity, laser pulse width, etc. [2], [9]. Light-thermal conversion, batteries, superconductors, electrocatalytic electrodes, sensor devices, energy harvesters, photoelectric devices, and other applications use laser-induced nanomaterials and nanostructures [10–14]. Aside from its use in energy conversion and storage, laser microfabrication technology has played an important role in biomedical applications over the last decade [15]. Recent growth in laser synthesis and laser microfabrication for nanomaterials in various energy sectors will be reviewed.

1.2   LASER METHODS IN MATERIAL SYNTHESIS

Recent breakthroughs in the creation of various lasers have opened up new opportunities for the solid-state or solution-state synthesis of materials. The photothermal reaction, photochemical reaction, and photothermal-chemical reaction mediated by an irradiation laser are caused by the generation of a highly controllable localised electromagnetic field [16]. With lasers, this form of material creation has been utilised for quite some time. Carbon-based materials occupied the first row in laser-induced nanomaterial synthesis, followed by non-carbon materials [2]. The section that follows will provide an exhaustive summary of current developments in the laser synthesis of carbon-based and non-carbon-based nanomaterials.

1.2.1   Carbon Nanomaterials

Conventional annealing and heat treatment procedures were used in the synthesis of a significant percentage of carbon nanomaterials. Polymer precursors are frequently thermally treated at high temperatures in furnaces or ovens, but since the sample dimension is less than the heated volume, the process experiences a slow cooling rate and substantial loss of energy. Additionally, it was challenging to inherit the shape and structure of the precursors or develop new nanostructures. Due to the localised thermal effect and absence of interference from the surrounding materials, it is possible to produce various nanomaterials by employing laser synthesis to address the aforementioned problems [17, 18]. Carbon-based nanomaterials, particularly graphene [19], are gaining popularity due to their availability and their potential to be used in light, wearable, and flexible electronics, enabling paradigms like the Internet of Things [20, 21]. The twenty-first century's most extensively studied nanomaterial is graphene. Nowadays, laser-assisted synthesis methods allow inexpensive and environmentally friendly strategies for graphene preparation from various precursors like graphene oxide, polymer, CH_4, SiC, etc. [22] In addition to graphene, laser processing produces diamond-like, glassy, as well as heteroatom-doped carbon [23]. In order to fabricate different electronic devices, methods were employed based on laser, for the preparation, reduction, alteration, cutting, and micro-patterning of graphene and put to use. The controlled growth of graphene and the reduction of its oxides, graphene oxide (GO), are the subject of extensive investigation in material chemistry [24]. Nowadays, a wide variety of applications are possible for GO due to its outstanding water dispersibility and tunability of its physical and chemical characteristics through control of the number of groups that contain oxygen relying on the extent of reduction [20]. The notable applications that demonstrate GO's potential as a dominant player in these industries include electronics, power production, sensors, wearables, catalysis, robotics, chemical protection, etc [25].

By employing a direct laser reduction technique, Zhang et al. reported the synthesis of graphene microcircuits on the films of GO. A 100x objective lens having a large numerical aperture of 1.4 focused a femtosecond (fs) laser pulse with a wavelength of 790 nm, a pulse width of 120 fs, and a repetition rate of 80 MHz onto the GO films [26]. The output power of the laser has a significant impact on the electrical conductivity of as-reduced graphene. The linear nature of the current-voltage curves in graphene microcircuits denoted stable ohmic conductivities. In a study by Wong et al. they converted electro-sprayed thin films of GO into reduced graphene oxide (rGO) and employed a nanosecond laser with a wavelength of 355 nm to ablate the unwanted regions [27]. Additionally, through the incomplete reduction of GO, Qu et al. prepared asymmetric graphene/graphene oxide (G/GO) fibre with the aid of the laser method [28]. The asymmetric G/GO fibre produced was a promising material to utilise as a moisture-sensitive fibre actuator. Furthermore, this group also studied a laser-triggered spontaneous reduction of GO by an aerogel approach [29]. High-power laser sources have been used to reduce GO suspended in ammonia. Reduced graphene oxide (rGO) was synthesised using a process involving 248 nm excimer laser irradiation and ammonia [30]. The GO sheets were likewise reduced by Ghadim et al. using a nanosecond pulsed laser in an ammonia solution at ambient temperature [31].

Polymers are an important class of non-graphitic carbon source for the fabrication of laser-induced graphene (LIG) [32]. It can be obtained from a variety of polymers, particularly polyimide (PI). Laser scribing of polymers enables the synthesis of porous graphene having a greater specific surface area as compared to the 2D sheet-like graphene obtained from GO. The pulsed laser irradiation in polymers induces the photothermal effect, followed by the rearrangement of sp^3 hybridised carbon arrangement to sp^2 hybridised carbon arrangement [2]. The first study, which employed PI as a source material, was published in 2014 [33]. The authors employed a 10.6 μm wavelength CO_2 laser to produce patterns on PI. The loss of oxygen and formation of graphene were confirmed by X-ray Photoelectron Spectroscopy (XPS) and Raman spectroscopy, respectively. The same procedure was performed on polyetherimide (PEI); however, the quality of the rGO was lower than that on PI [33]. Additionally, PEI was selectively converted into graphene when a mixture of PEI and polycarbonate (PC) was subjected to a CO_2 laser [34]. In order to produce sulphur-doped porous graphene structures, a sulfonated polymer was used as a precursor. Lamberti et al. produced LIG by irradiating sulfonated poly (ether ketone) (SPEEK) with a CO_2 laser. It resulted in the formation of a flexible conductive material, which can be utilised as an electrode for supercapacitors [35]. LIG can be produced via laser synthesis (LS) in addition to polymers by using natural sources. Ye et al. reported this for the first time by applying a CO_2 laser on various types of wood in a controlled reducing environment that was rich in H_2. XPS and Raman spectroscopy revealed the formation of graphene [36]. The high lignin content of the starting material is essential for successful graphene formation. Chyan et al. applied LS to food sources including potato skins, coconut shells, or other natural materials like cork [37]. The LS process can be carried out in an environment with an ambient atmosphere because of the greater lignin content. A hard, amorphous carbon film called diamond-like carbon (DLC) has a significant proportion of sp^3 hybridised carbon atoms as well as a considerable amount of hydrogen [38]. Stock et al., using a high-purity graphite target, conducted the synthesis of DLC thin films with the application of a pulsed laser deposition (PLD) method. It should be noted that an environment with high-vacuum and also with a residual pressure lower than 10^{-8} mbar was necessary [39]. The initial atomic layers of the thin-film structure of DLC were then modified through surface annealing by UV laser, utilising the resulting DLC as the substrate to form graphene-like layers with good conductivity. Han et al. created a method of laser-assisted synthesis for DLC sheets by irradiating the single-crystal Si substrates which were immersed in cyclohexane liquid with a KrF excimer laser. The deposition process was carried out in an open atmosphere at ambient temperature with a peak laser power density of about 10^8 W/cm^2. Scanning electron microscopy (SEM), XPS and Raman spectroscopy analysis of the deposited films revealed the diamond-like characteristics [40].

The greater localised temperature generated by the laser's photothermal effects during the laser processing was crucial in the material preparation therefore; the laser can act as a heat source. The heating synthesis of ordinary materials will benefit from this. Similar to this, laser synthesis can also be used to prepare heteroatom-doped carbon, enabling large-scale patterning [41]. Alshareef et al. synthesised the nitrogen-atom-doped graphene, where the precursor was urea containing polyimide [42]. A single-step laser-based transformation was used to manufacture anodes that are conductive and free of additives as well as binders for Na-ion batteries. Peng et al. developed porous boron-doped graphene having excellent electrochemical performance by incorporating boric acid (H_3BO_3) into PI [43]. In order to produce the graphene patterns doped with sulphur and nitrogen on a substrate of polyethylene terephthalate (PET) as well as on glass, Li et al. utilised a starting material based on organic polybenzimidazole (PBI) ink. By using laser processing, other heteroatom-doped graphene forms besides nitrogen were also synthesised [44]. Ruoff et al. suggested a method in which graphene covered with fluoropolymer was the precursor for the preparation of graphene that was F-doped [45].

1.2.2 NON-CARBON MATERIALS

The production of non-carbon compounds, such as metal carbides, metal disulphide, metal oxides, and so on, can be done on a massive scale using a fabrication technique called laser synthesis [46]. The photothermal and/or photochemical processes are the key mechanisms that are utilised in the manufacture of non-carbon nanomaterials through the utilisation of laser synthesis. Laser synthesis allows the creation of nanomaterials to be carried out in a variety of conditions, including non-aqueous habitats, liquid environments, and environments that do not include water. The synthesis method that takes place in a liquid environment is one of the most common methods used for the development of a variety of colloidal structures. The techniques known as laser ablation in liquids (LAL) and laser fragmentation in liquids (LFL) are the ones that are utilised most frequently in this category [47]. The production of complex NMs can be accomplished through either the "top-down" synthesis of pulsed laser ablation of a solid target in liquids (PLAL) or the "bottom-up" methodology of pulsed laser irradiation of colloidal NPs in liquids (PLICN). Both of these techniques fall under the category of laser-assisted liquid synthesis (LAL). A schematic representation of a pulsed laser deposition system is given in figure 1.1 In most cases, the bare exposed nanoparticles of gold were produced by ablating Au nanoparticles that were already present in the solution during the process of synthesising metals and alloys using PLAL [48, 49]. These nanoparticles play an important function in the catalytic processes of several applications. Mukherjee et al. proposed a method of laser ablation that utilises a Co target in a solution of K_2PtCl_4 in order to build a nanoalloy of PtCo that is included in CoO_x matrices [50]. This method might be used to create a nanoalloy of PtCo that is incorporated in CoO_x matrices. Zhang et al. created Au-added ZnO nanospheres (NSs) by irradiating liquids with a laser. These nanospheres are called Au-ZnO NSs [51].

In contrast to the "top-down" synthesis of PLAL that makes use of bulk targets, the bottom-up technique of PLICN is also an outstanding liquid-based laser strategy for the design of nanomaterials. This strategy involves the use of a laser. Recently, the versatility of PLICN in the synthesis of graphene materials modified with functional nanoparticles was demonstrated by Liu et al. in their report on a laser technique for the concomitant reduction and alteration of graphene oxide by using a variety of NPs such as Pt, Pt-Pd alloys, RuO_2, and MnO_x [53]. Pulsed laser irradiation was used by

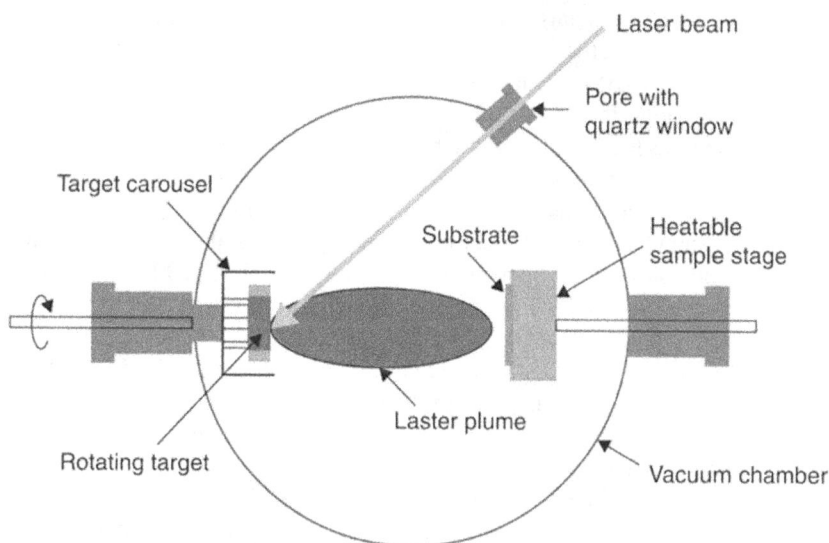

FIGURE 1.1 Schematic representation of a pulsed laser deposition system. Reprinted with permission from [52].

Guo et al. to synthesise carbon dots in chlorobenzene, and these carbon dots were then embedded in perovskite films [54]. The LFL-based preparation of nanomaterials occurs due to the laser energy absorption via coulomb explosion and photothermal vaporisation. Photothermal vaporisation and coulomb explosion are induced by lasers with pulse widths in the nanoscale range and ultrashort lasers with fs, respectively [47]. Zhou et al. revealed the use of laser fragmentation to produce a highly active Co_3O_4 catalyst. The laser irradiation resulted in the formation of numerous oxygen vacancies in the nanoparticle and enhanced the electrical conductivity [55]. Li et al. presented a one-step process free of any chemical reduction agent, the laser ablation in liquid (LAL) for the preparation of the heterostructure of reduced TiO_2-graphene oxide. The production rate of photocatalytic H_2 was found to be increased by using this heterostructure [56]. Numerous studies have been published on the synthesis of metal oxides, metal disulfides, and metal carbides in non-aqueous environments, in addition to the fabrication of non-carbon compounds in aqueous environments. Large-scale production can be achieved using laser technology. It occurs as a result of laser irradiation and subsequent photothermal and/or photochemical effects [2].

By using a series of metal as well as metal oxide nanoparticles synthesised by the laser technique, the energy storage and energy conversion industries have experienced rapid growth. The advantages of laser ablation were proved by its versatility for the preparation of different metal oxides, as well as by its large-scale production and fast synthesis procedure [57, 58]. Using laser ablation on the suitable substrates of metals, Ou et al. presented a facile method for producing a variety of hierarchically nanostructured metal oxides (MO_x, M= Ti, Mn, Fe, Co, Ni, Cu, Mo, Ag, Sn, W and NiFe) with potential electrocatalytic property [59]. The NiO nanocrystal on the corresponding Ni plate was found to be suitable as a bi-functional electrode for both hydrogen evolution reaction (HER) and oxygen evolution reaction (OER). It could be explained by NiO nanocrystals' huge surface area, rich defects, and strong hydrophilicity. Furthermore, when the $BiVO_4$ target was subjected to laser ablation, Zheng et al. found the growth of $BiVO_4$ that was selectively [001]-oriented on FTO substrates [60]. The laser synthesis of transition metal dichalcogenides (TMDs) is a promising strategy owing to its high thermal efficiency and less time-consuming nature. Most of the TMDs-based nanomaterials were prepared via laser thinning and laser annealing. For example, Steel et al. created a laser thinning method for the preparation of single-layer Molybdenum disulphide (MoS_2) from multilayer MoS_2 flakes [61]. By growing MoS_2 films and followed by laser thinning, Sow et al. fabricated MoS_2 monolayer domain encased by multilayer film [62]. Furthermore, by directly writing onto polyimide foils covered in a layer of MoS_2 dispersion, Lamberti et al. demonstrated a facile method for fabricating MoS_2-based laser-induced graphene (MoS_2-LIG) [63].

Zang et al. suggested a direct-writing strategy of laser patterning to generate conductive molybdenum carbide-graphene (MCG) composites directly on paper substrates. The mechanical stability and electrochemical activity of the MCG composites in their as-prepared state made them suitable for a range of possible purposes, including in supercapacitors, energy harvesters, and electrochemical ion detectors and gas sensors. It was also discovered to be useful in paper-based electronics [64].

1.3 APPLICATION OF LASER TECHNOLOGY IN MICROFABRICATION

The laser microfabrication method derives energy from a laser to irradiate on preferred spots as well as for selective laser patterning. The technique consists of laser synthesis and patterning. The superiority of this method could be attributed to its cost-effectiveness and relatively higher processing quality [65, 66]. This section will point out the numerous applications of the laser microfabrication technique (figure 1.2) in EES devices such as sensors, batteries, superconductors, electrodes, etc.

Potentially, green and sustainable technologies are greatly dependent on the efficiency of photothermal conversion. Recently, light-thermal conversion candidates like carbon-based materials and polymers, for example, have sparked significant attention due to their high conversion efficiency. In addition to the favourable optical and thermal properties offered by the light-thermal conversion materials, they should be cost-effective and allow for large-scale manufacturing. The laser processing approach is a facile and economical way for large-scale production of NMs [67, 68] through photothermal conversions. Optothermally-gated photon nudging (OPN) is an all-optical technology developed by Li et al. for attaining nanoscale accuracy [69]. OPN uses a tiny surfactant layer to optothermally control the particle-substrate interaction, permitting colloidal particles on solid substrates to be regulated with optical scattering force. This contactless nanomanipulation is applicable in nano photonics, nanofabrication, and colloidal sciences. Through the use of laser 3D printing technology, Qu et al. developed an ideal material platform of three-dimensional solar steam generation (SSG) [70]. Fan et al. proposed a unique methodology of thermal oxidation for the fabrication of semiconductor oxide nanowire arrays on a Cu surface [71]. In this study, the precursor micro/nano structures were introduced via ultrafast laser patterning. Moreover, Fan et al. also developed a hierarchical surface nanostructure in the shape of a cauliflower by using a laser directly writing on a

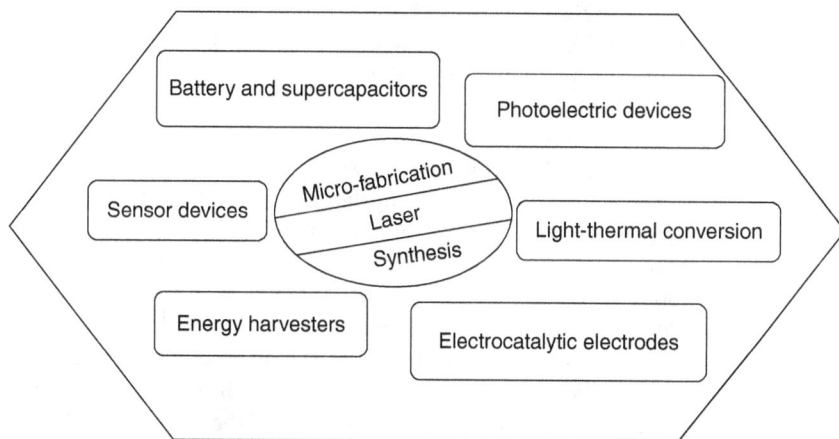

FIGURE 1.2 Applications of laser as a synthetic and microfabrication technology in different functional devices.

copper surface. The nanostructure as-fabricated showed a high water evaporation efficiency [72].

1.3.1 Battery and Supercapacitors

There is a need to store energy in order to compensate for the intermittent nature of renewable sources. Batteries and supercapacitors have witnessed a remarkable development among all electrochemical energy storage (EES) systems. Both batteries and supercapacitors serve similar functions of energy storing; however, there is a significant difference in the inner workings of these devices. Among these two, supercapacitors offer long cycle life, high power density and act as ideal storage technology where power bursts are required [73]. The application of the laser microfabrication method is an ideal solution to overcome poor mechanical performance and high cost. Furthermore, it will provide the devices with a longer life cycle.

Flexible solid-state supercapacitors (SCs) can meet the requirements of the lightweight, multifunctional wearable electronics that have lately gained attention thanks to their promising safety performance, long life span and intrinsically efficient charging-discharging characteristics. Numerous studies on the use of laser technology for SCs, particularly for carbon-based materials, have been conducted during the past decade [74, 75]. For instance, by laser reduction followed by the patterning of graphite oxide sheets, Gao et al. showed how a novel form of monolithic, all-carbon supercapacitor could be produced on a large scale [74]. In contrast to existing thin-film supercapacitors, the developed micro-supercapacitor devices offered superior cycle stability and energy storage capacities. Through a one-step process, Kaner et al. carried out the laser reduction for graphite oxide films to convert into laser-scribed graphene (LSG) [76]. This can be used for the fabrication of SC. Cai et al. conducted the fabrication of flexible all-solid-state carbon micro-supercapacitors (MSCs) via direct laser writing PI films, where a 405 nm blue-violet laser in an inert environment was employed [77]. In this study, the influence of laser irradiation atmosphere was explored by conducting the same procedure in the air and inert environment of Ar; the latter showed improved capacitive performance. Furthermore, after heat treatment, He et al. developed a flexible planar polyaniline (PANI) MSC with remarkable electrochemical properties using a laser printing lithography technique [78]. The fabrication of EES systems is driven by the rising need for energy resources. Recent advancements in laser microfabrication technology, which have attracted researchers, have been regarded as promising for lowering fabrication costs, extending battery life, and improving power density performance. A cathode catalyst based on MnO_2-(LIG) and dual polymer gel electrolyte (DPGE) were used by Ren et al. to develop a method for fabricating $Li-O_2$ batteries [79]. Veliscek et al. investigated the viability of silicon (Si) produced by Laser-assisted Chemical Vapour Pyrolysis (LaCVP) to act as anode material in lithium batteries [80]. Similarly, Munao et al. applied LaCVP for the preparation of Si-based nanocomposite anodes [81]. The as-prepared Si-based anode was subjected to electrochemical measurements, and the specific capacity was found to be the maximum. Zhang et al. proposed a method based on CO_2 laser to form hard carbon for anodes in Sodium (Na) ion batteries [82]. The application of

CO_2 laser to irradiate electrospun carbon nanofiber (CNF) films will improve their structural organisation. Furthermore, Zhang et al. explored the versatility of the laser-scribing process by fabricating an N-doped 3D graphene anode from urea containing PI [42].

1.3.2 SENSOR DEVICES VIA LASER TECHNIQUE

The surface of materials is modified into porous structures upon laser treatment. These properties make them suitable for sensors. Additionally, the laser microfabrication technology allows the easy fabrication of pattern array devices. As a result, the majority of devices manufactured via laser technology can be employed as sensors [83]. The remarkable advancements in the field of force sensors and gas sensors are discussed in this section. A force sensor is a sensor that assists in determining how much force has been applied to an object. Force sensors are extensively employed in a variety of industries, especially in wearable electronics, and smart textiles [84]. Carbon-based materials are widely used to fabricate force sensors. By directly writing PI with a laser, Tao et al. developed a wearable, cost-effective, and one-step LIG artificial throat that is capable of producing and detecting sound [85]. This LIG-based artificial throat operates radically differently from traditional acoustic transducers. If it operates as a sound source, a wide-band sound will be generated from LIG artificial throat. In contrast to that, if it acts as a sound detector it is capable of sensing different sound levels. LIG can accurately recognise different levels of tones and volumes and detect hum and cough. Additionally, it is equipped to comprehend words and sentences. Rahimi et al. fabricated a strain sensor which is unidirectional and similar to this [86]. This was achieved by embedding and transferring carbonised patterns created by laser microfabrication of thermoset polymers, allowing for real-time finger motion measurement. Nag et al. made use of a conductive layer of multi-walled carbon nanotube (MWCNT) composite and polydimethylsiloxane (PDMS) as substrate to enable monitoring of respiration and limb motions [87]. Moreover, Nag et al. explored using aluminium (Al) film to conduct on a substrate based on PI. Aluminium interdigitated electrodes for tactile sensors developed by laser microfabrication were responsive to finger pressure [88].

The detection of toxic gases by gas sensors has recently emerged as a crucial topic of research. Gas sensors have expanded significantly and are now widely applied in medicine, explosive detection, dangerous material detection, and environmental monitoring. The most common toxic gases such as carbon dioxide (CO_2), and carbon monoxide (CO), are odourless and colourless. Nevertheless, the detection of these gases is only feasible if the concentration is high enough to detect. In this context, researchers have contributed significantly to the development of portable sensors which have notably enhanced the safety of human lives [89]. Chang et al. proposed an ultrafast laser ablation technique to construct potential electrodes for a gas sensor with the added advantage of wireless circuits [90]. A wireless circuit was coupled with the graphene-based electrode to develop a gas sensor module. The efficiency of sensor modules to detect CO gas even at a lower concentration of parts per million (ppm) was found. Additionally, the developed sensor possessed high detection sensitivity to

H_2O, and air. The electrical resistance of the multilayer showed an appreciable change upon interaction with CO gas and revealed the effective performance of the sensor. In order to establish wearable monitoring, Park et al. pasted a humidity sensor on a fingernail using the interlocked rGO produced by a facile and single-step laser treatment [91]. Furthermore, Drmosha et al. fabricated a gas sensor consisting of a platinum (Pt)-based composite of rGO/ZnO, which was capable enough to detect the concentration of hydrogen at parts per million (ppm) levels of approximately 400 ppm by an efficiency of 99%. It outperformed the Pt-free composite by five times [92].

1.3.3 ELECTROCATALYTIC MATERIALS AND ELECTRODES

Transition metal oxides as well as hydroxides are the most prominent oxygen evolution reaction (OER) electrocatalysts. Laser ablation synthesis of OER electrocatalysts is viable when employing a laser as the power source in an environment with ambient air. Laser ablation in liquid is an effective method for the synthesis of OER. For instance, Blakemore et al. prepared a cobalt-based nanoparticle, cobalt oxide (Co_3O_4) through a pulsed laser ablation method (PLAM). The as-prepared catalyst exhibited high electrocatalytic oxygen evolution activity [93]. The morphology of Co_3O_4 NPs was studied by transmission electron microscopy (TEM) imaging as given in figure 1.3. Nishi et al. proposed the fs laser ablation method for the synthesis of $CoO–Co_2O_3–Co(OH)_2$ multiphase nanoparticles [94]. Cai et al. synthesised a low-cost and effective OER, called Ni-doped Fe_3O_4 nanoparticle clusters [95]. The promising electrocatalytic activity of this OER could be attributed to the hierarchical and porous architecture of the cluster. Moreover, the Ni doping induced an optimised

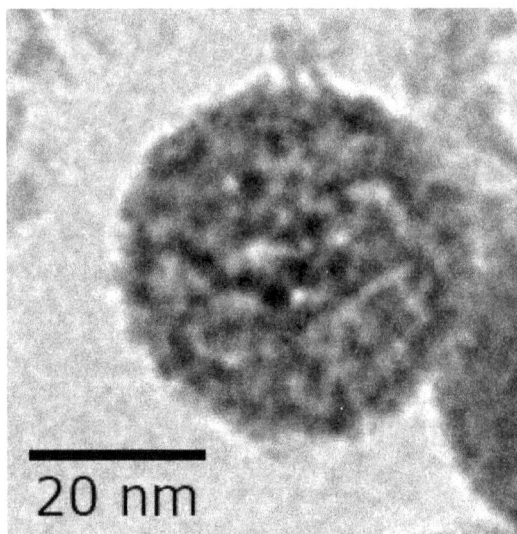

FIGURE 1.3 TEM image of Co_3O_4 NPs prepared via PLAM. Reprinted with permission from [93].

electronic configuration, which further enhanced the electrocatalytic activity. In addition to metal oxides and hydroxides, Zhang et al. reported the preparation of a metal-free catalyst by a laser process. The catalyst was oxidised laser-induced graphene (LIG-O), where the oxidation was conducted in an O_2 plasma atmosphere [96].

For electrochemical water splitting to be feasible, electrocatalysts should possess high durability and remarkable hydrogen evolution reaction (HER) performance. Li et al. presented a laser ablation technique for producing silver catalysts with greater activity and durability in an acid media [97]. The laser generated stacking faults which enhanced the catalytic activity. Chen et al. reported a preparation method for the synthesis of RuAu single-atom alloy (SAA) via laser ablation [98]. The higher catalytic activity of as-prepared catalyst could be ascribed to the Au active sites and relay catalysis. By increasing the surface area of Ni, Rauscher et al. made use of surface fabrication of Ni electrodes with the aid of a femtosecond pulse laser [99].

The LIG process is an ideal method for the synthesis of electrocatalyst to conduct the overall water splitting, thereby producing H_2 and O_2. Zhang et al. developed a procedure for the fabrication of an overall water-splitting device, where the opposing faces were occupied by HER and OER on a PI sheet [100]. The LIG electrodes offer long-term stability. The same group also outlined a method for the development of porous graphene from wood via CO_2 laser scribing [36]. The production of high-quality graphene is more advantageous with higher lignin content. In addition to LIG method, porous $Co_{0.75}Ni_{0.25}(OH)_2$ nanosheets were fabricated using a laser synthesis by Wang et al. to catalyze the total water splitting. According to experimental and theoretical findings, an ideal electronic structure for enhancing the OER and HER was developed by the abundant Co^{3+} ions on the nanosheets and the favourable atomic structure surrounding the Co^{3+} ions [101].

1.4 APPLICATIONS OF LASER-ASSISTED MICROFABRICATION METHODS IN OTHER FIELDS

Corrosion occurs as a result of the deterioration of material when it interacts with the environment. Fog, humidity, seawater, and alkaline or acidic soils are a few of the environments that promote corrosion. Every year, metal corrosion leads to tremendous economic losses and natural hazards. Recently, graphene has emerged as an ideal alternative as it is endowed with numerous advantages such as chemical inertness, and impermeability [2]. Even though graphene has considerable commercial value, growing it on carbon steel is difficult. As a result, it is generally grown on copper or nickel substrates. Ye et al. explored the anti-corrosion properties of graphene on carbon steel by developing a new method based on laser technology. They facilitated the growth of graphene by a laser alloying process in which Ni was incorporated into carbon steel to synthesise a catalyst based on Ni/Fe alloy. Graphene-coated carbon steel showed a corrosion rate of approximately 0.05 millimetres per year as compared to 0.09 millimetres per year for stainless steel [102]. A multitude of biomimetic designs were inspired by the intriguing properties of naturally occurring superhydrophobic and superhydrophilic surfaces. Kostal et al. proposed a unique three-step fabrication approach that mimicked the elytra of a beetle to improve the fog-collection performance of glasses [103]. In this study, a double hierarchical surface structure was developed with the aid of fs laser. The surface turned super hydrophilic (water contact angle= 10°) as a result of laser structuring, which improved the surface's intrinsic wetting property. In order to make the wetting state superhydrophobic, which has a water contact angle greater than 150°, a Teflon coating was used. Following this step, a selective laser ablation was conducted to uncover the superhydrophilicity pattern. The fog-collection efficiency of thus obtained micropatterns was 60% higher than that of blank glass. Furthermore, Chen et al. employed a femtosecond laser to construct hierarchical micropillars on a shape memory polymer (SMP). As-fabricated micropillars were capable of showing a switchable wetting state [104].

It has been a scientific and technological challenge to induce a mass flow from a laser beam. However, Wang et al. made use of gold nanoparticles (Au NPs) with the aid of a pulsed laser to induce a steady-state water flow [105]. The laser can regulate the direction, speed and size of the flow. This work was based on the optofluidic concept. According to that, as a result of pulsed laser irradiation and subsequent photoacoustic effect ultrasonic waves were generated from plasmonic NPs. Following this, a laser streaming will be generated in the neighbouring area. Similarly, Yue et al. demonstrated a revolutionary microfluidic pump based on photoacoustic laser streaming. It consisted of a quartz plate with Au particles implanted, and every location on the plate served as a micro pump. A plethora of microfluidic applications will be achievable with this laser-driven optofluidic approach [106].

1.5   OUTLOOK AND FUTURE CHALLENGES

The use of laser techniques that can operate in either pulsed or continuous wave modes has been put to use in the process of microfabricating various types of materials. Because the conversion, as well as the crystallisation of precursor materials, is driven by the photothermal or photochemical effects caused by the absorption of laser radiation, the required tuning can be achieved by varying both the scanning rate and the laser intensity. This is because the photothermal or photochemical effects are caused by the absorption of laser radiation. By doing this, it is possible to fabricate materials from the materials that served as their precursors by selecting the laser wavelength with great care. During the process of fabricating nanostructures with the assistance of a laser, two distinct types of manufacturing processes are utilised. One method involves producing nanostructures by means of a laser-induced photothermal effect, which is then followed by the localised production of the materials of interest. The sputtering etching effect is the primary factor on which the other method, known as the cold microfabrication process, relies when it comes to the creation of the suitable nanostructure pattern. The first technique of production is comparable to the laser synthesis process; however, this method is heavily reliant on the laser process parameters. The second method, on the other hand, calls for intensively concentrated radiation in order to avoid the burrs that are created by photothermal effects. Utilising an ultrafast laser is one method that can be used to lessen the impact of thermal effects and non-linear absorption. Because of their high intensity and ultrashort pulse width, this method can be utilised to create patterned nanostructures. This results in a minimal loss of heat throughout the manufacturing phase.

When compared to the traditional methods, laser processing has several advantages, including the following: (i) the processing is noticeably quick and saves time; (ii) it provides accurate control over the processing sites (the cold microfabrication method); (iii) it is compatible with a wide variety of materials, it can be manufactured in roll-to-roll mode, and it can adjust to variable size requirements; and (iv) it retains the precursor material's morphology through meticulous tuning of the laser focus Therefore, laser synthesis and microfabrication can be regarded as a potent technique for the production of nanomaterials [2]. In spite of the fact that this method provides many benefits, those advantages are somewhat overshadowed by the fact that the laser punch treatment is laborious and time-consuming when line-by-line scanning is taken into consideration. Additionally, laser processing makes it challenging to fabricate nanoparticles in three dimensions. By utilising either a patterned laser source or numerous laser beams, it is possible to solve the two issues that were discussed earlier to a significant degree. In contrast to wet chemical methods, the laser processing technique possesses limitations in the fabrication of 3-D nanostructures. However, a potential suggestion for expanding the scope of the use of the laser process is the in-situ alteration of previously created nanostructures via the thermal effects of an unfocused laser. This would allow for the laser process to be used for a wider variety of applications.

In addition to these uses, laser processing can also be broadened to include applications such as the controlled synthesis of nitrides, sulphides, carbides, and oxides under a variety of atmospheres, such as NH_3, N_2, H_2S, CH_4, and CO_2 respectively [2]. Additionally, the synthesis may be efficiently managed and the resolution can be enhanced by integrating a variety of optical techniques with the technology of laser processing. Therefore, the laser processing method can be improved in ways that go beyond the synthesis of nanomaterials by integrating with a variety of systems that are compatible with one another for a wide variety of applications.

1.6 CONCLUSION

Laser technology has emerged as a potential tool for the site-specific synthesis of nanomaterials. It has numerous applications in energy storage, conversion, production and in other significant areas such as anti-corrosion, and electro-catalysis. Laser technique as a synthetic method and microfabrication method offers a lot of advantages as it saves time and retains the morphology of the starting material. Precise tunability and controllability of various properties of the laser such as the spot size, pulse width, repetition rate, and peak power enables one to achieve this goal. Of various materials explored, carbon-based materials are regarded as the most useful candidates for the laser-assisted synthesis methods. Among all the carbon materials, graphene underwent a plethora of studies and was considered one of the prominent materials of endless possibilities.

REFERENCES

[1] N. Baig, I. Kammakakam, and W. Falath, "Nanomaterials: A review of synthesis methods, properties, recent progress, and challenges," *Mater. Adv.*, vol. 2, no. 6, pp. 1821–1871, 2021.

[2] L. Zhao, Z. Liu, D. Chen, F. Liu, Z. Yang, X. Li, H. Yu, H. Liu, and W. Zhou, "Laser synthesis and microfabrication of micro/nanostructured materials toward energy conversion and storage," *Nano-Micro Lett.*, vol. 13, no. 1, pp. 49, 2021.

[3] H. Palneedi, J. H. Park, D. Maurya, M. Peddigari, G.-T. Hwang, V. Annapureddy, J.-W. Kim, J.-J. Choi, B.-D. Hahn, S. Priya, K. J. Lee, and J. Ryu, "Laser irradiation of metal oxide films and nanostructures: applications and advances," *Adv. Mater.*, vol. 30, no. 14, pp. 1705148, Apr. 2018.

[4] S. Hong, H. Lee, J. Yeo, and S. H. Ko, "Digital selective laser methods for nanomaterials: From synthesis to processing," *Nano Today*, vol. 11, no. 5, pp. 547–564, 2016.

[5] J. Bian, L. Zhou, X. Wan, C. Zhu, B. Yang, and Y. Huang, "Laser transfer, printing, and assembly techniques for flexible electronics," *Adv. Electron. Mater.*, vol. 5, no. 7, pp. 1800900, Jul. 2019.

[6] A. Alex, S. Raj, S. K. Sugunan, and G. George, "Nanotechnology in soil remediation," in *Nanotechnology for Environmental Remediation*, Sabu Thomas, Merin Sara Thomas, Laly A Pothen, Eds. 2022, pp. 27–43. https://doi.org/10.1002/9783527834 143.ch3, https://onlinelibrary.wiley.com/doi/abs/10.1002/9783527834143.ch3

[7] Y. B. Pottathara, Y. Grohens, V. Kokol, N. Kalarikkal, and S. Thomas, "Chapter 1 – Synthesis and processing of emerging two-dimensional nanomaterials," in *Micro and Nano Technologies*, Y. Beeran Pottathara, S. Thomas, N. Kalarikkal, Y. Grohens, and V. B. T.-N. S. Kokol, Eds. Elsevier, 2019, pp. 1–25.

[8] M. Huston, M. DeBella, M. DiBella, and A. Gupta, "Green synthesis of nanomaterials," *Nanomaterials*, vol. 11, no. 8. 2021.

[9] C. Yang, Y. Huang, H. Cheng, L. Jiang, and L. Qu, "Rollable, stretchable, and reconfigurable graphene hygroelectric generators," *Adv. Mater.*, vol. 31, no. 2, pp. 1805705, Jan. 2019.

[10] H. Wang, D. Tran, J. Qian, F. Ding, and D. Losic, "MoS2/graphene composites as promising materials for energy storage and conversion applications," *Adv. Mater. Interfaces*, vol. 6, no. 20, pp. 1900915, Oct. 2019.

[11] T.-S. Kim, Y. Lee, W. Xu, Y. H. Kim, M. Kim, S.-Y. Min, T. H. Kim, H. W. Jang, and T.-W. Lee, "Direct-printed nanoscale metal-oxide-wire electronics," *Nano Energy*, vol. 58, pp. 437–446, 2019.

[12] M. T. Chorsi, E. J. Curry, H. T. Chorsi, R. Das, J. Baroody, P. K. Purohit, H. Ilies, and T. D. Nguyen, "Piezoelectric biomaterials for sensors and actuators," *Adv. Mater.*, vol. 31, no. 1, pp. 1802084, Jan. 2019.

[13] W. Dong, H. Liu, J. K. Behera, L. Lu, R. J. H. Ng, K. V. Sreekanth, X. Zhou, J. K. W. Yang, and R. E. Simpson, "Wide bandgap phase change material tuned visible photonics," *Adv. Funct. Mater.*, vol. 29, no. 6, pp. 1806181, Feb. 2019.

[14] D. J. Joe, S. Kim, J. H. Park, D. Y. Park, H. E. Lee, T. H. Im, I. Choi, R. S. Ruoff, and K. J. Lee, "Laser–material interactions for flexible applications," *Adv. Mater.*, vol. 29, no. 26, pp. 1606586, Jul. 2017.

[15] M. Cutroneo, L. Torrisi, and C. Scolaro, "Laser applications in bio-medical field," vol. 2010, Jul. 2012. DOI: 10.1285/i9788883050886p144, http://siba-ese.unisalento. it/index.php/psba2/article/view/11959

[16] A. Hatef, S. Fortin-Deschênes, E. Boulais, F. Lesage, and M. Meunier, "Photothermal response of hollow gold nanoshell to laser irradiation: Continuous wave, short and ultrashort pulse," *Int. J. Heat Mass Transf.*, vol. 89, pp. 866–871, 2015.

[17] K. Gao, B. Wang, L. Tao, B. V Cunning, Z. Zhang, S. Wang, R. S. Ruoff, and L. Qu, "Efficient metal-free electrocatalysts from n-doped carbon nanomaterials: Mono-doping and co-doping," *Adv. Mater.*, vol. 31, no. 13, pp. 1805121, Mar. 2019.

[18] R. Paul, F. Du, L. Dai, Y. Ding, Z. L. Wang, F. Wei, and A. Roy, "3D heteroatom-doped carbon nanomaterials as multifunctional metal-free catalysts for integrated energy devices," *Adv. Mater.*, vol. 31, no. 13, pp. 1805598, Mar. 2019.

[19] K. S. Novoselov, V. I. Fal'ko, L. Colombo, P. R. Gellert, M. G. Schwab, and K. Kim, "A roadmap for graphene," *Nature*, vol. 490, no. 7419, pp. 192–200, 2012.

[20] S. Sajjad, S. A. Khan Leghari, and A. Iqbal, "Study of graphene oxide structural features for catalytic, antibacterial, gas sensing, and metals decontamination environmental applications," *ACS Appl. Mater. Interfaces*, vol. 9, no. 50, pp. 43393–43414, Dec. 2017.

[21] H. Huang, H. Shi, P. Das, J. Qin, Y. Li, X. Wang, F. Su, P. Wen, S. Li, P. Lu, F. Liu, Y. Li, Y. Zhang, Y. Wang, Z.-S. Wu, and H.-M. Cheng, "The chemistry and promising applications of graphene and porous graphene materials," *Adv. Funct. Mater.*, vol. 30, no. 41, pp. 1909035, Oct. 2020.

[22] R. Kumar, R. K. Singh, D. P. Singh, E. Joanni, R. M. Yadav, and S. A. Moshkalev, "Laser-assisted synthesis, reduction and micro-patterning of graphene: Recent progress and applications," *Coord. Chem. Rev.*, vol. 342, pp. 34–79, 2017.

[23] R. D. Rodriguez, A. Khalelov, P. S. Postnikov, A. Lipovka, E. Dorozhko, I. Amin, G. V Murastov, J.-J. Chen, W. Sheng, M. E. Trusova, M. M. Chehimi, and E. Sheremet, "Beyond graphene oxide: Laser engineering functionalized graphene for flexible electronics," *Mater. Horizons*, vol. 7, no. 4, pp. 1030–1041, 2020.

[24] K. K. H. De Silva, H.-H. Huang, R. K. Joshi, and M. Yoshimura, "Chemical reduction of graphene oxide using green reductants," *Carbon N. Y.*, vol. 119, pp. 190–199, 2017.

[25] V. Scardaci, "Laser synthesized graphene and its applications," *Appl. Sci.*, vol. 11, no. 14, pp. 6304, 2021. https://doi.org/10.3390/app11146304, www.mdpi.com/2076-3417/11/14/6304, Received: 6 June 2021/Revised: 24 June 2021/Accepted: 1 July 2021/Published: 8 July 2021

[26] Y. Zhang, L. Guo, S. Wei, Y. He, H. Xia, Q. Chen, H.-B. Sun, and F.-S. Xiao, "Direct imprinting of microcircuits on graphene oxides film by femtosecond laser reduction," *Nano Today*, vol. 5, no. 1, pp. 15–20, 2010.

[27] B. Xie, Y. Wang, W. Lai, W. Lin, Z. Lin, Z. Zhang, P. Zou, Y. Xu, S. Zhou, C. Yang, F. Kang, and C.-P. Wong, "Laser-processed graphene based micro-supercapacitors for ultrathin, rollable, compact and designable energy storage components," *Nano Energy*, vol. 26, pp. 276–285, 2016.

[28] H. Cheng, J. Liu, Y. Zhao, C. Hu, Z. Zhang, N. Chen, L. Jiang, and L. Qu, "Graphene fibers with predetermined deformation as moisture-triggered actuators and robots," *Angew. Chemie Int. Ed.*, vol. 52, no. 40, pp. 10482–10486, Sep. 2013.

[29] H. Cheng, M. Ye, F. Zhao, C. Hu, Y. Zhao, Y. Liang, N. Chen, S. Chen, L. Jiang, and L. Qu, "A general and extremely simple remote approach toward graphene bulks with in situ multifunctionalization," *Adv. Mater.*, vol. 28, no. 17, pp. 3305–3312, May 2016.

[30] L. Huang, Y. Liu, L.-C. Ji, Y.-Q. Xie, T. Wang, and W.-Z. Shi, "Pulsed laser assisted reduction of graphene oxide," *Carbon N. Y.*, vol. 49, no. 7, pp. 2431–2436, 2011.

[31] E. E. Ghadim, N. Rashidi, S. Kimiagar, O. Akhavan, F. Manouchehri, and E. Ghaderi, "Pulsed laser irradiation for environment friendly reduction of graphene oxide suspensions," *Appl. Surf. Sci.*, vol. 301, pp. 183–188, 2014.

[32] L. Cao, S. Zhu, B. Pan, X. Dai, W. Zhao, Y. Liu, W. Xie, Y. Kuang, and X. Liu, "Stable and durable laser-induced graphene patterns embedded in polymer substrates," *Carbon N. Y.*, vol. 163, pp. 85–94, 2020.

[33] J. Lin, Z. Peng, Y. Liu, F. Ruiz-Zepeda, R. Ye, E. L. G. Samuel, M. J. Yacaman, B. I. Yakobson, and J. M. Tour, "Laser-induced porous graphene films from commercial polymers," *Nat. Commun.*, vol. 5, no. 1, pp. 5714, 2014.

[34] A. Z. Yazdi, I. O. Navas, A. Abouelmagd, and U. Sundararaj, "Direct creation of highly conductive laser-induced graphene nanocomposites from polymer blends," *Macromol. Rapid Commun.*, vol. 38, no. 17, pp. 1700176, Sep. 2017.

[35] A. Lamberti, M. Serrapede, G. Ferraro, M. Fontana, F. Perrucci, S. Bianco, A. Chiolerio, and S. Bocchini, "All-SPEEK flexible supercapacitor exploiting laser-induced graphenization," *2D Mater.*, vol. 4, no. 3, pp. 35012, 2017.

[36] R. Ye, Y. Chyan, J. Zhang, Y. Li, X. Han, C. Kittrell, and J. M. Tour, "Laser-induced graphene formation on wood," *Adv. Mater.*, vol. 29, no. 37, pp. 1702211, Oct. 2017.

[37] Y. Chyan, R. Ye, Y. Li, S. P. Singh, C. J. Arnusch, and J. M. Tour, "Laser-induced graphene by multiple lasing: Toward electronics on cloth, paper, and food," *ACS Nano*, vol. 12, no. 3, pp. 2176–2183, Mar. 2018.

[38] J. Robertson, "Diamond-like carbon films, properties and applications," *Compr. Hard Mater.*, vol. 3, pp. 101–139, Mar. 2014.

[39] F. Stock, F. Antoni, L. Diebold, C. Chowde Gowda, S. Hajjar-Garreau, D. Aubel, N. Boubiche, F. Le Normand, and D. Muller, "UV laser annealing of diamond-like carbon layers obtained by pulsed laser deposition for optical and photovoltaic applications," *Appl. Surf. Sci.*, vol. 464, pp. 562–566, 2019.

[40] Y. X. Han, H. Ling, Y. F. Lu, M. J. O'Keefe, and T. McKindra, "Laser-assisted synthesis of diamond-like carbon from cyclohexane liquid," in *Proc. SPIE*, vol. 6107, pp. 61070P, Feb. 2006.

[41] J. Cai, C. Lv, C. Hu, J. Luo, S. Liu, J. Song, Y. Shi, C. Chen, Z. Zhang, S. Ogawa, E. Aoyagi, and A. Watanabe, "Laser direct writing of heteroatom-doped porous carbon for high-performance micro-supercapacitors," *Energy Storage Mater.*, vol. 25, pp. 404–415, 2020.

[42] F. Zhang, E. Alhajji, Y. Lei, N. Kurra, and H. N. Alshareef, "Highly doped 3D graphene Na-Ion battery anode by laser scribing polyimide films in nitrogen ambient," *Adv. Energy Mater.*, vol. 8, no. 23, pp. 1800353, Aug. 2018.

[43] Z. Peng, R. Ye, J. A. Mann, D. Zakhidov, Y. Li, P. R. Smalley, J. Lin, and J. M. Tour, "Flexible boron-doped laser-induced graphene microsupercapacitors," *ACS Nano*, vol. 9, no. 6, pp. 5868–5875, Jun. 2015.

[44] Y. Huang, L. Zeng, C. Liu, D. Zeng, Z. Liu, X. Liu, X. Zhong, W. Guo, and L. Li, "Laser direct writing of heteroatom (n and s)-doped graphene from a polybenzimidazole ink donor on polyethylene terephthalate polymer and glass substrates," *Small*, vol. 14, no. 44, pp. 1803143, Nov. 2018.

[45] W. H. Lee, J. W. Suk, H. Chou, J. Lee, Y. Hao, Y. Wu, R. Piner, D. Akinwande, K. S. Kim, and R. S. Ruoff, "Selective-Area fluorination of graphene with fluoropolymer and laser irradiation," *Nano Lett.*, vol. 12, no. 5, pp. 2374–2378, May 2012.

[46] F. Davodi, E. Mühlhausen, D. Settipani, E.-L. Rautama, A.-P. Honkanen, S. Huotari, G. Marzun, P. Taskinen, and T. Kallio, "Comprehensive study to design advanced metal-carbide@garaphene and metal-carbide@iron oxide nanoparticles with tunable structure by the laser ablation in liquid," *J. Colloid Interface Sci.*, vol. 556, pp. 180–192, 2019.

[47] D. Zhang, B. Gökce, and S. Barcikowski, "Laser synthesis and processing of colloids: Fundamentals and applications," *Chem. Rev.*, vol. 117, no. 5, pp. 3990–4103, Mar. 2017.

[48] S. Hebié, Y. Holade, K. Maximova, M. Sentis, P. Delaporte, K. B. Kokoh, T. W. Napporn, and A. V Kabashin, "Advanced electrocatalysts on the basis of bare Au nanomaterials for biofuel cell applications," *ACS Catal.*, vol. 5, no. 11, pp. 6489–6496, Nov. 2015.

[49] V. Piotto, L. Litti, and M. Meneghetti, "Synthesis and shape manipulation of aniso-tropic gold nanoparticles by laser ablation in solution," *J. Phys. Chem. C*, vol. 124, no. 8, pp. 4820–4826, Feb. 2020.

[50] S. Hu, G. Goenaga, C. Melton, T. A. Zawodzinski, and D. Mukherjee, "PtCo/CoOx nanocomposites: Bifunctional electrocatalysts for oxygen reduction and evolution reactions synthesized via tandem laser ablation synthesis in solution-galvanic replace-ment reactions," *Appl. Catal. B Environ.*, vol. 182, pp. 286–296, 2016.

[51] H. Zhang, S. Wu, J. Liu, Y. Cai, and C. Liang, "Laser irradiation-induced Au–ZnO nanospheres with enhanced sensitivity and stability for ethanol sensing," *Phys. Chem. Chem. Phys.*, vol. 18, no. 32, pp. 22503–22508, 2016.

[52] S. M. Shang and W. Zeng, "4 – Conductive nanofibres and nanocoatings for smart textiles," in *Woodhead Publishing Series in Textiles*, T. B. T.-M. K.-H. for S.-T. D. Kirstein, Ed. Woodhead Publishing, 2013, pp. 92–128.

[53] Y. Peng, J. Cao, J. Yang, W. Yang, C. Zhang, X. Li, R. A. W. Dryfe, L. Li, I. A. Kinloch, and Z. Liu, "Laser assisted solution synthesis of high performance graphene supported electrocatalysts," *Adv. Funct. Mater.*, vol. 30, no. 32, pp. 2001756, Aug. 2020.

[54] P. Guo, X. Yang, Q. Ye, J. Zhang, H. Wang, H. Yu, W. Zhao, C. Liu, H. Yang, and H. Wang, "Laser-Generated nanocrystals in perovskite: universal embedding of ligand-free and sub-10 nm nanocrystals in solution-processed metal halide perovskite films for effectively modulated optoelectronic performance," *Adv. Energy Mater.*, vol. 9, no. 35, pp. 1901341, Sep. 2019.

[55] Y. Zhou, C.-K. Dong, L. Han, J. Yang, and X.-W. Du, "Top-Down preparation of active cobalt oxide catalyst," *ACS Catal.*, vol. 6, no. 10, pp. 6699–6703, Oct. 2016.

[56] L. Li, L. Yu, Z. Lin, and G. Yang, "Reduced TiO2-graphene oxide heterostructure as broad spectrum-driven efficient water-splitting photocatalysts," *ACS Appl. Mater. Interfaces*, vol. 8, no. 13, pp. 8536–8545, Apr. 2016.

[57] L. Peng, P. Xiong, L. Ma, Y. Yuan, Y. Zhu, D. Chen, X. Luo, J. Lu, K. Amine, and G. Yu, "Holey two-dimensional transition metal oxide nanosheets for efficient energy storage," *Nat. Commun.*, vol. 8, no. 1, p. 15139, 2017.

[58] D. Gao, R. Liu, J. Biskupek, U. Kaiser, Y.-F. Song, and C. Streb, "Modular design of noble-metal-free mixed metal oxide electrocatalysts for complete water splitting," *Angew. Chemie Int. Ed.*, vol. 58, no. 14, pp. 4644–4648, Mar. 2019.

[59] G. Ou, P. Fan, H. Zhang, K. Huang, C. Yang, W. Yu, H. Wei, M. Zhong, H. Wu, and Y. Li, "Large-scale hierarchical oxide nanostructures for high-performance electrocatalytic water splitting," *Nano Energy*, vol. 35, pp. 207–214, 2017.

[60] H. S. Han, S. Shin, D. H. Kim, I. J. Park, J. S. Kim, P.-S. Huang, J.-K. Lee, I. S. Cho, and X. Zheng, "Boosting the solar water oxidation performance of a BiVO4 photoanode by crystallographic orientation control," *Energy Environ. Sci.*, vol. 11, no. 5, pp. 1299–1306, 2018.

[61] A. Castellanos-Gomez, M. Barkelid, A. M. Goossens, V. E. Calado, H. S. J. van der Zant, and G. A. Steele, "Laser-Thinning of MoS2: On demand generation of a single-layer semiconductor," *Nano Lett.*, vol. 12, no. 6, pp. 3187–3192, Jun. 2012.

[62] J. Lu, J. H. Lu, H. Liu, B. Liu, K. X. Chan, J. Lin, W. Chen, K. P. Loh, and C. H. Sow, "Improved photoelectrical properties of MoS2 films after laser micromachining," *ACS Nano*, vol. 8, no. 6, pp. 6334–6343, Jun. 2014.

[63] F. Clerici, M. Fontana, S. Bianco, M. Serrapede, F. Perrucci, S. Ferrero, E. Tresso, and A. Lamberti, "In situ MoS2 decoration of laser-induced graphene as flexible supercapacitor electrodes," *ACS Appl. Mater. Interfaces*, vol. 8, no. 16, pp. 10459–10465, Apr. 2016.

[64] X. Zang, C. Shen, Y. Chu, B. Li, M. Wei, J. Zhong, M. Sanghadasa, and L. Lin, "Laser-Induced molybdenum carbide–graphene composites for 3D foldable paper electronics," *Adv. Mater.*, vol. 30, no. 26, pp. 1800062, Jun. 2018.

[65] E. H. Penilla, L. F. Devia-Cruz, A. T. Wieg, P. Martinez-Torres, N. Cuando-Espitia, P. Sellappan, Y. Kodera, G. Aguilar, and J. E. Garay, "Ultrafast laser welding of ceramics," *Science*, vol. 365, no. 6455, pp. 803–808, Aug. 2019.

[66] H. Luo, C. Wang, C. Linghu, K. Yu, C. Wang, and J. Song, "Laser-driven programmable non-contact transfer printing of objects onto arbitrary receivers via an active elastomeric microstructured stamp," *Natl. Sci. Rev.*, vol. 7, no. 2, pp. 296–304, Feb. 2020.

[67] M. Gao, L. Zhu, C. K. Peh, and G. W. Ho, "Solar absorber material and system designs for photothermal water vaporization towards clean water and energy production," *Energy Environ. Sci.*, vol. 12, no. 3, pp. 841–864, 2019.

[68] G. Liu, J. Xu, and K. Wang, "Solar water evaporation by black photothermal sheets," *Nano Energy*, vol. 41, pp. 269–284, 2017.

[69] J. Li, Y. Liu, L. Lin, M. Wang, T. Jiang, J. Guo, H. Ding, P. S. Kollipara, Y. Inoue, D. Fan, B. A. Korgel, and Y. Zheng, "Optical nanomanipulation on solid substrates via optothermally-gated photon nudging," *Nat. Commun.*, vol. 10, no. 1, pp. 5672, 2019.

[70] P. Zhang, Q. Liao, H. Yao, H. Cheng, Y. Huang, C. Yang, L. Jiang, and L. Qu, "Three-dimensional water evaporation on a macroporous vertically aligned graphene pillar array under one sun," *J. Mater. Chem. A*, vol. 6, no. 31, pp. 15303–15309, 2018.

[71] P. Fan, B. Bai, J. Long, D. Jiang, G. Jin, H. Zhang, and M. Zhong, "Broadband high-performance infrared antireflection nanowires facilely grown on ultrafast laser structured cu surface," *Nano Lett.*, vol. 15, no. 9, pp. 5988–5994, Sep. 2015.

[72] P. Fan, H. Wu, M. Zhong, H. Zhang, B. Bai, and G. Jin, "Large-scale cauliflower-shaped hierarchical copper nanostructures for efficient photothermal conversion," *Nanoscale*, vol. 8, no. 30, pp. 14617–14624, 2016.

[73] B. E. Conway, "Similarities and differences between supercapacitors and batteries for storing electrical energy BT – electrochemical supercapacitors: scientific fundamentals and technological applications," B. E. Conway, Ed. Boston, MA: Springer US, 1999, pp. 11–31.

[74] W. Gao, N. Singh, L. Song, Z. Liu, A. L. M. Reddy, L. Ci, R. Vajtai, Q. Zhang, B. Wei, and P. M. Ajayan, "Direct laser writing of micro-supercapacitors on hydrated graphite oxide films," *Nat. Nanotechnol.*, vol. 6, no. 8, pp. 496–500, 2011.

[75] J. Liang, C. Jiang, and W. Wu, "Toward fiber-, paper-, and foam-based flexible solid-state supercapacitors: Electrode materials and device designs," *Nanoscale*, vol. 11, no. 15, pp. 7041–7061, 2019.

[76] M. F. El-Kady, V. Strong, S. Dubin, and R. B. Kaner, "laser scribing of high-performance and flexible graphene-based electrochemical capacitors," *Science*, vol. 335, no. 6074, pp. 1326–1330, Mar. 2012.

[77] J. Cai, C. Lv, and A. Watanabe, "Laser direct writing of high-performance flexible all-solid-state carbon micro-supercapacitors for an on-chip self-powered photodetection system," *Nano Energy*, vol. 30, pp. 790–800, 2016.

[78] W. He, R. Ma, and D. J. Kang, "High-performance, flexible planar microsupercapacitors based on crosslinked polyaniline using laser printing lithography," *Carbon N. Y.*, vol. 161, pp. 117–122, 2020.

[79] M. Ren, J. Zhang, C. Zhang, M. G. Stanford, Y. Chyan, Y. Yao, and J. M. Tour, "Quasi-solid-state Li–O2 batteries with laser-induced graphene cathode catalysts," *ACS Appl. Energy Mater.*, vol. 3, no. 2, pp. 1702–1709, Feb. 2020.

[80] Z. Veliscek, L. S. Perse, R. Dominko, E. Kelder, and M. Gaberscek, "Preparation, characterisation and optimisation of lithium battery anodes consisting of silicon synthesised using laser assisted chemical vapour pyrolysis," *J. Power Sources*, vol. 273, pp. 380–388, 2015.

[81] D. Munaò, M. Valvo, J. van Erven, E. M. Kelder, J. Hassoun, and S. Panero, "Silicon-based nanocomposite for advanced thin film anodes in lithium-ion batteries," *J. Mater. Chem.*, vol. 22, no. 4, pp. 1556–1561, 2012.

[82] B. Zhang, M. Deschamps, M.-R. Ammar, E. Raymundo-Piñero, L. Hennet, D. Batuk, and J.-M. Tarascon, "Laser synthesis of hard carbon for anodes in na-ion battery," *Adv. Mater. Technol.*, vol. 2, no. 3, pp. 1600227, Mar. 2017.

[83] T. Han, A. Nag, N. Afsarimanesh, S. C. Mukhopadhyay, S. Kundu, and Y. Xu, "Laser-Assisted printed flexible sensors: A review," *Sensors*, vol. 19, no. 6. 2019.

[84] X. Wang, Z. Liu, and T. Zhang, "Flexible sensing electronics for wearable/attachable health monitoring," *Small*, vol. 13, no. 25, pp. 1602790, Jul. 2017.

[85] L.-Q. Tao, H. Tian, Y. Liu, Z.-Y. Ju, Y. Pang, Y.-Q. Chen, D.-Y. Wang, X.-G. Tian, J.-C. Yan, N.-Q. Deng, Y. Yang, and T.-L. Ren, "An intelligent artificial throat with sound-sensing ability based on laser induced graphene," *Nat. Commun.*, vol. 8, no. 1, pp. 14579, 2017.

[86] R. Rahimi, M. Ochoa, W. Yu, and B. Ziaie, "Highly stretchable and sensitive unidirectional strain sensor via laser carbonization," *ACS Appl. Mater. Interfaces*, vol. 7, no. 8, pp. 4463–4470, Mar. 2015.

[87] A. Nag, S. C. Mukhopadhyay, and J. Kosel, "Flexible carbon nanotube nanocomposite sensor for multiple physiological parameter monitoring," *Sensors Actuators A Phys.*, vol. 251, pp. 148–155, 2016.

[88] A. Nag, S. C. Mukhopadhyay, and J. Kosel, "Tactile sensing from laser-ablated metallized PET films," *IEEE Sens. J.*, vol. 17, no. 1, pp. 7–13, 2017.

[89] G. Dubourg and M. Radović, "Multifunctional screen-printed TiO2 nanoparticles tuned by laser irradiation for a flexible and scalable uv detector and room-temperature ethanol sensor," *ACS Appl. Mater. Interfaces*, vol. 11, no. 6, pp. 6257–6266, Feb. 2019.

[90] T.-L. Chang, C.-Y. Chou, C.-P. Wang, T.-C. Teng, and H.-C. Han, "Picosecond laser-direct fabrication of graphene-based electrodes for a gas sensor module with wireless circuits," *Microelectron. Eng.*, vol. 210, pp. 19–26, 2019.

[91] R. Park, H. Kim, S. Lone, S. Jeon, Y. W. Kwon, B. Shin, and S. W. Hong, "One-step laser patterned highly uniform reduced graphene oxide thin films for circuit-enabled tattoo and flexible humidity sensor application," *Sensors*, vol. 18, no. 6, pp. 19, 2018.

[92] Q. A. Drmosh, Z. H. Yamani, A. H. Hendi, M. A. Gondal, R. A. Moqbel, T. A. Saleh, and M. Y. Khan, "A novel approach to fabricating a ternary rGO/ZnO/Pt system for high-performance hydrogen sensor at low operating temperatures," *Appl. Surf. Sci.*, vol. 464, pp. 616–626, 2019.

[93] J. D. Blakemore, H. B. Gray, J. R. Winkler, and A. M. Müller, "Co3O4 nanoparticle water-oxidation catalysts made by pulsed-laser ablation in liquids," *ACS Catal.*, vol. 3, no. 11, pp. 2497–2500, Nov. 2013.

[94] T. Nishi, Y. Hayasaka, T. M. Suzuki, S. Sato, N. Isomura, N. Takahashi, S. Kosaka, T. Nakamura, S. Sato, and T. Morikawa, "Electrochemical water oxidation catalysed by CoO-Co2O3-Co(OH)2 multiphase-nanoparticles prepared by femtosecond laser ablation in water," *ChemistrySelect*, vol. 3, no. 17, pp. 4979–4984, May 2018.

[95] M. Cai, R. Pan, W. Liu, X. Luo, C. Chen, H. Zhang, and M. Zhong, "Laser-assisted doping and architecture engineering of Fe3O4 nanoparticles for highly enhanced oxygen evolution reaction," *ChemSusChem*, vol. 12, no. 15, pp. 3562–3570, Aug. 2019.

[96] J. Zhang, M. Ren, L. Wang, Y. Li, B. I. Yakobson, and J. M. Tour, "Oxidized laser-induced graphene for efficient oxygen electrocatalysis," *Adv. Mater.*, vol. 30, no. 21, pp. 1707319, May 2018.

[97] Z. Li, J.-Y. Fu, Y. Feng, C.-K. Dong, H. Liu, and X.-W. Du, "A silver catalyst activated by stacking faults for the hydrogen evolution reaction," *Nat. Catal.*, vol. 2, no. 12, pp. 1107–1114, 2019.

[98] C.-H. Chen, D. Wu, Z. Li, R. Zhang, C.-G. Kuai, X.-R. Zhao, C.-K. Dong, S.-Z. Qiao, H. Liu, and X.-W. Du, "Ruthenium-based single-atom alloy with high electrocatalytic activity for hydrogen evolution," *Adv. Energy Mater.*, vol. 9, no. 20, pp. 1803913, May 2019.

[99] T. Rauscher, C. I. Müller, A. Gabler, T. Gimpel, M. Köhring, B. Kieback, W. Schade, and L. Röntzsch, "Femtosecond-laser structuring of Ni electrodes for highly active hydrogen evolution," *Electrochim. Acta*, vol. 247, pp. 1130–1139, 2017.

[100] J. Zhang, C. Zhang, J. Sha, H. Fei, Y. Li, and J. M. Tour, "Efficient water-splitting electrodes based on laser-induced graphene," *ACS Appl. Mater. Interfaces*, vol. 9, no. 32, pp. 26840–26847, Aug. 2017.

[101] X. Wang, Z. Li, D.-Y. Wu, G.-R. Shen, C. Zou, Y. Feng, H. Liu, C.-K. Dong, and X.-W. Du, "Porous cobalt–nickel hydroxide nanosheets with active cobalt ions for overall water splitting," *Small*, vol. 15, no. 8, pp. 1804832, Feb. 2019.

[102] X. Ye, J. Long, Z. Lin, H. Zhang, H. Zhu, and M. Zhong, "Direct laser fabrication of large-area and patterned graphene at room temperature," *Carbon N. Y.*, vol. 68, pp. 784–790, 2014.

[103] E. Kostal, S. Stroj, S. Kasemann, V. Matylitsky, and M. Domke, "Fabrication of biomimetic fog-collecting superhydrophilic–superhydrophobic surface micropatterns using femtosecond lasers," *Langmuir*, vol. 34, no. 9, pp. 2933–2941, Mar. 2018.

[104] X. Bai, Q. Yang, Y. Fang, J. Zhang, J. Yong, X. Hou, and F. Chen, "Superhydrophobicity-memory surfaces prepared by a femtosecond laser," *Chem. Eng. J.*, vol. 383, pp. 123143, 2020.

[105] Y. Wang, Q. Zhang, Z. Zhu, F. Lin, J. Deng, G. Ku, S. Dong, S. Song, M. K. Alam, D. Liu, Z. Wang, and J. Bao, "Laser streaming: Turning a laser beam into a flow of liquid," *Sci. Adv.*, vol. 3, no. 9, pp. e1700555, Nov. 2022.

[106] S. Yue, F. Lin, Q. Zhang, N. Epie, S. Dong, X. Shan, D. Liu, W.-K. Chu, Z. Wang, and J. Bao, "Gold-implanted plasmonic quartz plate as a launch pad for laser-driven photoacoustic microfluidic pumps," *Proc. Natl. Acad. Sci.*, vol. 116, no. 14, pp. 6580–6585, Apr. 2019.

2 Phase Change Materials for Energy Harvesting and Management for Energy Production, Storage, and Conversion

Sajjad Moradpour and Fathollah Pourfayaz

2.1 INTRODUCTION

All human activities on the globe are based on energy. Due to the high energy consumption in various fields, there has been a considerable rise in the need for energy sources in recent years. Fossil fuels have been accessible to mankind as a cheap and accessible source of energy from the distant past. However, these fuels contribute significantly to pollution and do a lot of environmental harm, such as producing greenhouse gases [1]. In addition to environmental concerns, the price of fossil fuels has increased and has been trending upward in recent years.

The demand for sustainable energy sources and energy security has increased as a result of population growth and economic development of countries. Therefore, it is inevitable that efforts will be made to develop new energy production technologies or optimize the current energy systems in order to address energy-related environmental issues, resource scarcity, and high production prices. The development and designing of energy storage systems (ESS) is one of the most efficient ways to enhance the utilization of energy resources, as well as their protection and recovery. Energy storage systems are an optimal way to store energy when they are available and use this energy when needed [2, 3].

In this chapter, the thermal type of energy storage (TES) systems with the help of phase change materials (PCMs) has been investigated and focused on. Also, in the rest of this chapter, various types of phase change materials, their thermal improvement methods, governing equations and theories and their applications have been introduced.

DOI: 10.1201/9781003318859-2

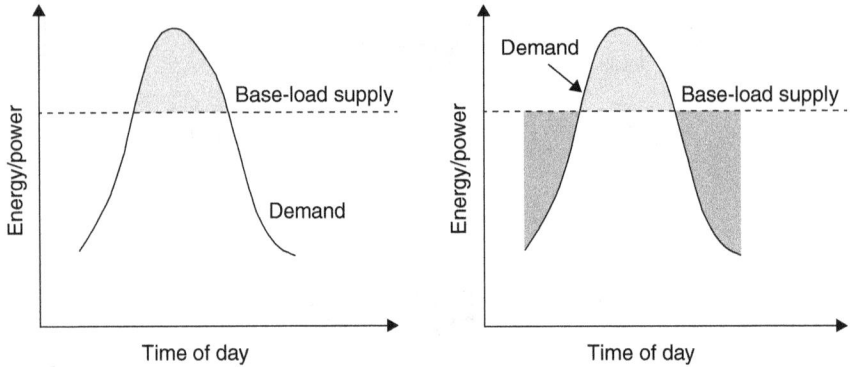

FIGURE 2.1   ESS performance. [Reprinted with permission from Springer Nature].
Source: [4].

2.1.1 ENERGY STORAGE SYSTEMS

Energy storage systems not only reduce the mismatch and help to close the gap between energy supply and demand (in fact, they make the load factor[1] closer to one) but also increase the reliability of energy systems and play a significant role in saving energy and reducing costs. According to Figure 2.1, which shows a conventional energy demand diagram, at the time of low demand and increased energy production (green part), the storage system is charged, and at the time of high demand and lack of energy production (red part), the storage system is discharged.

2.2   ENERGY STORAGE SYSTEMS DESIGN CRITERIA

To design a suitable energy storage system, there are three main criteria of technical specifications, economic efficiency and environmental effects that should be taken into consideration [5]. Thermal energy storage should be done with high capacity to reduce system volume and increase efficiency. Also, there must be a suitable heat transfer rate between the heat transfer fluid (HTF) and the storage material so that the energy storage and release process occurs at a suitable speed and in the shortest possible time. The energy storage material must have good mechanical and chemical stability so that its thermal and thermos-physical properties do not decline after several thermal cycles.

   Economic efficiency determines the return period of the investment cost. The price of a thermal energy storage system is mostly determined by three components: the storage materials, the heat exchanger, and the land used to install the system, which largely depends on the technical specifications mentioned above.

2.3   TYPES OF ENERGY STORAGE TECHNOLOGIES

Energy storage is done in different ways according to the required form of energy and its application and technical feasibility. Electro-chemical, mechanical and thermal

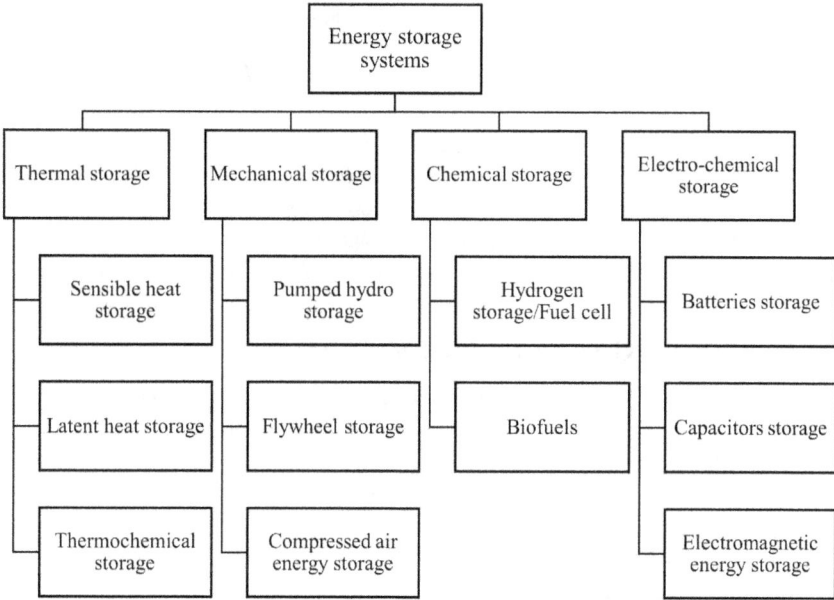

FIGURE 2.2   Energy storage mechanism.

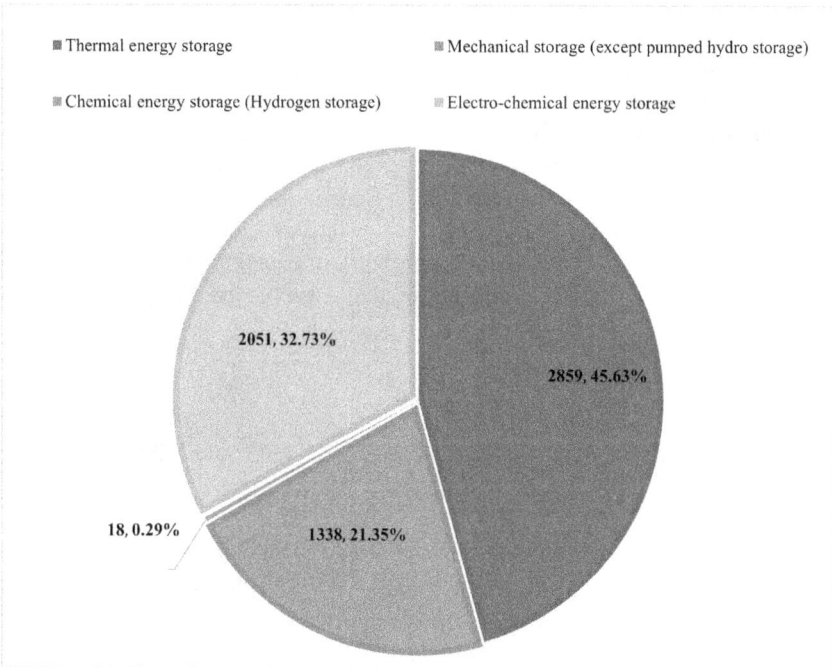

FIGURE 2.3   Distribution of Grid-connected installed capacities of all other energy storage mechanism (MW).

storage can be mentioned among the most important storage methods [6]. A type of classification of these methods is shown in Figure 2.2.

About 96% of the capacity of energy storage systems installed worldwide is dedicated to pumped hydro mechanical storage (169.557 GW) [6]. The distribution of other storage systems in the world is according to Figure 2.3, of which about 45% is the share of thermal systems (latent and sensible) and about one-third is the share of electro-chemical systems (batteries), of which the growth of its use has been fast especially after the introduction of lithium-ion batteries instead of lead-acid batteries [4, 6]. The focus of this chapter is on the thermal (latent) type of storage systems.

2.4   THERMAL ENERGY STORAGE SYSTEMS

In this section, thermal energy storage systems are introduced, which have been the focus of many researchers in the last two decades, and the number of articles published in this field has increased year by year. Figure 2.4 shows the number of articles published in the WOS[2] database on thermal energy storage systems, which has grown significantly since 2006.

2.5   TYPES OF THERMAL ENERGY STORAGE (TES) SYSTEMS

In general, thermal energy storage methods are classified into two categories: physical and chemical processes. The physical processes themselves include sensible heat and latent heat storage methods. In sensible energy storage, thermal energy is stored in a solid or liquid object with an increase in temperature, and subsequently, with a decrease in temperature, the stored energy is released, the amount of which is a function of temperature, specific heat capacity, and the amount of the object [7]. Latent heat storage is based on the phase change of the energy storage material [8]; That is, for example, by converting from a solid phase to a liquid phase, thermal energy is stored in the material, and then by converting from a liquid phase to a solid phase, energy is released. In the chemical reaction method, certain materials are subjected to a chemical reaction and according to the endothermic and exothermic nature of the reaction, thermal energy is stored and discharged. Figure 2.5 shows the diagram of what has been said.

2.5.1   SENSIBLE ENERGY STORAGE SYSTEMS

One of the common methods for thermal energy storage is sensible heat storage. According to Figure 2.6, the storage of sensible heat increases the temperature of the substance.

The amount of stored heat is directly related to the amount of storage material, specific heat capacity and temperature change and is calculated using the following equation [7]:

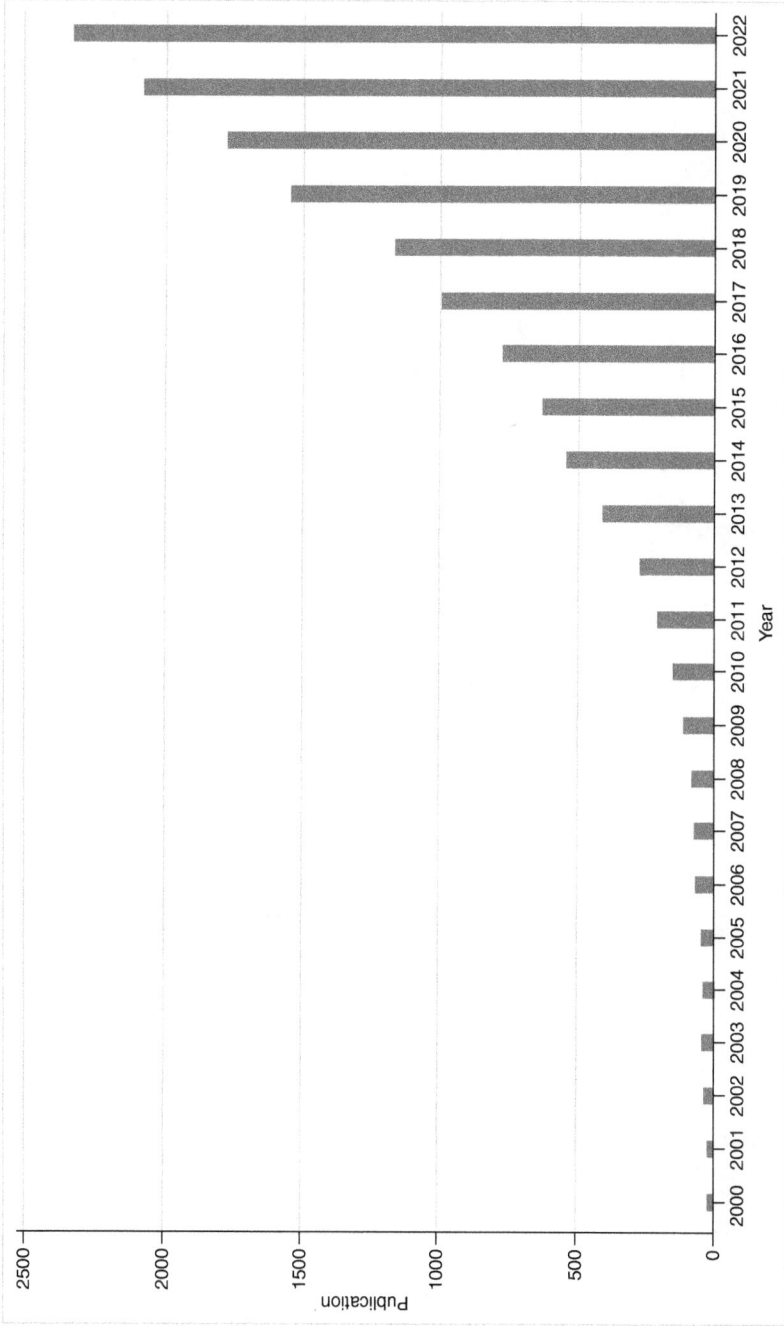

FIGURE 2.4 Publication of "thermal energy storage" in WOS.

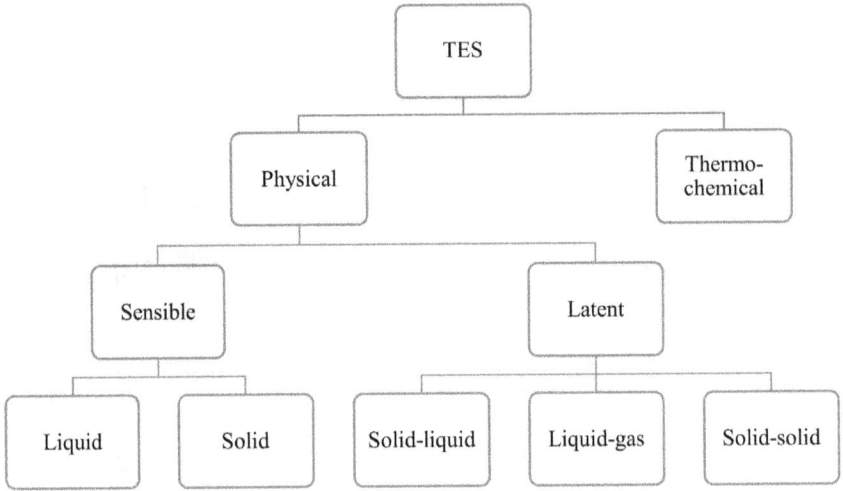

FIGURE 2.5   Types of TES systems.

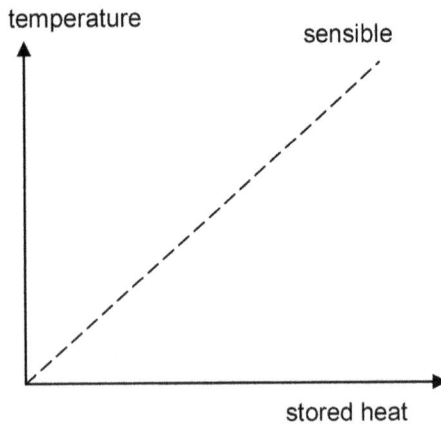

FIGURE 2.6   Sensible energy storage diagram. [Reprinted with permission from Springer Nature].

Source: [9].

$$Q = \int_{T_i}^{T_f} mC_p \, dT = mC_p \left(T_f - T_i \right) \tag{2.1}$$

where Q (kJ) is the stored heat, m (kg) is the mass of the material, C_p (kJ/kg°C) is the specific heat capacity, T_i (°C) is the initial temperature and T_f (°C) is the final temperature.

TABLE 2.1
Materials Used for Sensible Energy Storage

Material	Working temperature (C)		Density (kg/m³)	Thermal conductivity (W/(m K))	Specific heat (kJ/(kg C))
Sand-rock minerals	200–300	Solid	1700	1.0	1.30
Reinforced concrete	200–400		2200	1.5	0.85
Cast iron	200–400		7200	37.0	0.56
Silica fire bricks	200–700		1820	1.5	1.00
HitecXL solar salt	120–500	Liquid	1992	0.52	
Mineral oil	200–300		770	0.12	
Synthetic oil	250–350		900	0.11	

Source: [5].

Examples of energy storage materials in the form of sensible heat are shown in Table 2.1. In general, the advantages of sensible heat include high thermal conductivity and low price of storage materials. Its biggest disadvantage is the low thermal capacity of the material, which causes the volume of the storage unit to become too large. Water is one of the best storage materials, because firstly, it is very cheap and secondly, it has a high heat capacity. Gases are not used due to their very low heat capacity.

2.5.2 LATENT ENERGY STORAGE SYSTEMS

In latent heat storage, the phase of energy storage material must be changed. The solid–liquid phase change by the process of melting and freezing can store and release a large amount of thermal energy. During melting, the heat is transferred to the storage material and the temperature of the material remains constant, which is called the phase change temperature. In Figure 2.7 the process of stored energy in latent form is shown. First, the energy is stored sensibly and linearly with the change of temperature in the material. After reaching the phase change temperature, with the temperature remaining constant, the energy is latently stored in the material and finally, the energy is stored again in a sensible form [9].

In general, the sensible and latent stored thermal energy is obtained with the help of the following relationship [7]:

$$Q = \int_{T_i}^{T_m} mc_p \, dT + ma_m \Delta h_m + \int_{T_m}^{T_f} mc_p \, dT \tag{2.2}$$

where Q (kJ) is stored heat, m (kg), mass of material, C_p (kJ/kg°C), specific heat capacity, T_i (°C), initial temperature, T_m (°C), melting temperature, T_f (°C), final temperature, Δh_m (kJ/kg) is the enthalpy change per unit mass and a_m is the melting ratio.

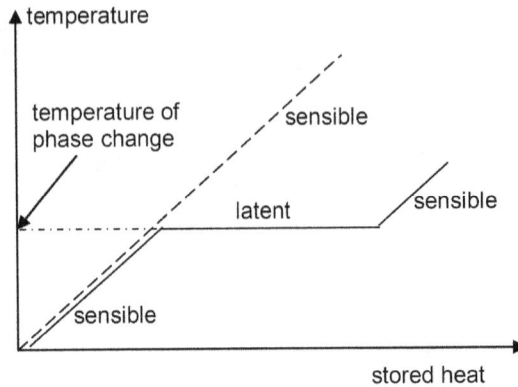

FIGURE 2.7 Latent energy storage diagram. [Reprinted with permission from Springer Nature].

Source: [9].

In solid–solid phase change, when a substance changes from one crystal to another, the heat is stored and released. The solid–solid phase change has a smaller latent heat and a smaller volume change than the solid–liquid phase change. Solid–vapour and liquid–vapour phase changes have higher latent heat, but their volume changes are also greater, which causes problems for the energy storage tank. The vapour–liquid phase change done by the evaporation and condensation process is highly dependent on the boundary conditions [9]:

- In closed systems with constant volume, evaporation causes an increase in vapour pressure, and as a result of an increase in vapour pressure, the phase change temperature also increases, so it is not useful for thermal energy storage.
- In closed systems with constant pressure, the evaporation increases the volume, which is also not suitable for energy storage.
- Open systems in ambient pressure are a third option. Here, the temperature of the phase change does not change significantly, but when the substance evaporates due to the open system, the substance enters the environment and we will have a loss.

According to the above-mentioned cases, the solid–liquid phase change is very suitable and practical in latent heat storage, and materials used to store latent heat are known as phase change materials (PCM).

One of the advantages of latent heat compared to sensible heat is the high phase change enthalpy (100 to 200 times larger). Also, the energy density is higher in the latent heat method, which makes the volume of the system smaller. However, the higher price of energy storage materials and the lower thermal conductivity are the disadvantages of latent heat compared to sensible heat. In the following sections, different methods for improving the thermal conductivity of phase change materials are discussed.

TABLE 2.2
Materials Used for Thermo-chemical Storage

Material	Temperature range (C)	Enthalpy change during reaction	Chemical reaction
Iron carbonate [10]	180	2.6 GJ/m³	$FeCO_3 \leftrightarrow FeO + CO_2$
Hydroxides [11]	500	3 GJ/m³	$Ca(OH)_2 \leftrightarrow CaO + H_2O$
Calcium carbonate [11]	800–900	4.4 GJ/m³	$CaCO_3 \leftrightarrow CaO + CO_2$

Source: [5].

2.5.3 THERMO-CHEMICAL ENERGY STORAGE SYSTEMS

When a chemical reaction occurs, the enthalpy of the substance at the beginning of the process is different from the enthalpy of the substance at the end of the process, and this enthalpy change is known as the enthalpy of reaction [9]. If the reaction is endothermic, it absorbs heat and if the reaction is exothermic, it releases heat [9]. Any chemical process with high reaction enthalpy can be used for an energy storage system under two conditions: First, the reaction products should be storable, and second, the process should be reversible so that the absorbed heat is released.

The amount of heat stored in a chemical reaction can be obtained with the help of the following equation [9]:

$$\Delta Q = \Delta H = m.\Delta h \tag{2.3}$$

Where; the enthalpy change per unit mass is Δh (kJ/kg), the mass of the material is m (kg), and the amount of stored heat is ΔQ (kJ).

When an energy storage system is designed with the help of a chemical reaction, three basic criteria should be taken into consideration, which are chemical reversibility, appropriate chemical enthalpy change, and simple chemical reaction conditions [5]. In this method, the energy density is much higher compared to the latent heat method, however, the chemical reaction method has not been widely considered so far. Among the basic problems in this method, we can mention the need for complex reactors for some chemical reactions, poor chemical reversibility and poor chemical stability.

2.6 WHY SOLID–LIQUID LATENT HEAT (RATHER THAN SENSIBLE, THERMO-CHEMICAL OR OTHER PHASE CHANGE)?

In the previous section, it was explained that sensible heat storage has good thermal conductivity and is more cost-effective than other thermal energy storage mechanisms, but due to lower heat capacity and lower energy density, the volume of the system increases significantly. The chemical reaction method has a higher energy density compared to other methods but there are problems such as poor reversibility, low stability, and complexity of the reaction.

Latent heat storage is one of the most optimal methods for thermal energy storage. This method works on the basis of phase change, that is, during energy storage or discharge, the temperature is almost constant, which is very desirable and makes it suitable for constant temperature applications. Also, due to the high energy density in latent heat storage, the volume of the system becomes smaller. In latent heat storage, the solid–liquid phase change during melting and freezing can store or release a large amount of heat. This type of phase change has been widely studied, researched and used. Although the latent heat of this phase change is less than the latent heat of liquid-gas and gas-solid phase change, it does not have the problems of volumetric expansion and dependence on boundary conditions.

2.7   PHASE CHANGE MATERIALS (PCMS)

Phase change material refers to a material that stores latent heat. First, by increasing the temperature of the material, the heat is stored in a sensible form, and then when the phase change temperature is reached, the heat is stored in a latent form at a constant temperature. During the release of thermal energy, this process is done in reverse [12, 13]. Phase change material in latent thermal energy storage (LTES) systems is the most efficient method to store thermal energy. Phase change materials were first discovered in the early 1900s through the work of Yale University's Alan Tower Waterman [14].

2.8   PROPERTIES

When deciding if a particular PCM is suitable and desirable for LTES, we must consider its thermal, kinetic, and chemical properties [14].

PCMs used in thermal energy storage systems must have the following properties [8]:

Thermal properties:

- Appropriate phase change temperature according to system working temperature
- Suitable phase change enthalpy to reduce system volume
- Appropriate heat transfer to reduce energy storage (charge) and release (discharge) time. (The thermal conductivity coefficient of most phase change materials is low ranging from 0.1 to 0.6 W/mK [15])

Physical properties:

- Proper phase equilibrium
- High density in order to reduce the volume of the chamber containing PCM
- Small volume change during phase change
- Low vapour pressure

Kinetic properties:

- Low Subcooling (also called supercooling) temperature (conditions where the temperature must be a few degrees lower than the freezing temperature for the PCM to start freezing)
- Appropriate crystallization rate

TABLE 2.3
PCM Selection Criteria

Criteria	Items to be considered
Thermal criteria	Suitable phase change temperature
	High enthalpy of fusion per volume
	High thermal conductivity of melted/solid PCM
	Preferably suitable specific heat capacity for sensible heat transfer
Chemical criteria	Being non-toxic
	non-corrosive
	High stability and lower rate of cycle degradation
	Non-explosive
	Non-flammable
	No decomposition of composite materials
Physical criteria	High density to increase enthalpy per volume
	Low vapor pressure to reduce sealing problems
	Little volume change during phase change
Kinetic criteria	Low subcooling when discharging
Economic criteria	Inexpensive or economical
	Easy accessibility
	Recyclability

Source: [7, 8, 14].

Chemical properties:

- High chemical stability (for longer PCM operation)
- Compatibility with other materials used in the energy storage system
- Non-toxic and non-flammable

Economic properties:

- To be abundant
- Being available
- Economic efficiency

2.9   TYPES OF PCM

There are many different phase change materials available in a wide range of melting/solidification and working temperatures. In a general classification, PCMs are classified into three general categories: organic materials, inorganic materials and eutectic compounds (see Figure 2.8) [8, 16].

Organic materials include paraffins and non-paraffins, which have the ability to be stored and released frequently without phase separation (which happens when PCM has several components) and reducing melting enthalpy. They have a low sub cooling temperature and are generally not corrosive. Melting enthalpy per unit volume of organic materials is almost half of that of inorganics [8] (see Figure 2.9).

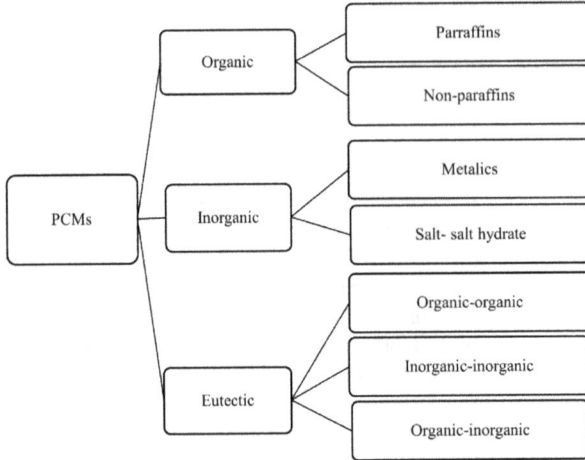

FIGURE 2.8   Types of PCMs.

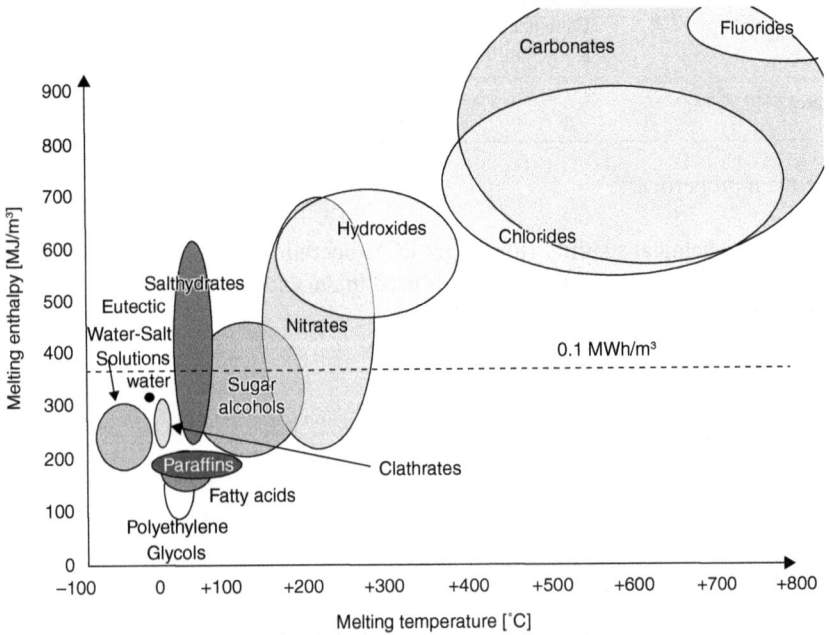

FIGURE 2.9   Distribution of PCMs based on their melting enthalpy and melting temperature.
[Reprinted with permission from Springer Nature].

Source: [9].

TABLE 2.4
Examples of PCMs

Material	Phase change temps (C)	Density (kg/m³)	Thermal conductivity (W/(m K))	Latent heat (kJ/kg)
RT100 (paraffin)	100	880	0.20	124
A164 (organic)	164	1500	-	306
KNO₃	333	2110	0.5	226
Na₂CO₃	854	2533	2	275.7
48%CaCO3–45%KNO3–7% NaNO₃	130	-	-	-
MgCl2–KCl–NaCl	380	2044	0.5	149.7

Source: [5].

Inorganic materials include metals, hydrated salts and salts (above 150°C) [7, 9] and cover a wide range of working temperatures. They do not have the problem of decreasing the saturation temperature over time, but there are problems of segregation and supercoiling [17]. In eutectic compounds, their compounds are used in order to reach the desired melting temperatures and correct the problems of organic and inorganic materials.

Examples of all three types of PCMs are given in Table 2.4. More examples for each type individually can be found in [7].

2.10 PHASE CHANGE THEORY

A melting–freezing process for a pure ideal substance is shown in Figure 2.10 at constant atmospheric pressure in a phase diagram. During the process (freezing/solidification), the temperature and heat flux change as Figure 2.11. But for a mixture (here a binary mixture) the phase diagram is as Figure 2.12 which there is a region of the two-phase field (the liquid and solid solutions).

2.11 NUMERICAL STUDIES

Looking into the literature, to solve conservation equations governing PCM, including continuity equation, conservation of momentum and conservation of energy you can find two methods – the effective heat capacity method and the enthalpy method (as well as enthalpy-porosity in Fluent software) – that allow us to consider phase changes when dealing with the energy equation. These two methods have been widely studied in the context of solid–liquid interface boundary conditions (Stefan or moving boundary problem) [17–24].

Due to the fact that for the modeling of phase change materials in the molten state, the free convection flow should also be considered (especially in large chambers), the enthalpy-porosity method should be used, and the existing relationships for modeling are as follows:

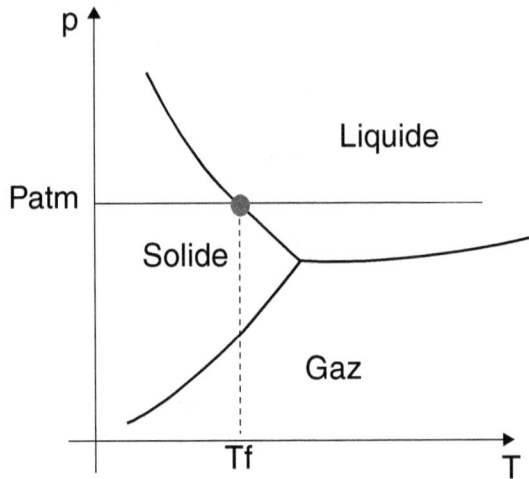

FIGURE 2.10   Melting/freezing process shown at phase diagram. [Reprinted with permission from Elsevier].

Source: [18].

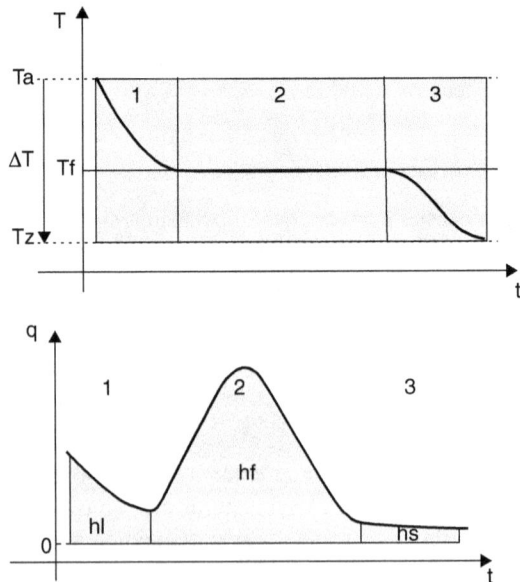

FIGURE 2.11   Temperature and heat flux of a pure material when freezing. [Reprinted with permission from Elsevier].

Source: [18].

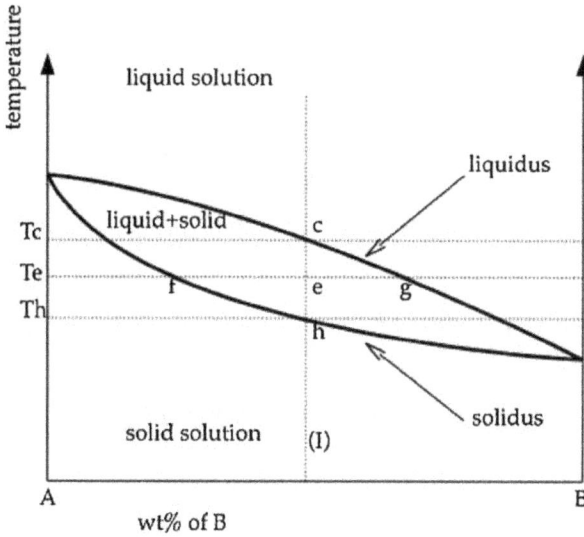

FIGURE 2.12 Binary phase diagram. [Reprinted with permission from Elsevier].
Source: [18].

- Conservation of mass equation [25]:

$$\frac{\partial \rho}{\partial t} + \nabla.(\rho \mathbf{v}) = S_m \tag{2.4}$$

Where S_m is the source (or sink) term. Here this term is zero due to no mass production inside the control volume.

- Equation of conservation of momentum:

To obtain this equation, Boussinesq approximation and the technique of considering the volume in the process of freezing or melting as porous have been used. The sentence related to the source term is obtained from the following formula [26]:

$$S = \frac{(1-\beta)^2}{(\beta^3 + \varepsilon)} A_{mush} (\mathbf{V} - \mathbf{V_p}) \tag{2.5}$$

In this equation, β is the volume fraction of liquid, which is calculated from equation (2.6) [26]:

$$
\begin{aligned}
&\beta = 0 \rightarrow if \rightarrow T < T_{solidus} \\
&\beta = 1 \rightarrow if \rightarrow T > T_{liquidus} \\
&\beta = \frac{T - T_{solidus}}{T_{liquidus} - T_{solidus}} \rightarrow if \rightarrow T_{solidus} < T < T_{liquidus}
\end{aligned}
\tag{2.6}
$$

ε is also a small number like 0.001 to avoid division by zero. V_p is the velocity of the frozen liquid. A_{mush} is the constant of the sink term and the larger its value, the higher the speed damping will be.

In Boussinesq approximation, the term related to density in the continuity equation and also in the first term on the left side of the momentum equation is considered constant, but in the direction of gravity, its changes are a function of temperature. As a result, the momentum equation for the melting-freezing process will be in the form of equation (2.7) [26, 27]:

$$\rho\frac{DV}{Dt} = -\nabla P + \nabla.\left(\bar{\tau}\right) - \rho g B\left(T - T_{ef}\right) + S \tag{2.7}$$

In the formula above, B is the coefficient of thermal expansion.

- Energy conservation equation [26]:

$$\frac{\partial}{\partial t}(\rho H) + \nabla.(\rho V H) = \nabla.(K \nabla T) + S \tag{2.8}$$

In the above equation, H is the enthalpy of the substance, which is equal to the sum of the sensible enthalpy and the latent heat of fusion [26]:

$$H = h + \Delta H$$
$$h = h_{ref} + \int_{T_{ref}}^{T} c_p dT \tag{2.9}$$
$$\Delta H = \beta L$$

In this equation, β is the liquid volume fraction and L is the latent heat of melting (freezing).

2.12 PCM INCORPORATION

Different techniques for using and integrating PCMs, especially in buildings, are as follows [17]:

- Direct incorporation
- Impregnation
- Encapsulation
- Shape-stabilized PCMs

2.13 HEAT TRANSFER ENHANCEMENT TECHNIQUES

To overcome the low thermal conductivity of PCMs and enhance heat transfer between PCMs and HTF there are different techniques, like using extended surfaces (fins); thermal conductivity enhancement; using multi-PCMs (cascade); encapsulation; and embedding of heat pipe [15].

2.13.1 FINS

Fins are an important tool in providing extra heat transfer surfaces. The contribution of alternative fin configurations in improving latent thermal energy storage systems' performance has been thoroughly investigated by numerous researchers.

2.13.2 THERMAL CONDUCTIVITY ENHANCEMENT

To increase thermal conductivity, the use of materials with high thermal conductivity has been tested. Placing these materials in the storage system increases thermal conductivity:

The effective thermal conductivity can be increased when high-conductivity porous material (either naturally occurring porous materials like graphite or metal matrices composed of aluminium, copper, etc.) is impregnated with PCM. Also placing metal structures or carbon fibers into PCMs can be helpful for increasing the effective thermal conductivity [15].

2.13.3 MICROENCAPSULATION OF PCM

It is also suggested that using microencapsulated PCMs will help increase the rate of heat transfer between PCM and its source/sink [15].

2.13.4 CASCADE LATENT HEAT STORAGE

The combination of several phase change materials with different melting temperatures is known as cascade storage (see Figure 2.13).

One of the problems with storage with a phase change material is that when storing thermal energy, the fluid passing through the system (HTF) quickly transfers heat to the phase change material; therefore, the temperature of the HTF decreases and as a

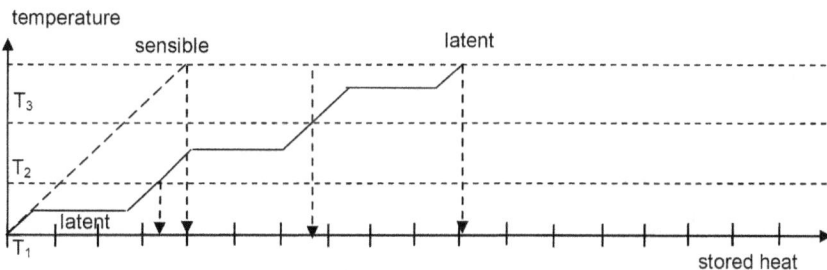

FIGURE 2.13   Cascade latent storage. [Reprinted with permission from Springer Nature].

Source: [9].

result, the temperature difference between the phase change material and the HTF decreases, which causes a weaker heat transfer to occur at the end of the storage process. In other words, at the beginning of the entry of the HTF into the system, the phase change material melts faster, but at the end of the storage, when the HTF is leaving, the phase change material melts more slowly. By using cascade storage, this problem can be solved. In this way, at the end of the storage system, which melts occur more slowly, another phase change material with a different melting temperature is replaced in order to maintain the temperature difference between the HTF and the PCM and thus the melting occurs at higher speed.

2.13.5 HEAT PIPE

Another way to overcome the low thermal conductivity of PCMs is to use heat pipes. With heat pipes, the heat transfer mechanism uses the liquid–gas phase change process. They are devices with a very high thermal conductivity coefficient [28, 29] and can transfer heat from HTF to PCMs and vice versa at a high rate by properly placing them in phase change materials [30, 31].

2.14 APPLICATION

PCMs are now being used in many industries such as construction, textiles, automotive, and solar energy installation. Plus, with the recent advancements in technology, PCMs are being utilized in electronics and medical fields too [7].

2.15 PHASE CHANGE MATERIALS FOR BUILDING

With the rising costs of fossil fuels and for the sake of our environment, using TES for heating and cooling buildings is becoming more and more essential. PCMs in passive[3] or active form can reduce the cost of energy consumption as well as temperature fluctuations during the day and night and move the peak consumption to an off-peak time [13, 32].

2.16 PHASE CHANGE MATERIALS FOR HEATING OR COOLING WATER

It was shown in the literature that using PCM can be helpful for water heating. For example [33] demonstrated that adding a PCM module to a water tank for household hot water delivery allowed for the provision of hot water for a longer period of time, even in the absence of an external energy source.

2.17 PHASE CHANGE MATERIALS FOR SOLAR ENERGY

In thermosolar energy systems, where heat must be stored throughout the day for use at night, significant attempts have been made in recent years to utilize LHS materials [34, 35].

2.18 PHASE CHANGE MATERIALS FOR THE TEXTILE INDUSTRY

NASA created the technology to include PCM microcapsules in fibres in the early 1980s to enhance their functionality for controlling the thermal barrier qualities of clothing textiles, particularly space suits with the goal of making it easier for astronauts to handle drastic temperature changes when on space missions [36, 37]. Effort for adding PCMs to fabrics to control temperature condition has been developing.

2.19 PHASE CHANGE MATERIALS FOR BIOMEDICAL

For several biomedical applications that require thermal protection, such as specialized bandages or burn wound dressings, PCMs are now regarded as promising materials [38, 39].

2.20 PHASE CHANGE MATERIALS FOR ELECTRONICS

Designing electronic devices that aren't used continuously for extended periods of time has become more complex, so a PCM-based cooling system could be a great solution for thermal management [40, 41].

2.21 PHASE CHANGE MATERIALS FOR THE AUTOMOTIVE INDUSTRY

In the automotive sector, PCMs are utilized for internal combustion engines, engine cooling, improving passenger thermal comfort, and preheating catalytic converters [42, 43].

2.22 PHASE CHANGE MATERIALS FOR SPACE APPLICATIONS

In space applications, PCMs are very important in order to store energy for the required time [44].

2.23 PHASE CHANGE MATERIALS FOR THE FOOD INDUSTRY

PCMs are also used in the food industry. Applications like increasing of drying rate or uniformity of food temperature are given in the references [45, 46].

2.24 CONCLUSION

Phase change materials (PCMs) are substances that exhibit the unique ability to absorb and release large amounts of thermal energy during the process of charging and discharging. Due to their potential for energy storage and thermal management applications, PCMs have gained considerable attention in recent years. When compared to traditional thermal storage systems, PCMs offer several advantages such as high enthalpy per volume, smaller volume required and the ability to maintain a constant temperature during the phase change process. As a result, these materials have found wide-ranging applications in different industries including building construction, transportation, electronics cooling, and renewable energy systems. Throughout this chapter, we have discussed briefly the most important aspects related to PCMs so that the readers can gain a basic knowledge about them.

NOTES

1 Load factor is defined as the average load divided by the peak load in a specified time period.
2 www.webofscience.com
3 Passive heating/cooling techniques use the inherent qualities of the building and its surroundings to maintain a pleasant indoor temperature without the use of mechanical heating/cooling systems, such as fans.

REFERENCES

[1] P. Breeze, "Chapter 2 – The carbon cycle and atmospheric warming," in *Electricity Generation and the Environment*, P. Breeze Ed.: Academic Press, 2017, pp. 13–21.

[2] M. A. Hamdan and F. A. Elwerr, "Thermal energy storage using a phase change material," *Solar Energy*, vol. 56, no. 2, pp. 183–189, 1996/02/01/ 1996. doi: https://doi.org/10.1016/0038-092X(95)00090-E

[3] B. Zalba, J. M. Marín, L. F. Cabeza, and H. Mehling, "Free-cooling of buildings with phase change materials," *International Journal of Refrigeration*, vol. 27, no. 8, pp. 839–849, 2004/12/01/ 2004. doi: https://doi.org/10.1016/j.ijrefrig.2004.03.015

[4] A. H. Alami, *Mechanical Energy Storage for Renewable and Sustainable Energy Resources*. Springer International Publishing, 2020.

[5] Y. Tian and C. Y. Zhao, "A review of solar collectors and thermal energy storage in solar thermal applications," *Applied Energy*, vol. 104, pp. 538–553, 2013/04/01/ 2013. doi: https://doi.org/10.1016/j.apenergy.2012.11.051

[6] N. Khan, S. Dilshad, R. Khalid, A. R. Kalair, and N. Abas, "Review of energy storage and transportation of energy," *Energy Storage*, vol. 1, no. 3, pp. e49, 2019/06/01 2019. https://doi.org/10.1002/est2.49 doi: https://doi.org/10.1002/est2.49.

[7] K. Pielichowska and K. Pielichowski, "Phase change materials for thermal energy storage," *Progress in Materials Science*, vol. 65, pp. 67–123, 2014/08/01/ 2014. doi: https://doi.org/10.1016/j.pmatsci.2014.03.005

[8] A. Sharma, V. V. Tyagi, C. R. Chen, and D. Buddhi, "Review on thermal energy storage with phase change materials and applications," *Renewable and Sustainable Energy Reviews*, vol. 13, no. 2, pp. 318–345, 2009/02/01/ 2009. doi: https://doi.org/10.1016/j.rser.2007.10.005

[9] H. Mehling and L. F. Cabeza, *Heat and Cold Storage with PCM: An up to Date Introduction into Basics and Applications*. Springer Berlin Heidelberg, 2008.

[10] S. Kalaiselvam and R. Parameshwaran, "Chapter 6 - Thermochemical Energy Storage," in Thermal Energy Storage Technologies for Sustainability. Boston: Academic Press, 2014, pp. 127–144. https://doi.org/10.1016/B978-0-12-417291-3.00006-

[11] E. Hahne, "Thermal energy storage some views on some problems," in *International Heat Transfer Conference Digital Library*. San Francisco, USA, August 1986: Begel House Inc, 1986. pp. 279–292, doi: 10.1615/IHTC8.2490.

[12] B. Abdoos, M. Ghazvini, F. Pourfayaz, M. H. Ahmadi, and A. Nouralishahi, "A comprehensive review of nano-phase change materials with a focus on the effects of influential factors," *Environmental Progress & Sustainable Energy*, vol. 41, no. 2, pp. e13808, 2022/03/01 2022. doi: https://doi.org/10.1002/ep.13808

[13] A. Kasaeian, L. bahrami, F. Pourfayaz, E. Khodabandeh, and W.-M. Yan, "Experimental studies on the applications of PCMs and nano-PCMs in buildings: A critical review," *Energy and Buildings*, vol. 154, pp. 96–112, 2017/11/01/ 2017. doi: https://doi.org/10.1016/j.enbuild.2017.08.037

[14] M. Mhadhbi, *Phase Change Materials and Their Applications*. IntechOpen, 2018.

[15] S. Jegadheeswaran and S. D. Pohekar, "Performance enhancement in latent heat thermal storage system: A review," *Renewable and Sustainable Energy Reviews*, vol. 13, no. 9, pp. 2225–2244, 2009/12/01/ 2009. doi: https://doi.org/10.1016/j.rser.2009.06.024

[16] F. P. M. Taheri, R. Habibi, A. Maleki, "Exergy analysis of charge and discharge processes of thermal energy storage system with various phase change materials: A comprehensive comparison," *Accepted in Journal of Thermal Science*, in press. Doi: https://doi.org/10.1007/s11630-023-1859-y

[17] B. Lamrani, K. Johannes, and F. Kuznik, "Phase change materials integrated into building walls: An updated review," *Renewable and Sustainable Energy Reviews,* vol. 140, pp. 110751, 2021/04/01/ 2021. doi: https://doi.org/10.1016/j.rser.2021.110751

[18] F. Kuznik, D. David, K. Johannes, and J.-J. Roux, "A review on phase change materials integrated in building walls," *Renewable and Sustainable Energy Reviews*, vol. 15, no. 1, pp. 379–391, 2011/01/01/ 2011. doi: https://doi.org/10.1016/j.rser.2010.08.019

[19] S. N. Al-Saadi and Z. Zhai, "Modeling phase change materials embedded in building enclosure: A review," *Renewable and Sustainable Energy Reviews,* vol. 21, pp. 659–673, 2013/05/01/ 2013. doi: https://doi.org/10.1016/j.rser.2013.01.024

[20] A. W. Date, "A strong enthalpy formulation for the Stefan problem," *International Journal of Heat and Mass Transfer*, vol. 34, no. 9, pp. 2231–2235, 1991/09/01/ 1991. doi: https://doi.org/10.1016/0017-9310(91)90049-K

[21] L. E. Goodrich, "Efficient numerical technique for one-dimensional thermal problems with phase change," *International Journal of Heat and Mass Transfer*, vol. 21, no. 5, pp. 615–621, 1978/05/01/ 1978. doi: https://doi.org/10.1016/0017-9310(78)90058-3

[22] F. Kuznik *et al.*, "Impact of the enthalpy function on the simulation of a building with phase change material wall," *Energy and Buildings*, vol. 126, pp. 220–229, 2016/08/15/ 2016. doi: https://doi.org/10.1016/j.enbuild.2016.05.046

[23] V. Voller and M. Cross, "Accurate solutions of moving boundary problems using the enthalpy method," *International Journal of Heat and Mass Transfer*, vol. 24, no. 3, pp. 545–556, 1981/03/01/ 1981. doi: https://doi.org/10.1016/0017-9310(81)90062-4

[24] M. Yao and A. Chait, "An alternative formulation of the apparent heat capacity method for phase-change problems," *Numerical Heat Transfer, Part B: Fundamentals*, vol. 24, no. 3, pp. 279–300, 1993/10/01 1993. doi: 10.1080/10407799308955894

[25] F. M. White, *Fluid Mechanics*. Mcgraw-Hill Kogakusha, 1979.

[26] S. Seddegh, X. Wang, and A. D. Henderson, "A comparative study of thermal behaviour of a horizontal and vertical shell-and-tube energy storage using phase change materials," *Applied Thermal Engineering*, vol. 93, pp. 348–358, 2016/01/25/ 2016, doi: https://doi.org/10.1016/j.applthermaleng.2015.09.

[27] L. M. Jiji, *Heat Convection*. Springer, 2006.

[28] A. Faghri, *Heat Pipe Science and Technology*. Global Digital Press, 2016.

[29] D. Reay, R. McGlen, and P. Kew, *Heat Pipes: Theory, Design and Applications*. Elsevier Science, 2013.

[30] C. W. Robak, T. L. Bergman, and A. Faghri, "Enhancement of latent heat energy storage using embedded heat pipes," *International Journal of Heat and Mass Transfer*, vol. 54, no. 15, pp. 3476–3484, 2011/07/01/ 2011. doi: https://doi.org/10.1016/j.ijheatmasstransfer.2011.03.038

[31] S. Tiari, S. Qiu, and M. Mahdavi, "Numerical study of finned heat pipe-assisted thermal energy storage system with high temperature phase change material," *Energy Conversion and Management*, vol. 89, pp. 833–842, 2015/01/01/ 2015. doi: https://doi.org/10.1016/j.enconman.2014.10.053

[32] S. Nekoonam and F. Pourfayaz, "An energy storage system utilization for performance enhancement of wind catcher cooling," (in en), *Iranian Journal of Chemistry and Chemical Engineering*, vol. 41, no. 6, pp. 2087–2099, 2022. doi: 10.30492/ijcce.2022.523765.4549

[33] C. Solé, M. Medrano, A. Castell, M. Nogués, H. Mehling, and L. F. Cabeza, "Energetic and exergetic analysis of a domestic water tank with phase change material," *International Journal of Energy Research*, vol. 32, no. 3, pp. 204–214, 2008/03/10 2008. https://doi.org/10.1002/er.1341 doi: https://doi.org/10.1002/er.1341

[34] Y. Maleki, F. Pourfayaz, and M. Mehrpooya, "Experimental study of a novel hybrid photovoltaic/thermal and thermoelectric generators system with dual phase change materials," *Renewable Energy*, vol. 201, pp. 202–215, 2022/12/01/ 2022. doi: https://doi.org/10.1016/j.renene.2022.11.037

[35] P. Pourmoghadam, M. Farighi, F. Pourfayaz, and A. Kasaeian, "Annual transient analysis of energetic, exergetic, and economic performances of solar cascade organic Rankine cycles integrated with PCM-based thermal energy storage systems," *Case Studies in Thermal Engineering*, vol. 28, pp. 101388, 2021/12/01/ 2021. doi: https://doi.org/10.1016/j.csite.2021.101388

[36] S. Mondal, "Phase change materials for smart textiles – An overview," *Applied Thermal Engineering*, vol. 28, no. 11, pp. 1536–1550, 2008/08/01/ 2008. doi: https://doi.org/10.1016/j.applthermaleng.2007.08.009

[37] G. Nelson, "Application of microencapsulation in textiles," *International Journal of Pharmaceutics*, vol. 242, no. 1, pp. 55–62, 2002/08/21/ 2002. doi: https://doi.org/10.1016/S0378-5173(02)00141-2

[38] R. De Santis, V. Ambrogi, C. Carfagna, L. Ambrosio, and L. Nicolais, "Effect of microencapsulated phase change materials on the thermo-mechanical properties of poly(methyl-methacrylate) based biomaterials," *Journal of Materials Science: Materials in Medicine*, vol. 17, no. 12, pp. 1219–1226, 2006/12/01 2006. doi: 10.1007/s10856-006-0595-7

[39] Y. Lv, Y. Zou, and L. Yang, "Feasibility study for thermal protection by microencapsulated phase change micro/nanoparticles during cryosurgery," *Chemical Engineering Science*, vol. 66, no. 17, pp. 3941–3953, 2011/09/01/ 2011. doi: https://doi.org/10.1016/j.ces.2011.05.031

[40] R. Akhilesh, A. Narasimhan, and C. Balaji, "Method to improve geometry for heat transfer enhancement in PCM composite heat sinks," *International Journal of Heat and Mass Transfer*, vol. 48, no. 13, pp. 2759–2770, 2005/06/01/ 2005. doi: https://doi.org/10.1016/j.ijheatmasstransfer.2005.01.032

[41] F. L. Tan and C. P. Tso, "Cooling of mobile electronic devices using phase change materials," *Applied Thermal Engineering*, vol. 24, no. 2, pp. 159–169, 2004/02/01/ 2004. doi: https://doi.org/10.1016/j.applthermaleng.2003.09.005

[42] K.-b. Kim, K.-w. Choi, Y.-j. Kim, K.-h. Lee, and K.-s. Lee, "Feasibility study on a novel cooling technique using a phase change material in an automotive engine," *Energy*, vol. 35, no. 1, pp. 478–484, 2010/01/01/ 2010. doi: https://doi.org/10.1016/j.energy.2009.10.015

[43] L. L. Vasiliev, V. S. Burak, A. G. Kulakov, D. A. Mishkinis, and P. V. Bohan, "Latent heat storage modules for preheating internal combustion engines: application to a bus

petrol engine," *Applied Thermal Engineering,* vol. 20, no. 10, pp. 913–923, 2000/07/01/ 2000. doi: https://doi.org/10.1016/S1359-4311(99)00061-7

[44] B. Yimer and M. Adami, "Parametric study of phase change thermal energy storage systems for space application," *Energy Conversion and Management,* vol. 38, no. 3, pp. 253–262, 1997/02/01/ 1997. doi: https://doi.org/10.1016/S0196-8904(96)00045-3

[45] S. Devahastin and S. Pitaksuriyarat, "Use of latent heat storage to conserve energy during drying and its effect on drying kinetics of a food product," *Applied Thermal Engineering,* vol. 26, no. 14, pp. 1705–1713, 2006/10/01/ 2006. doi: https://doi.org/10.1016/j.applthermaleng.2005.11.007

[46] Y. L. Lu, W. H. Zhang, P. Yuan, M. D. Xue, Z. G. Qu, and W. Q. Tao, "Experimental study of heat transfer intensification by using a novel combined shelf in food refrigerated display cabinets (Experimental study of a novel cabinets)," *Applied Thermal Engineering,* vol. 30, no. 2, pp. 85–91, 2010/02/01/ 2010. doi: https://doi.org/10.1016/j.applthermaleng.2008.10.003

3 Porous Carbon-based Materials for Energy Production, Storage, and Conversion

Sebastian Torres, Jesús Iniesta, and Alicia Gomis-Berenguer

3.1 INTRODUCTION

The ever-increasing demand for energy along with the high price of fossil fuels like natural gas or oil which are attributed to a supply reduction, have caused serious concerns about the fast depletion of world fossil fuel production in the near future. As far as the environmental implications, the weight of fossil fuels is still too high despite the incentives promoted by public and private organizations for the use of renewable energy with the aim of mitigating the environmental impact of using traditional energy resources. In this context, it is estimated (according to the World Energy Council) [1] that by 2050 the world will be required to double its energy supply.

Accordingly, even though fossil fuel energy consumption is still above 80%, it is important to develop new types of clean, environment-friendly and sustainable energy storage and conversion technologies. Some examples lead to the use of electrochemistry methodology through integrated energy systems coupled with nuclear or renewable energy. In this regard, electrochemical supercapacitors, batteries, fuel cells and water electrolysis reactors are the most representative. Additionally, the use of water and sunlight as sustainable energy generation and conversion (e.g. photoelectrochemical water splitting to release hydrogen and oxygen gases, or photodegradation of pollutants to reach the decontamination of residual waters) have been used for several sustainable applications having medium or high research technology level. The design and development of novel (electro)catalysts or electrode materials is vital for the process performance. Example of catalysts research for the degradation of environmental contaminants is focus on the improved harvesting features across the visible region; electrocatalysts research is centred on the oxygen reduction reaction (ORR) which is taking place at the cathode of fuel cell devices, oxygen and hydrogen evolution reaction (OER and HER, respectively) both taking place at the electrolysis of water, CO_2 reduction reaction, (CO2RR) for the conversion of high added value compounds and fuels, as well as the mitigation of greenhouse gas, among others [2,3].

DOI: 10.1201/9781003318859-3

New, challenging frontiers are being opened by the recent technology which offers new materials and methodologies for energy generation, storage and conversion devices. In particular, carbon-based materials (metal-free materials) like graphene, carbon nanosheets, carbon gels or activated carbons play a significant role in highly efficient devices, avoiding the use of metallic catalysts based on precious metals or metal oxides [2,3]. Specifically, porous carbon materials have attained considerable attention due to their availability, good electrical conductivity, high surface area and tuneable pore size distribution, structure, and surface chemistry [3]. They are prepared by various methods including conventional, named physical and chemical activations, and novel processes, such as nanocasting, self-activation, sol-gel approaches or hydrothermal synthesis as most representatives. Additionally, several strategies are considered to introduce heteroatoms (e.g. O-, N-, S-, B- or P-containing groups, or even the co-incorporation of; for example B and N, N and P, or B and P) in the carbonaceous network, including oxidation or functionalization by acidic/basic treatments or gasification [4,5]. This chapter concentrates on the use of porous carbon materials for energetic applications, by considering the influence of the BET (Brunauer-Emmer-Teller) surface area and its active surface area (ASA), the pore size and the balance of micro- and mesoporosity, the type of heteroatom and how functionalization influences hydrophobic and hydrophilic properties, without deepening, to the extent possible, on the synthetic procedures which are well reported in the literature, including this book [6].

3.2 ELECTROCHEMICAL ENERGY STORAGE

This chapter section discusses the role of nanoporous carbon materials in energy storage applications. It is not the goal to point at a much wider vision of the role of activated carbons, as reported recently in a review by Bandosz [7], not to go through comparative studies or provide extensive data in terms of capacitive charge, energy density and, finally, nor to accurately show the synthesis methodologies and characterization of the nanoporous carbons used in the field of energy storage. This chapter section focuses on how the pore dimensions and their distribution, the differences between the BET surface and the active surface area, the surface chemistry, and the presence of defects influence the performance of the carbonaceous electrodes in energy storage devices.

3.2.1 BATTERIES

Metal ion batteries (MIB), metal ion capacitors (MIC) or electrical double layer capacitors (EDCL) correspond nowadays to strategies which will meet the future needs for the storage of energy and its myriad of applications such as wearable sensors and biosensors, medical devices, environmental, food industry, robotics, and mainly electric vehicles [8].

The fundamentals and mechanisms of MIB and MIC have been extensively treated in [7,9,10]. The key component of MIB or MIC devices is the negative electrode which is made of carbon materials. At high voltage, MIB or MIC might suffer from the decomposition of the electrolyte, so thicker electrodes will provide higher energy density and power density, wherein charging discharging is a function of ions mobility, the distribution of the active electrode material and processes involving intercalation and de-intercalation of the electrolyte (Figure 3.1A). As far as the anode of MIB is concerned, a plethora of carbon materials have been used which includes graphite, carbon nanotubes or graphenes and its derivatives from the physical or chemical modification [3,11]. Even though porous carbons may go for a high reversible capacity associated with the intercalation of M at low potential and a reduction of the solid electrolyte interface development [11], the anode material for MIB applications is limited to nonporous carbon materials with low active surface area [12] instead of the commonly used activated carbons with high BET surface area, in order to reduce the irreversible capacity of the battery [13]. Moreover, the use of nanoporous carbon materials in MIB is an inappropriate strategy, not only because of its high BET surface area but also for exhibiting high ASA (even when those porous carbon materials are thermally treated under inert gas). However, conventional porous carbons or biomass-derived porous carbon materials are being widely used in devices like Li-S batteries as an alternative to the MIB (for a comprehensive review see reference [14]) or in metal redox flow batteries [15]. The latter review is more concentrated on the use of carbon or graphitic felts and on the use of C-C composites, but nothing is mentioned about the use of turbostratic nanoporous carbons. The above reviews, nonetheless, put on the table how crucial the textural and surface chemistry, the presence and type of heteroatoms, and the presence of defects and pores size distribution on the performance of batteries is in terms of capacity, energy density and power energy. Finally, not only has the research led to the 3D turbostratic porous carbon, but it has also emphasized the synthesis of 2D graphene porous nanosheets to be employed in storage applications like Li-, Na-, K-, Zn- or Li-S-ion batteries [16].

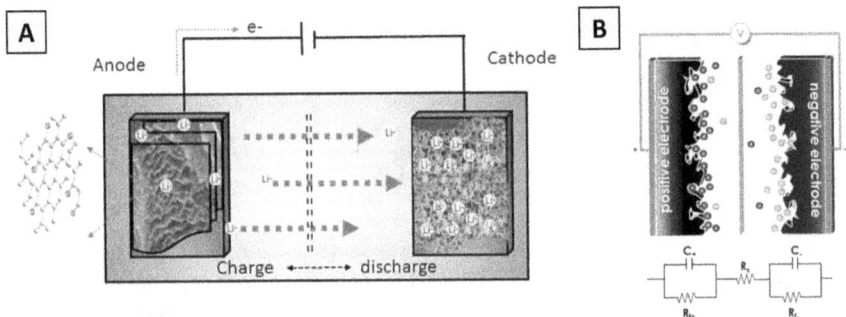

FIGURE 3.1 (A) Lithium-sulphur battery working principle. Reproduced from [7] with permission from Elsevier. Copyright 2017. (B) Scheme of charged state of a symmetric electrical double layer capacitor using porous carbon electrode and its simplified equivalent circuit. Reprinted from [17] with permission from Elsevier. Copyright 2019.

3.2.2 SUPERCAPACITORS

EDLCs are common devices for energy storage wherein the use of carbon materials plays the leading role (Figure 3.1B) [17]. The fundamentals and mechanisms of EDLCs have been extensively treated in [18–20]. The advantage of EDLCs is that they benefit from a high power density compared to MIB devices, whereas the former exhibits moderate specific energy density, the requirements of which high surface area carbon materials in EDLCs must accomplish, namely, the higher the surface area the higher the electrification of the double layer within the micro- and mesopores. Activated carbons represent one of the most commercial materials for this specific use. As well as the specific surface area, the control of porosity has to be balanced when the EDLC works at high voltage under non-aqueous solutions, as reported by Béguin et al. [21].

Irrespective of the synthesis methodology used, activated carbons with high BET surface area are obtained generally in the order of 1000 to 2000 m^2/g. Most EDLC devices point at a cell voltage of *ca.* 2.7 V in a non-aqueous solution and specific capacitance over 100 F/g [18]. Obviously, the cell voltage will depend on the stability of the electrolyte and how stability is linked to the presence of functional groups, e.g. oxygen or nitrogen functionalities but also on the presence of defects, both of them denoting the presence of active sites in the nanoporous carbon (*vide infra*). By looking at the three dimensionality of the carbon material, not only is a higher surface area a decisive factor influencing the performance of the EDLC, but also the existence of micro- and mesoporosity, and their connectivity, thereby favouring diffusivity of the electrolyte and speeding the charge transfer process [22]. Poorly connected microporosity will hinder the accessibility of the electrolyte. To overcome this crucial point larger porous (mesopores) have to be incorporated by keeping a high specific surface area, and therefore creating novel wide pathways. The above strategy has been reported by Fernandez et al. [23] where the authors used poly(vinyl alcohol) as a precursor for the synthesis of mesoporous carbons, though critically, a balance of micro- and mesoporosity must be taken into account when evaluating the volumetric specific capacitance, the energy density and power density. However, it is also worth highlighting the ordering or disordering of the network of the porous carbons. In this regard, the use of ordered mesoporous carbons has brought significant results in terms of specific capacitance [24]. Furthermore, the activation of mesoporous carbons can be performed by chemical or physical procedures in order to balance the meso-to-micro ratio, and porosity and therefore to increase the specific capacity of the so-called hierarchical nanoporous carbons [25].

A comprehensive study of the use of gels (aerogels, cryogels and xerogels) has been extensively compiled by Kraiwattanawong [26] where gels exhibit highly porous properties with very controlled pore size distribution leading to valuable nanoporous carbon materials as electrodes for EDLCs. The above work is again pointing out how electronic and ion conductivity, the tortuosity of the pore channel, micro-to-meso ratio, pore size distribution, and sort of electrolyte and solvent are crucial for the design of suitable electrodes in EDLCs.

Nanoporous carbons rich with functional groups is also of paramount importance that must be taken for the synthesis of carbonaceous materials, and therefore, the type of functionalization and its content will determine the performance of the electrode in a supercapacitor. As far as the above goal, to profit from using biomass, N-doped mesoporous carbons have been synthesized by using milk powder through one step including pyrolysis of the biomass and KOH as an activating agent, leading to high specific capacitance and high stability after thousands of cycles [27]. On the other side, the presence of O-containing groups could be detrimental to the performance of flexible supercapacitors based on carbon textile electrodes stated by Barczak and Bandosz [28] due to lower electric conductivity and lower wettability of the electrolyte. Moreover, the above authors reported in their work that O-functional groups reduce the effective micropore volume thereby decreasing the diffusion of electrolyte species. Thus, an utter knowledge of the surface chemistry is required (e.g. the presence of -COH, -COO and -COOH groups within the nanoporous carbon network is rather preferred, instead of -OH or -C=O groups [29]). On the contrary, the introduction of S- and N-containing species (in their reduced form) does not block the microporosity way of obtaining those with higher values in capacitance.

The presence of defects on the surface of porous carbons will also influence and determine the efficiency of the EDLC device. A recent report has led to the synthesis of B/S-co-doped porous carbon derived from the use of biological wastes towards supercapacitor applications [30]; rich B and S doping porous carbons exhibited high specific capacitance. In other work, P-doped porous carbon derived from walnut shell biomass [31] has been synthesized to be applied as an anode in Zn-ion hybrid capacitors (MIC). Generally, this hybrid system is founded on the reversible process on the metal (in this case Zn) and physical adsorption/desorption of ions in the porous carbon electrode, with benefits in both high energy density and power density [32]. The advantage of using P as a doping atom relies on the creation of defects associated with the longer bond distance of P-C and a better redistribution of charge between P and C atoms because of the lower electronegativity of P atom, thereby accelerating the diffusion of metal ions and achieving excellent energy storage. The use of B/P- co-doping has been also used in porous carbons for Zn-ion capacitors with excellent specific capacity retention [33]. For the above two works, it is worth noting that the low grade of functionalization within nanoporous carbons is vital for improving the performance of specific capacity and lifespan of carbonaceous electrodes. Additionally, this chapter section brings to light supercapacitors made of 3D nanoporous carbons using hierarchical porous carbons which exhibit a controllable connection of the pore structure and BET surface area with amelioration of diffusion pathways for ion transport [34,35].

3.3 ENERGY CONVERSION

Beyond energy storage, a wide variety of porous carbon materials have been studied for their application in energy conversion processes mainly based in the catalysis field. This chapter section follows the use of nanoporous carbons in electrocatalytic and photocatalytic processes with the aim of having a general view of the effect of the carbon features in both applications.

3.3.1 ELECTROCATALYSIS

Electrocatalysis plays a relevant role in industrial processes since it allows the reduction of the reaction energy barrier, thereby accelerating reaction rates and improving energy efficiency. Among the electrochemical energy conversion processes, hydrogen production is considered one of the most promising options for clean energy generation that could replace the use of fossil fuels. Hydrogen production uses water as a green source by the HER [36]. Moreover, ORR [37,38] and OER [39] are essential processes for renewable energy conversion technologies such as fuel cells and water splitting. Another technology for obtaining sustainable energy is the use of electrodes capable of favouring CO_2 reduction reactions towards, for instance CO, which is part of the synthesis gas and therefore the building block to generate fuels. Chemical products of high industrial demand can also be obtained through CO2RR [40,41].

To carry out the main aforementioned reactions, electrodes based on noble metals have been applied mainly for energy conversion using Pt, considered the most active electrocatalyst for HER and ORR [42–44]. Nevertheless, several studies showed that noble metal-based catalysts, such as Ir, Rh, etc., are also highly active for the above reactions [45,46]. Although noble metal-based catalysts are highly effective electrocatalysts, these electrodes have a high cost, poor stability, considerable environmental toxicity, and high demand in industrial processes. In this sense, metal-free catalysts based on carbon materials are presented as a friendlier alternative since they exhibit high stability, good electrical conductivity and high surface area that favour the availability of active sites. Metal-free porous carbons have great versatility in terms of structure, electron distribution, and composition that allow them to adapt their electrochemical properties. Moreover, the presence of porosity could improve the activity and selectivity by accelerating the electron transfer and mass transport of the reactants which affects the kinetics of the reaction. A structure with high surface area and pores in the range of micro- and mesoporosity can increase the number of catalytic sites exposed to the electroactive species and can facilitate electron transfer and, consequently, increase the efficiency of reactions [41,47,48]. However, pristine carbons are considered poorly active materials for electrocatalytic reactions, therefore, several strategies have been explored to improve the activity of carbon materials including its surface chemistry modification by doping with heteroatoms of electronegativities different from that of the carbon atom. The heteroatoms can act as electron acceptors or donor groups and form defects in the structure [49,50]. Additionally, it has been shown that certain heteroatom-doped nanoporous carbons possess multiple active sites for electrocatalysis.

Upon doping carbon materials with different heteroatoms (e.g. N, S, B, P, F and O), the C atoms adjacent to the heteroatom dopants can become either negatively or positively charged, depending on their electronegativity differences and relative positions [51]. Among heteroatoms, N- has been the most extensively explored due to its similar electronegativity and electroaffinity to carbon. Although the local structure of active sites of N-doped materials is under debate, it seems that C atoms next to N-pyridinic with Lewis basicity are the active sites for ORR when an acidic electrolyte is used (Figure 3.2A) [52–54]. Furthermore, to improve the Faradaic efficiency and yield of CO2RR, N-pyridinic doping can provide basic sites to capture CO_2 which is a Lewis acid, to stabilize the CO_2-intermediate and to enhance the activity of CO2RR. Indeed, it seems that N-pyridinic is the most selective site for CO2RR over HER [55]. Research has been carried out to obtain nanoporous carbons with the presence of structural defects, which have been proposed as active sites. As far as CO2RR, N-doped mesoporous carbon materials were first prepared which, through various thermal treatments, progressively eliminated part of the nitrogenous species [56]. Defect-rich mesoporous carbons generated by the elimination of the nitrogenous species exhibited a higher CO generation compared to the non-annealed carbons. In addition, suppression of HER was observed, which is a reaction that competes for active sites of N-species. With an applied overpotential of 0.49 V *vs.* RHE, the formation of CO from CO_2 was obtained with a Faradaic efficiency close to 80% and a partial current density for CO of -2.9 mA/cm². Density functional theory (DFT) calculations further revealed that the active sites were the defects generated by N removal, which lowered the energy barriers to CO2RR.

Song et al. [57] reported the synthesis of nanoporous carbon by nanocasting method with uniform micropores (~0.84 nm in diameter) in the walls of ordered mesopore channels with hexagonal geometry (~7.4 nm in diameter). Authors studied the obtained material for ethanol production via CO2RR. It seems that the micropores induce desolvation to accumulate electrolytic ions and allow high local electrical potentials, so that the active sites on the surface, N-pyrrolic and N-pyridine, can generate the activation of CO_2 molecules and the C-C coupling of intermediates, resulting in the formation of ethanol with high production rate and Faradaic efficiency of 78% at -0.56 V *vs.* RHE.

Prabu and co-workers [58] reported the synthesis of porous carbon from dried palm plant waste obtaining graphene-like hierarchical porous nanosheets after pyrolysis and then basic activation. This carbonaceous material presented a high surface area, with a hierarchical micro/mesoporous architecture (S^{BET}: 1297 m²/g, Vtotal: 0.6 cm³/g) and presence of N-pyridine, N-pyrrole, and N-graphitic (coming from the organic components of the biomass), properties that are seeking to promote HER catalysis. Indeed, in an acid medium, the electrode showed a minimum overpotential of -0.33 V vs. RHE at 10 mA/cm², a small Tafel slope value of 63 mV/dec, and high long-term stability (Figure 3.2B).

Beyond N-, other heteroatoms have been used for dope nanoporous carbons [59]. For example, Wang et al. [60] reported the synthesis of ordered mesoporous carbon with a sulphur content of 1.5 wt.%. This material showed a high catalytic activity for ORR, mainly ascribed to the presence of sulphide groups. Taking into account the influence of doping with heteroatoms and the presence of defects in the structure of porous carbons, recent research has focused on improving the electrocatalytic activity of the materials to be used, using as a strategy the doping with multiple heteroatoms, looking for the synergistic effects that can be generated. The bi- (e.g. N/S-, N/P-, N/B-) and tri- (e.g. N/S/O-) co-doping in porous carbons improve catalytic activity and selectivity, and the porous structure exposes more active sites to reaction species and facilitates electrolyte transport, leading to a significant increase in catalytic activities [61]. For ORR and CO2RR, electrodes based on carbons with a hierarchical porosity and N/S co-doped, presented a high activity towards ORR in acid and alkaline electrolytes, and for CO2RR to CO with a maximum Faraday efficiency of 87.8% [62]. N/S- co-doping could produce more active sites for catalysis than only N- or S-doping alone due to the enhancement of spin maximum density at carbon [63]. In another study, looking forward to improving the efficiency in the reduction of CO_2 to CO, a N/S- co-doped carbon electrode was prepared, which reached a high Faradaic efficiency of 92% at an overpotential of 0.49 V vs RHE. The authors found that the incorporation of S leads to the generation of a higher density of N-pyridinic

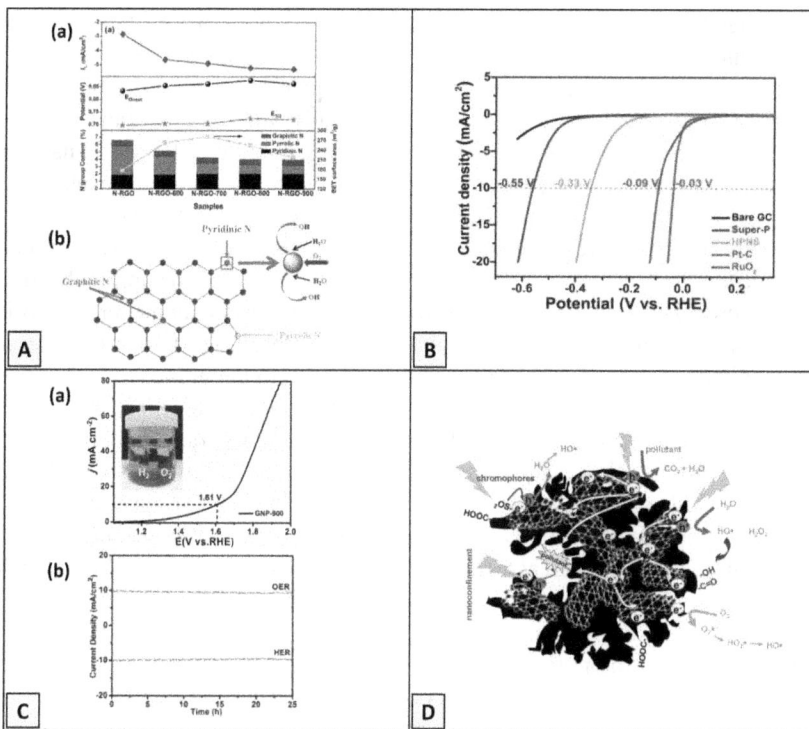

FIGURE 3.2 (A.a) Relations between E_{onset}, $E_{1/2}$, IL and (A.b) contents of different N-groups and BET surface area scheme mechanism of N-pyridinic group in ORR. Reproduced from [54] with permission from Elsevier. Copyright 2017; (B) Comparison of linear sweep voltammetry polarization curves for HER at 20 mV/s in acidic medium of synthetized carbon material with other commercial catalysts. Reproduced from [58] with permission from Elsevier. Copyright 2019; (C.a) Polarization curve of water splitting and insert image showing H_2 and O_2 bubbles, (C.b) current-time curve at 179 and 299 mV in 1 M NaOH electrolyte for 25 h Reproduced from [65] with permission from Elsevier. Copyright 2021. (D) Scheme of the mechanism proposed for the exciton formation and fate upon irradiation of nanoporous carbon. Reprinted from [68] with permission form Elsevier. Copyright 2016.

sites wherein CO_2 is activated [64]. Liu and co-workers [65] reported the synthesis of carbon material tri- co-doped with N/P/O- with a total pore volume of 0.53 cm^3/g. They showed a high catalytic response of this material for HER, OER, and ORR, and they reported its bifunctional use for water splitting reaction, delivering 10 mA/cm^2 with a potential of 1.61 V vs RHE (Figure 3.2C).

3.3.2 PHOTOCATALYSIS

Photochemical reactions (light-initiated reactions) are particularly useful because the excitation of electronic molecular states, using adequate energies, may induce chemical bond breaking, offering interesting opportunities in environmental chemistry, and energy production and conversion. Some of the most noteworthy examples are as follows: the photoelectrochemical hydrogen generation, the photoreduction of CO_2 (artificial photosynthesis), and the photocatalytic degradation of environmental contaminants [66–68].

The most common materials used in photochemical reactions are semiconductors (e.g., TiO_2, WO_3, Bi_2WO_6, etc.); however, generally, their major drawback is low photonic efficiency under sunlight. For this reason, numerous approaches have been established to improve the photocatalytic activity, of which some examples are already known, such as organic metal halides, perovskites, transition metal oxides and sulphides, non-noble metal-based catalysts, hybrid nanostructures, or graphitic carbon nitrides [69–73]. Another significant strategy is based on the immobilization of the photoactive material on porous substrates, e.g. carbon/semiconductor composites. In this case, the enhancement of the photocatalytic response of such carbon/semiconductor composites has been traditionally attributed to the improved mass transfer due to the presence of the carbon matrix and/or the electron acceptor role of the carbon material that enhances the splitting of the photogenerated excitons [74,75]. Since those early studies, many researchers have investigated the key role of carbon matrices coupled to all types of photoactive materials [76,77]. The enhanced performance and increased photoactivity of the composites compared to the semiconductor alone are discussed by different mechanisms that depend on various factors including (i) the mixture method of both phases [78,79], (ii) the interaction (or not) between the semiconductor and carbonaceous phase [79,80], and (iii) the carbon matrix nature [80,81]. Centred on the porosity of the carbon material, it seems that in the case of nanoporous carbons, the porous network allows the mass transfer of the adsorbed species (target pollutant to be degraded in the case of heterogeneous photocatalysis, CO_2 in the case of CO2RR or H_2O for its photooxidation) from the bulk solution to the photoactive semiconductor phase through the interface between the two catalyst components.

Likewise, nanoporous carbon materials have been employed as additives to semiconductors to fabricate photoanodes with the objective of studying water photooxidation for energy production [82,83]. For example, composites WO_3/nanoporous carbon revealed superior photocurrent and incident photon-to-current efficiency values for hybrid electrodes compared with the WO_3 film alone, indicating that the porosity decreases the path length for the photogenerated charge carriers to reach the surface and react with electron acceptors [83]. Interestingly, different studies reported, in 2010, the self-photochemical activity of some nanoporous carbons under UV-Vis irradiation in the absence of a semiconductor, an ability that until that time, had been neglected and disregarded [84,85]. These studies showed an improvement in photooxidation of phenol in aqueous solution by irradiation of an activated carbon demonstrating the effect of the carbon material beyond the synergistic effect of the porosity. Numerous works have been published on this topic since then, exploring the photooxidation of various pollutants in water using nanoporous carbons of varied nature as photocatalysts [68,85–88].

The origin of the photochemical response of the semiconductor-free nanoporous carbons could be explained by several possible scenarios: (i) adsorption of the target pollutant on the porosity of the nanoporous carbon causing an enhancement in the conversion due to the effect of the pollutant pre-concentration, (ii) occurrence of carbon/light interactions that would provoke the degradation of the compounds adsorbed, (iii) formation of radical species upon illumination of the carbon material [68].

With the aim of confirming the effect of the porosity, several works studied the photochemical reactions in the confined porosity of nanoporous carbons [89–91]. In those works, the carbons were preloaded with a target pollutant and then irradiated in an aqueous suspension, after irradiation the remaining compounds, retained inside the porosity, were extracted and analyzed. Following that protocol, the degradation of pollutant caused by secondary reactions (such as adsorption or photolysis) was avoided. These studies revealed that both the porosity and surface chemistry of the nanoporous carbon directly affect the performance reaction. It seems that tighter confinement of target molecule favours its degradation due to the higher host-guest interaction inside the porous network [90,91]. On the other side, the surface chemistry has the role of facilitating the interaction between the target molecule and carbon material affecting the affinity between both components and, as consequence, modifying the degradation yield (i.e. the higher affinity, the better degradation performance). Additionally, it is known that some groups such as N-, S- and O- can act as chromophores, photogenerating excitons that also participate in the observed photochemical reactions (Figure 3.2D) [90,92,93]. Porous carbon materials were also investigated as photoanodes for water splitting [71,89,94–96]. An anodic photocurrent of up to 0.45 mA/cm^2 was recorded associated to water oxidation [92]. Light exposure causes the photogeneration of charge carriers (electrons and holes) on the chromophore-like moieties (e.g. O-, N- and S-containing groups), the porosity facilitates the adsorption of water molecules in the pores close to the chromophores, enhancing the electron transfer from oxygen in water molecules to the vacancies (holes).

3.4 CONCLUSIONS

With a scenario pointing out growing emissions of CO_2 in the atmosphere together with the decrease in energy supply and high energy prices, novel strategies and materials are being developed linked to energy production, storage and conversion.

Free-metal catalysts porous carbon materials have been extensively used so far in storage applications (supercapacitor, metal ion batteries and their hybrid devices), the production of energy (ORR involved in fuel cells) or the energy conversion by using water electrolyzers (HER or OER) or (photo)electrochemical or photochemical reactors (CO2RR or the elimination of pollutants in air or water). On the one hand, recent efforts are still focused on porous textures, high surface area, proper electrical conductivity and presence of heteroatoms, such as variables remarkably determining for example (i) the specific capacitance, energy density and power density as well as cyclic stability and lifespan for storage devices, (ii) the Faradaic efficiency, stability and selectivity for the electrochemical reduction of oxygen and CO_2, (iii) the production rate of water splitting via (photo)electrochemical processes, or (iv) the photointrinsic activity or quantum yield towards the degradation of hazardous pollutants confined into microreactor like nanopores. On the other hand, the chapter has brought out how relevant is the development of deeper understanding about a well-balanced micro/mesopores in carbon samples in order to increase mass transfer, the presence of active sites or defects which appears during the synthesis process or by the incorporation of heteroatoms (N, P, O and or B), the exact role of the active surface area, and number and kind of surface functionalization which can provide amelioration in wettability or simply the control of hydrophobic/hydrophilic features of the porous carbons. All the above deep knowledge is required for a rational design of porous-based carbon electrodes and catalysts leading to high outcome in energy production, storage and conversion. Finally, new trends must go for simple and low-cost electrode preparations with feasible scale up of the process and efficient deployment of the above photochemical and electrochemical methodologies in the market.

ACKNOWLEDGEMENTS

ST thanks the Santiago Grisolía program for the fellowship 2021-51136H0435. JI thanks the Spanish Ministry of Science, Innovation and Universities (MICINN, www.ciencia.gob.es/) with grant PID2019-108136RB-C32 for financial support. AG-B thanks European Union NextGenerationEU (ZAMBRANO21-10) for the funding.

REFERENCES

[1] *Energy council.* Available at www.worldenergy.org/
[2] V. Meunier, C. Ania, A. Bianco, Y. Chen, G.B. Choi and Y.A. Kim et al., Carbon science perspective in 2022: Current research and future challenges, *Carbon* 195 (2022), pp. 272–291.
[3] C. Hu, Q. Dai and L. Dai, Multifunctional carbon-based metal-free catalysts for advanced energy conversion and storage, *Cell Reports Phys. Sci.* 2 (2021), pp. 100328.
[4] Y.J. Xu, G. Weinberg, X. Liu, O. Timpe, R. Schlögl and D.S. Su, Nanoarchitecturing of activated carbon: Facile strategy for chemical functionalization of the surface of activated carbon, *Adv. Funct. Mater.* 18 (2008), pp. 3613–3619.

[5] J.L. Figueiredo, Functionalization of porous carbons for catalytic applications, *J. Mater. Chem. A.* 1 (2013), pp. 9351–9364.

[6] H. Marsh and F. Rodríguez-Reinoso, *Activated Carbon*, Elsevier (2006).

[7] T.J. Bandosz and T.Z. Ren, Porous carbon modified with sulfur in energy related applications, *Carbon.* 118 (2017), pp. 561–577.

[8] J. Verma and D. Kumar, Metal-ion batteries for electric vehicles: Current state of the technology, issues and future perspectives, *Nanoscale Adv.* 3 (2021), pp. 3384–3394.

[9] Y. Liu, X. Li, L. Fan, S. Li, H. Maleki Kheimeh Sari and J. Qin, A review of carbon-based materials for safe lithium metal anodes, *Front. Chem.* 7 (2019), pp. 721.

[10] L. Wang, J. Han, D. Kong, Y. Tao and Q.H. Yang, Enhanced roles of carbon architectures in high-performance lithium-ion batteries, *Nano-Micro Letters* 11 (2019).

[11] C.O. Ania, P.A. Armstrong, T.J. Bandosz, F. Beguin, A.P. Carvalho and A. Celzard et al., Engaging nanoporous carbons in "beyond adsorption" applications: Characterization, challenges and performance, *Carbon* 164 (2020), pp. 69–84.

[12] F. Béguin, F. Chevallier, C. Vix, S. Saadallah, J.N. Rouzaud and E. Frackowiak, A better understanding of the irreversible lithium insertion mechanisms in disordered carbons, *J. Phys. Chem. Solids* 65 (2004), pp. 211–217.

[13] F. Béguin, F. Chevallier, C. Vix-Guterl, S. Saadallah, V. Bertagna and J.N. Rouzaud et al., Correlation of the irreversible lithium capacity with the active surface area of modified carbons, *Carbon* 43 (2005), pp. 2160–2167.

[14] Q. Li, Y. Liu, Y. Wang, Y. Chen, X. Guo and Z. Wu et al., Review of the application of biomass-derived porous carbon in lithium-sulfur batteries, *Ionics* 26 (2020), pp. 4765–4781.

[15] M. Schnucklake, M. Cheng, M. Maleki and C. Roth, A mini-review on decorating, templating of commercial and electrospinning of new porous carbon electrodes for vanadium redox flow batteries, *J. Phys. Mater.* 4 (2021), pp. 032007.

[16] S. Khan, M. Ul-Islam, M.W. Ahmad, M.S. Khan, M. Imran and S.H. Siyal et al., Synthetic methodologies and energy storage/conversion applications of porous carbon nanosheets: A systematic review, *Energy and Fuels* 36 (2022), pp. 3420–3442.

[17] P. Ratajczak, M.E. Suss, F. Kaasik and F. Béguin, Carbon electrodes for capacitive technologies, *Energy Storage Mater.* 16 (2019), pp. 126–145.

[18] R. Dubey and V. Guruviah, Review of carbon-based electrode materials for supercapacitor energy storage, *Ionics* 25 (2019), pp. 1419–1445.

[19] A. Wang, K. Sun, R. Xu, Y. Sun and J. Jiang, Cleanly synthesizing rotten potato-based activated carbon for supercapacitor by self-catalytic activation, *J. Clean. Prod.* 283 (2021), pp. 125385.

[20] X. Chen, R. Paul and L. Dai, Carbon-based supercapacitors for efficient energy storage, *Natl. Sci. Rev.* 4 (2017), pp. 453–489.

[21] R. Mysyk, E. Raymundo-Piñero and F. Béguin, Saturation of subnanometer pores in an electric double-layer capacitor, *Electrochem. Commun.* 11 (2009), pp. 554–556.

[22] L. Liu, H. Zhao and Y. Lei, Advances on three-dimensional electrodes for micro-supercapacitors: A mini-review, *InfoMat.* 1 (2019), pp. 74–84.

[23] J.A. Fernández, T. Morishita, M. Toyoda, M. Inagaki, F. Stoeckli and T.A. Centeno, Performance of mesoporous carbons derived from poly(vinyl alcohol) in electrochemical capacitors, *J. Power Sources.* 175 (2008), pp. 675–679.

[24] W. Li, J. Liu and D. Zhao, Mesoporous materials for energy conversion and storage devices. *Nat Rev Mater.* 1 (2016), 16023. https://doi.org/10.1038/natrevmats.2016.23.

[25] K. Xia, Q. Gao, J. Jiang and J. Hu, Hierarchical porous carbons with controlled micropores and mesopores for supercapacitor electrode materials, *Carbon* 46 (2008), pp. 1718–1726.

[26] K. Kraiwattanawong, A review on the development of a porous carbon-based as modeling materials for electric double layer capacitors, *Arab. J. Chem.* 15 (2022), pp. 103625.

[27] S. Jia, Y. Wang, G. Xin, S. Zhou, P. Tian and J. Zang, An efficient preparation of N-doped mesoporous carbon derived from milk powder for supercapacitors and fuel cells, *Electrochim. Acta* 196 (2016), pp. 527–534.

[28] M. Barczak and T.J. Bandosz, Evaluation of nitrogen- and sulfur-doped porous carbon textiles as electrode materials for flexible supercapacitors, *Electrochim. Acta* 305 (2019), pp. 125–136.

[29] T. Tojo, K. Sakurai, H. Muramatsu, T. Hayashi, K.S. Yang and Y.C. Jung et al., Electrochemical role of oxygen containing functional groups on activated carbon electrode, *RSC Adv.* 4 (2014), pp. 62678–62683.

[30] Y. Wang, D. Wang, Z. Li, Q. Su, S. Wei and S. Pang et al., Preparation of Boron/Sulfur- codoped porous carbon derived from biological wastes and its application in a supercapacitor, *Nanomaterials* 12 (2022), pp. 1182.

[31] H. Sun, C. Liu, D. Guo, S. Liang, W. Xie and S. Liu et al., P-doped porous carbon derived from walnut shell for zinc ion hybrid capacitors, *RSC Adv.* 12 (2022), pp. 24724–24733.

[32] T. Chen and L. Dai, Carbon nanomaterials for high-performance supercapacitors, *Mater. Today* 16 (2013), pp. 272–280.

[33] Y.G. Lee and G.H. An, synergistic effects of phosphorus and boron co-incorporated activated carbon for ultrafast zinc-ion hybrid supercapacitors, *ACS Appl. Mater. Interfaces* 12 (2020), pp. 41342–41349.

[34] D.V. Cuong, B.M. Matsagar, M. Lee, M.S.A. Hossain, Y. Yamauchi and M. Vithanage et al., A critical review on biochar-based engineered hierarchical porous carbon for capacitive charge storage, *Renew. Sustain. Energy Rev.* 145 (2021), pp. 111029.

[35] L. Liu, H. Zhao and Y. Lei, Advances on three-dimensional electrodes for micro-supercapacitors: A mini-review, *InfoMat* 1 (2019), pp. 74–84.

[36] S.E. Hosseini and M.A. Wahid, Hydrogen production from renewable and sustainable energy resources: Promising green energy carrier for clean development, *Renew. Sustain. Energy Rev.* 57 (2016), pp. 850–866.

[37] M. Shao, Q. Chang, J.P. Dodelet and R. Chenitz, Recent advances in electrocatalysts for oxygen reduction reaction, *Chem. Rev.* 116 (2016), pp. 3594–3657.

[38] A. Kulkarni, S. Siahrostami, A. Patel and J.K. Nørskov, Understanding catalytic activity trends in the oxygen reduction reaction, *Chem. Rev.* 118 (2018), pp. 2302–2312.

[39] M. Tahir, L. Pan, F. Idrees, X. Zhang, L. Wang and J.J. Zou et al., Electrocatalytic oxygen evolution reaction for energy conversion and storage: A comprehensive review, *Nano Energy* 37 (2017), pp. 136–157.

[40] J. Qiao, Y. Liu, F. Hong and J. Zhang, A review of catalysts for the electroreduction of carbon dioxide to produce low-carbon fuels, *Chem. Soc. Rev.* 43 (2013), pp. 631–675.

[41] X. Duan, J. Xu, Z. Wei, J. Ma, S. Guo and S. Wang et al., Metal-Free carbon materials for CO_2 electrochemical reduction, *Adv. Mater.* 29 (2017), pp. 1701784.

[42] N.M. Marković, T.J. Schmidt, V. Stamenković and P.N. Ross, Oxygen reduction reaction on Pt and Pt bimetallic surfaces: A selective review, *Fuel Cells* 1 (2001), pp. 105–116.

[43] C. Zhang, X. Shen, Y. Pan and Z. Peng, A review of Pt-based electrocatalysts for oxygen reduction reaction, *Front. Energy* 113 11 (2017), pp. 268–285.

[44] K.H. Liu, H.X. Zhong, S.J. Li, Y.X. Duan, M.M. Shi and X.B. Zhang et al., Advanced catalysts for sustainable hydrogen generation and storage via hydrogen evolution and carbon dioxide/nitrogen reduction reactions, *Prog. Mater. Sci.* 92 (2018), pp. 64–111.

[45] B. Şen, A. Aygün, A. Şavk, C. Yenikaya, S. Cevik and F. Şen, Metal-organic frameworks based on monodisperse palladiumcobalt nanohybrids as highly active and reusable nanocatalysts for hydrogen generation, *Int. J. Hydrogen Energy* 44 (2019), pp. 2988–2996.

[46] J. Li, P. Zhou, F. Li, J. Ma, Y. Liu and X. Zhang et al., Shape-controlled synthesis of Pd polyhedron supported on polyethyleneimine-reduced graphene oxide for enhancing the efficiency of hydrogen evolution reaction, *J. Power Sources* 302 (2016), pp. 343–351.

[47] C. Hu and L. Dai, Carbon-based metal-free catalysts for electrocatalysis beyond the ORR, *Angew. Chemie Int. Ed.* 55 (2016), pp. 11736–11758.

[48] J. Zhang, Z. Xia and L. Dai, Carbon-based electrocatalysts for advanced energy conversion and storage, *Sci. Adv.* 1 (2015).

[49] S. Wang, E. Iyyamperumal, A. Roy, Y. Xue, D. Yu and L. Dai, Vertically aligned BCN nanotubes as efficient metal-free electrocatalysts for the oxygen reduction reaction: A synergetic effect by co-doping with boron and nitrogen, *Angew. Chem. Int. Ed. Engl.* 50 (2011), pp. 11756–11760.

[50] Y. Jia, L. Zhang, L. Zhuang, H. Liu, X. Yan and X. Wang et al., Identification of active sites for acidic oxygen reduction on carbon catalysts with and without nitrogen doping, *Nat. Catal.* 2019 28 2 (2019), pp. 688–695.

[51] R. Paul, F. Du, L. Dai, Y. Ding, Z.L. Wang and F. Wei et al., 3D heteroatom-doped carbon nanomaterials as multifunctional metal-free catalysts for integrated energy devices, *Adv. Mater.* 31 (2019), pp. 1805598.

[52] S.K. Singh, K. Takeyasu and J. Nakamura, Active sites and mechanism of oxygen reduction reaction electrocatalysis on nitrogen-doped carbon materials, *Adv. Mater.* 31 (2019), pp. 1804297.

[53] D. Guo, R. Shibuya, C. Akiba, S. Saji, T. Kondo and J. Nakamura, Active sites of nitrogen-doped carbon materials for oxygen reduction reaction clarified using model catalysts, *Science* 351 (2016), pp. 361–365.

[54] H. Miao, S. Li, Z. Wang, S. Sun, M. Kuang and Z. Liu et al., Enhancing the pyridinic N content of Nitrogen-doped graphene and improving its catalytic activity for oxygen reduction reaction, *Int. J. Hydrogen Energy* 42 (2017), pp. 28298–28308.

[55] S. Siahrostami, K. Jiang, M. Karamad, K. Chan, H. Wang and J. Nørskov, Theoretical investigations into defected graphene for electrochemical reduction of CO_2, *ACS Sustain. Chem. Eng.* 5 (2017), pp. 11080–11085.

[56] R. Daiyan, X. Tan, R. Chen, W.H. Saputera, H.A. Tahini and E. Lovell et al., Electroreduction of CO_2 to CO on a mesoporous carbon catalyst with progressively removed nitrogen moieties, *ACS Energy Lett.* 3 (2018), pp. 2292–2298.

[57] Y. Song, S. Wang, W. Chen, S. Li, G. Feng and W. Wei et al., Enhanced ethanol production from CO_2 electroreduction at micropores in nitrogen-doped mesoporous carbon, *ChemSusChem* 13 (2020), pp. 293–297.

[58] N. Prabu, R.S.A. Saravanan, T. Kesavan, G. Maduraiveeran and M. Sasidharan, An efficient palm waste derived hierarchical porous carbon for electrocatalytic hydrogen evolution reaction, *Carbon* 152 (2019), pp. 188–197.

[59] B. Wang, B. Liu and L. Dai, Non-N-doped carbons as metal-free electrocatalysts, *Adv. Sustain. Syst.* 5 (2021), pp. 2000134.

[60] H. Wang, X. Bo, Y. Zhang and L. Guo, Sulfur-doped ordered mesoporous carbon with high electrocatalytic activity for oxygen reduction, *Electrochim. Acta* 108 (2013), pp. 404–411.

[61] K. Qu, Y. Zheng, X. Zhang, K. Davey, S. Dai and S.Z. Qiao, Promotion of electrocatalytic hydrogen evolution reaction on Nitrogen-doped carbon nanosheets with secondary heteroatoms, *ACS Nano* 11 (2017), pp. 7293–7300.

[62] R. Li, F. Liu, Y. Zhang, M. Guo and D. Liu, Nitrogen, Sulfur co-doped hierarchically porous carbon as a metal-free electrocatalyst for oxygen reduction and carbon dioxide reduction reaction, *ACS Appl. Mater. Interfaces* 12 (2020), pp. 44578–44587.

[63] J. Liang, Y. Jiao, M. Jaroniec and S.Z. Qiao, Sulfur and Nitrogen dual-doped mesoporous graphene electrocatalyst for oxygen reduction with synergistically enhanced performance, *Angew. Chemie Int. Ed.* 51 (2012), pp. 11496–11500.

[64] F. Pan, B. Li, W. Deng, Z. Du, Y. Gang and G. Wang et al., Promoting electrocatalytic CO_2 reduction on nitrogen-doped carbon with sulfur addition, *Appl. Catal. B Environ.* 252 (2019), pp. 240–249.

[65] Z. Liu, M. Wang, X. Luo, S. Li, S. Li and Q. Zhou et al., N-, P-, and O-doped porous carbon: A trifunctional metal-free electrocatalyst, *Appl. Surf. Sci.* 544 (2021), pp. 148912.

[66] J. Jian, G. Jiang, R. Van de Krol, B. Wei and H. Wang, Recent advances in rational engineering of multinary semiconductors for photoelectrochemical hydrogen generation, *Nano Energy* 51 (2018), pp. 457–450.

[67] A.D. Handoko, K. Li and J. Tang, Recent progress in artificial photosynthesis: CO_2 photoreduction to valuable chemicals in a heterogeneous system, *Curr. Opin. Chem, Eng.* 2 (2013), pp. 200–206.

[68] A. Gomis-Berenguer, L.F. Velasco, I. Velo-Gala and C.O. Ania, Photochemistry of nanoporous carbons: Perspectives in energy conversion and environmental remediation, *J. Colloid Interface Sci.* 490 (2017), pp. 879–901.

[69] M.A. Green, A. Ho-Baillie and H.J. Snaith, The emergence of perovskite solar cells, *Nat. Photonics* 8 (2014), pp. 506–514.

[70] O. Malinkiewicz, A. Yella, Y.H. Lee, G.M. Espallargas, M. Graetzel and M.K. Nazeeruddin et al., Perovskite solar cells employing organic charge-transport layers, *Nat. Photonics* 8 (2013), pp. 128–132.

[71] C. Hu, X. Chen, Q. Dai, M. Wang, L. Qu and L. Dai, Earth-abundant carbon catalysts for renewable generation of clean energy from sunlight and water, *Nano Energy* 41 (2017), pp. 367–376.

[72] J. Liang, Z. Jiang, P.K. Wong and C.S. Lee, Recent progress on carbon nitride and its hybrid photocatalysts for CO_2 reduction, *Sol. RRL* 5 (2021), pp. 2000478.

[73] G. Guan, E. Ye, M. You and Z. Li, Hybridized 2D nanomaterials toward highly efficient photocatalysis for degrading pollutants: Current status and future perspectives, *Small* 16 (2020), pp. 1907087.

[74] J. Matos, J. Laine and J.M. Herrmann, Synergy effect in the photocatalytic degradation of phenol on a suspended mixture of titania and activated carbon, *Appl. Catal. B Environ.* 18 (1998), pp. 281–291.

[75] B. Tryba, A.W. Morawski and M. Inagaki, Application of TiO2-mounted activated carbon to the removal of phenol from water, *Appl. Catal. B Environ.* 41 (2003), pp. 427–433.

[76] J.L. Faria and W. Wang, *Carbon Materials in Photocatalysis*, in *Carbon Materials for Catalysis*, John Wiley and Sons, (2008), pp. 1–579.

[77] R. Leary and A. Westwood, Carbonaceous nanomaterials for the enhancement of TiO2 photocatalysis, *Carbon* 49 (2011), pp. 741–772.

[78] R.J. Carmona, L.F. Velasco, M.C. Hidalgo, J.A. Navío and C.O. Ania, Boosting the visible-light photoactivity of Bi2WO6 using acidic carbon additives, *Appl. Catal. A Gen.* 505 (2015), pp. 467–477.

[79] G.D. Gesesse, *Photocatalytic and Photoelectrocatalytic Activity of Mixed Semiconductor/Nanoporous Carbon Materials: Application to the Degradation of Environmental Pollutants*, University of Orléans, France, (2021).

[80] R.J. Carmona, L.F. Velasco, E. Laurenti, V. Maurino and C.O. Ania, Carbon materials as additives to WO3 for an enhanced conversion of simulated solar light, *Front. Mater.* 3 (2016), pp. 9.

[81] A. Gomis-Berenguer, I. Eliani, V.F. Lourenço, R.J. Carmona, L.F. Velasco and C.O. Ania, Insights on the use of carbon additives as promoters of the visible-light photocatalytic activity of Bi2WO6, *Materials* 12 3 (2019), pp. 385. https://doi.org/10.3390/ma12030385

[82] M. Haro, L.F. Velasco and C.O. Ania, Carbon-mediated photoinduced reactions as a key factor in the photocatalytic performance of C/TiO2, *Catal. Sci. Technol.* 2 (2012), pp. 2264–2272.

[83] A. Gomis-Berenguer, V. Celorrio, J. Iniesta, D.J. Fermin and C.O. Ania, Nanoporous carbon/WO3 anodes for an enhanced water photooxidation, *Carbon* 108 (2016), pp. 471–479.

[84] L.F. Velasco, J.B. Parra and C.O. Ania, Role of activated carbon features on the photocatalytic degradation of phenol, *Appl. Surf. Sci.* 256 (2010), pp. 5254–5258.

[85] L.F. Velasco, I.M. Fonseca, J.B. Parra, J.C. Lima and C.O. Ania, Photochemical behaviour of activated carbons under UV irradiation, *Carbon* 50 (2012), pp. 249–258.

[86] I. Velo-Gala, J.J. López-Peñalver, M. Sánchez-Polo and J. Rivera-Utrilla, Activated carbon as photocatalyst of reactions in aqueous phase, *Appl. Catal. B Environ.* 142–143 (2013), pp. 694–704.

[87] L.F. Velasco, V. Maurino, E. Laurenti and C. Ania, Light-induced generation of radicals on semiconductor-free carbon photocatalysts, *Appl. Catal. A Gen.* 453 (2013), pp. 310–315.

[88] L.F. Velasco and J.C. Lima, C. Ania, Visible-light photochemical activity of nanoporous carbons under monochromatic light, *Angew. Chemie Int. Ed.* 53 (2014), pp. 4146–4148.

[89] L.F. Velasco, A. Gomis-Berenguer, J.C. Lima and C.O. Ania, Tuning the surface chemistry of nanoporous carbons for enhanced nanoconfined photochemical activity, *ChemCatChem* 7 (2015), pp. 3012–3019.

[90] A. Gomis-Berenguer, M. Seredych, J. Iniesta, J.C. Lima, T.J. Bandosz and C.O. Ania, Sulfur-mediated photochemical energy harvesting in nanoporous carbons, *Carbon* 104 (2016).

[91] A. Gomis-Berenguer, J. Iniesta, A. Moro, V. Maurino, J.C. Lima and C.O. Ania, Boosting visible light conversion in the confined pore space of nanoporous carbons, *Carbon* 96 (2016).

[92] M. Baca, G.E.O. Borgstahl, M. Boissinot, P.M. Burke, D.W.R. Williams and K.A. Slater et al., Complete chemical structure of photoactive yellow protein: Novel thioester-linked 4-hydroxycinnamyl chromophore and photocycle chemistry, *Biochemistry* 33 (1994), pp. 14369–14377.

[93] J.E. Beecher, T. Durst, J.M.J. Fréchet, A. Godt, A. Pangborn and D.R. Robello et al., New chromophores containing sulfonamide, sulfonate, or sulfoximide groups for second harmonic generation, *Adv. Mater.* 5 (1993), pp. 632–634.

[94] C.O. Ania, M. Seredych, E. Rodríguez-Castellón and T.J. Bandosz, Visible light driven photoelectrochemical water splitting on metal free nanoporous carbon promoted by chromophoric functional groups, *Carbon* 79 (2014), pp. 432–441.

[95] A. Gomis-Berenguer, I. Velo-Gala, E. Rodríguez-Castellón and C.O. Ania, Surface modification of a nanoporous carbon photoanode upon irradiation, *Molecules* 21 (2016), pp. 1611. 10.3390/molecules21111611

[96] Y. Peng, A. Rendón-Patiño, A. Franconetti, J. Albero, A. Primo and H. García, Photocatalytic overall water splitting activity of templateless structured graphitic nanoparticles obtained from cyclodextrins, *ACS Appl. Energy Mater.* 3 (2020), pp. 6623–6632.

4 Synthetic Strategies for the Preparation of Nanoporous Carbons

Ana Casanova, Encarnación Raymundo-Piñero,
Conchi O. Ania, and Alicia Gomis-Berenguer

4.1 INTRODUCTION

Developing and deploying clean energy technologies as part of the world's climate and clean energy transition policies has created significant pressure in the demand for raw critical materials needed for catalytic reactions, particularly in the fields of energy generation, conversion and storage. Overcoming such drawbacks with metal-free materials based on abundant renewable resources has become a priority towards sustainable energy generation/conversion/storage technologies (e.g., fuel cells, batteries, supercapacitors, water splitting, carbon dioxide conversion into fuels, nitrogen fixation) as promising solutions. Some examples include efficient materials for energy storage in supercapacitors and batteries, and catalysts for promoting the oxygen and hydrogen evolution reactions (OER, HER) for water splitting, hydrogen oxidation reaction (HOR), oxygen reduction reaction (ORR) in fuel cells, metal-air batteries, and CO_2 reduction reaction (CO_2RR) for the synthesis of solar fuels and in Li-CO_2 batteries. For most of these reactions, metallic catalysts based on precious metals are the most efficient solutions. To alleviate strains on critical materials, research efforts are directed towards the development of non-precious metal-based catalysts.

In this sense, metal-free porous carbons have attained considerable attention to replace noble (metal-based catalysts) in energy storage and conversion applications [1]. Owing to their availability, low cost, good electronic conductivity, high surface area and tuneable composition, pore structure, unique catalytic performance, reusability and stability have been reported for metal-free porous carbons in such fields of application [2–4]. As a few examples, metal-free porous semiconducting carbons have proved to be very appealing for hydrogen generation using solar light, although solar-to-hydrogen yields are still lower than those of metallic catalysts [5,6]. Nanostructured carbons have also demonstrated excellent ORR performance in alkaline electrolytes but is still challenging [7]. Other areas of growing interest are the CO_2 reduction reaction (CO_2RR) for direct conversion of CO_2 into fuels and Li-CO_2 batteries and the nitrogen reduction reaction (NRR) for the synthesis of NH_3 at ambient conditions. For CO_2RR, excellent performance has been reported for N-doped carbon catalysts in terms of stability and selectivity towards CO, and more reduced feedstocks

DOI: 10.1201/9781003318859-4

(e.g., C1 and C2 products) [8,9]. For NRR, the immobilization of ionic liquids in the nanopores of materials has been reported to increase the uptake of nitrogen gas, which is expected to improve the electrochemical activation of nitrogen [10].

In the case of energy storage in supercapacitors and batteries, the pore structure of the carbon electrodes is essential for the performance of the device. Generally, high surface areas are necessary, although the correlation with the electrochemical performance is not straightforward. An interconnected hierarchical micropore network is typically required for an efficient charge accommodation in the electrical double layer to ensure a continuous electron transport pathway and shorten ion/mass transport distances [2]. On the other hand, macropores and mesopores act as ion-buffering reservoirs, and channels for rapid ionic transport, respectively [11]. The electronic conductivity of porous carbons is also an important parameter in most energy storage and conversion applications. Although some carbons (e.g., graphite, graphene and its derivatives, carbon nanotubes) present electronic properties close to those of metallic electrodes, most nanoporous carbons display a low degree of structural order due to the high density of defects in the twisted graphitic layers upon the development of the nanopore network. As a result, the conductivity of nanoporous carbons is limited (typically 4–5 orders of magnitude lower than that of graphite or graphene) [12,13]. Hence increasing the conductivity of nanoporous carbons without compromising the porosity has become an important research topic.

Furthermore, unlike metal-based catalysts, carbons may present multiple catalytic active sites that can be modulated upon functionalization (e.g., doping, structural defects) and adequate pore control. These features are interesting for coupling different reactions in self-powered integrated systems, consisting for instance of a photo-electrochemical water-splitting unit and Zn-air battery for renewable generation of electricity from sunlight and water [4]. In this chapter, we summarize the most relevant synthetic methods for the preparation of nanoporous carbons covering current industrial practices at large scale, as well as other alternatives proposed in the abundant literature to control the pore structure while using green approaches (i.e., lowering the environmental fingerprint using low energy-demanding processes and low cost and abundant precursors). Figure 4.1 collects a summary of the synthetic routes described in this chapter, with details about the methodology and the impact on the porous structure of the resulting carbons. A special emphasis will be paid to the methods aiming to control the microporosity of carbons since micropores (e.g., pores with width below 2 nm) are considered the active sites for most energy conversion and storage applications.

4.2 CONVENTIONAL SYNTHESIS METHODS

Activated carbons are produced at an industrial scale by physical and chemical activation methods with the use of high-temperature furnaces under a controlled atmosphere. Most processes start with a carbonization (pyrolysis) under an inert atmosphere or an oxidation step (typically air) of the precursor, followed by the activation of the carbonized material in the presence of an activating agent to develop a porous structure. The preparation of activated carbons using conventional activation methods is a well-known process at the industrial scale, and extensive work has been done to

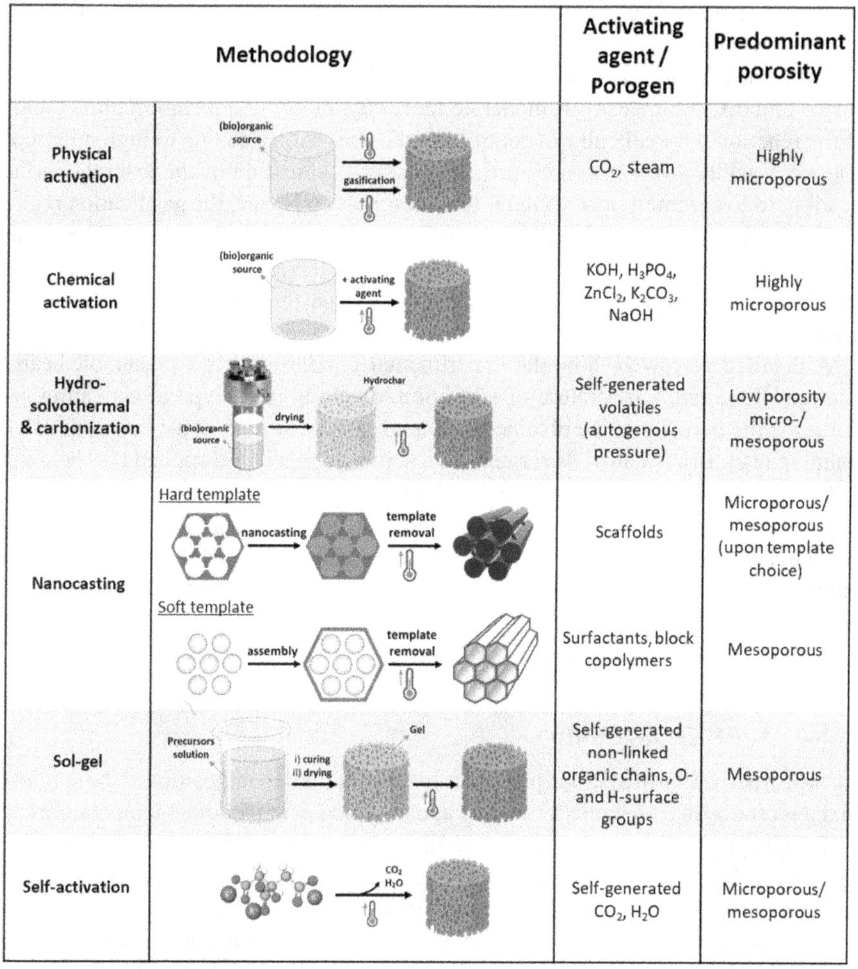

Methodology		Activating agent / Porogen	Predominant porosity
Physical activation	(bio)organic source → gasification	CO_2, steam	Highly microporous
Chemical activation	(bio)organic source → + activating agent	KOH, H_3PO_4, $ZnCl_2$, K_2CO_3, NaOH	Highly microporous
Hydro-solvothermal & carbonization	(bio)organic source → drying → Hydrochar	Self-generated volatiles (autogenous pressure)	Low porosity micro-/ mesoporous
Nanocasting	Hard template nanocasting → template removal	Scaffolds	Microporous/ mesoporous (upon template choice)
	Soft template assembly → template removal	Surfactants, block copolymers	Mesoporous
Sol-gel	Precursors solution → i) curing ii) drying → Gel	Self-generated non-linked organic chains, O- and H-surface groups	Mesoporous
Self-activation	→ CO_2 H_2O	Self-generated CO_2, H_2O	Microporous/ mesoporous

FIGURE 4.1 Illustration of most relevant synthesis methodologies to produce nanoporous carbons.

investigate the influence of the operating conditions, precursors and activating agents to enhance the porosity [14]. We herein summarize the most relevant aspects.

4.2.1 PHYSICAL ACTIVATION

Physical (or thermal) activation of a carbon precursor is commonly carried out in two consecutive heating steps, starting with the carbonization of the precursor to obtain a char and further activation of the char in the gas phase. The carbonization or pyrolysis involves the thermal transformation of the precursor under an inert atmosphere at moderate/high temperatures (typically 400–600 °C). The obtained char is an amorphous solid that exhibits an incipient porosity which will be enlarged during the activation step. The activation is typically carried out under steam, carbon dioxide or their mixtures, at elevated temperatures (between 750–1100 °C). It is important to point out that the gasification temperature depends on the reactivity of the precursor under CO_2 or steam. A moderate reactivity, assuring a diffuse regime control of the reaction, typically allows controlling the activation (leading to high micropore volumes), while a fast reactivity provokes a fast gasification of the external surface (leading to lower micropore volumes and mesopores). Hence, the gasification regime of the char must be determined by measuring the thermal profile under CO_2 or steam. Additionally, for certain precursors (e.g., coals, wool wastes) a pre-oxidation step is recommended before the carbonization to favour the formation of an incipient porosity [15], and a well-developed microporosity after the gasification.

A detailed review of how the experimental conditions (e.g., precursor, heating rate, gas flow rate, temperature of activation, dwelling time, type of activating gas) influence the porosity of the obtained activated carbon, as well as the reactions can be found in [14]. Briefly, slow flow rates of an activating agent are applied (*ca.* below 10 ml/min) to allow uniform gasification with the formation of micropores, as opposed to external gasification with a predominance of large macropores [16]. Regarding the activation extent (or burn-off), values lower than *ca.* 40–50 wt.% usually render microporous carbons (linear dependence of surface areas and pore volumes with the burn-off). Above 40–50 wt.% burn-off, a widening of the micropores takes place with the development of mesopores [17,18].

4.2.2 Chemical Activation

Chemical activation of a carbon precursor consists of the decomposition of the precursor in the presence of a chemical activating agent, typically at moderate temperatures and in a single stage (although an extensive final washing step of the activated carbon is required). The most common activating agents are $ZnCl_2$, H_3PO_4, alkaline hydroxides and carbonates, and the most used precursors are lignocellulosic and biomass-derived precursors [14]. Compared to physical activation, chemical activation allows the use of lower temperatures and dwelling time which leads to higher yields. In general, chemical activation is preferred when biomass and lignocellulosic materials are used as precursors [19], with $ZnCl_2$ and H_3PO_4 being applied at temperatures around 450–550 °C, while alkaline hydroxides and carbonates are applied at temperatures around 700 °C (between 600–800 °C).

The main operating parameters that affect the porosity of the activated carbon during chemical activation procedures are typically the ratio of activating agent: precursor, mixing procedure, activation temperature, heating rate, dwelling time and activation atmosphere. As a general rule, the impregnation of the activating agent from the solution is more efficient (in terms of porosity development) than a solid mixture of the reactants [20]. The amount of activating agent favours the development of high surface areas (at the expense of the overall yield) due to the formation of micropores, whereas high impregnation ratios may lead to the development of mesopores and macropores [19,21,22]. The activation mechanisms of physical and chemical activation methods are different, and so is the porosity of the resulting carbons. In general, physical activation yields predominantly narrow pores, i.e., microporous carbons, with a small contribution of mesoporosity, while chemical activation allows the formation of wider micropore size distributions, upon the activating agent used, and favours the development of mesoporosity.

4.3   NOVEL SYNTHESIS METHODS

Both physical and chemical activation procedures do not allow fine control of the porosity of the resulting activated carbon. Thus, many efforts have been directed to develop new synthetic approaches for the fabrication of nanoporous carbons with high porous features and well-defined pore architectures. In the following sections, we will summarize the most relevant methodologies to control the microporosity in the synthesis of nanoporous carbons.

4.3.1   HYDRO-SOLVOTHERMAL CARBONIZATION

The thermochemical transformation under subcritical water at moderate temperatures and pressures of wet/dry biomass feedstocks and carbohydrates has received much attention for the preparation of porous carbons [23–25]. The decomposition reaction (hydrolysis) leads to the formation of a carbonaceous solid (known as hydrochar). The reaction can also be carried out in other solvents such as alcohols (solvothermal carbonization). Both high (300–800 °C) and low temperature (<300 °C) conditions can be applied for the preparation of carbon materials, although the second one is more common for the preparation of hydrochars. The nature of the precursor and the hydrothermal carbonization operating conditions (e.g., temperature, residence time, pressure) have a strong influence on the properties of the obtained hydrochars and further activated carbons [26–28].

While most hydrochars obtained display poor textural parameters (e.g., surface areas rarely exceeding $ca.$ 50 m^2/g), their activation leads to porous carbons with interesting features. On the other hand, high surface areas hydrochars can be prepared by controlling the salinity of the solution to modulate the cross-linking and growth of the primary carbon particles during hydrothermal carbonization. For instance, hydrochars with surface areas of $ca.$ 673 m^2/g have been reported from the hydrothermal carbonization of glucose under hypersaline conditions (Figure 4.2A) [29]. The nature of the salt seems to be important, with high surface areas obtained for hygroscopic salt ions or eutectic salt melts (e.g., ZnCl$_2$, LiCl, NaCl, KCl). Another route to increase the surface area of hydrochar is the use of metallic nanoparticles to accelerate the dehydration, condensation, and carbonization of the precursor during hydrothermal carbonization [30]. Another alternative for the synthesis of hydrochars with a well-developed pore structure is based on post-synthetic activation procedures. The reactivity and composition of the hydrochar precursor are important indicators to estimate the porous development of the hydrochar upon further activation processes [28,29]. For instance, high surface area activated carbons ($ca.$ up to 3362 m^2/g) have been prepared by chemical and physical activation of hydrochars obtained from different precursors (e.g., furfural, glucose, rice husks, starch, cellulose, eucalyptus

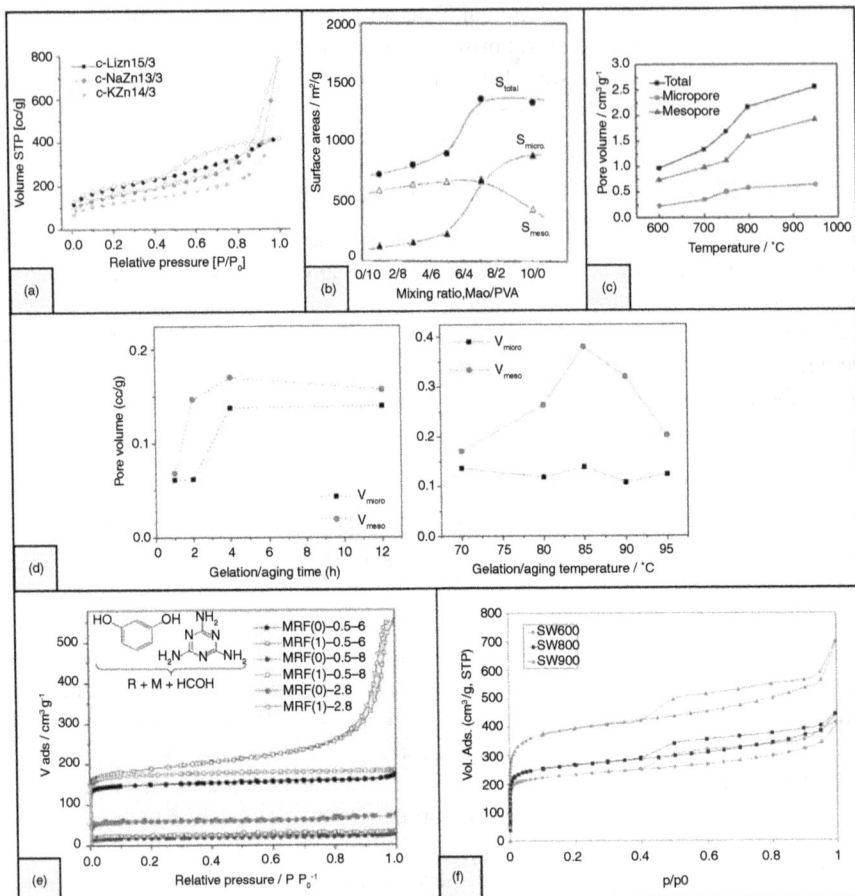

FIGURE 4.2 (A) N$_2$ adsorption/desorption isotherms at -196 °C for a series of hydrothermal activated carbons obtained at different temperatures. Reproduced from [29] with permission from the Royal Society of Chemistry. (B) Surface areas of nanoporous carbons prepared in Mg gluconate/PVA systems through powder mixing. Reproduced from [50] with permission from Elsevier. Copyright 2010. (C) Effect of temperature on pore volume upon the preparation of nanoporous carbon using ZnO nanoparticles as templates. Reprinted from [53] with permission from Elsevier. Copyright 2021. (D) Correlation between the gelation/aging time and temperature and the micro–mesopore volumes of porous carbons obtained by a sol-gel method. Reproduced from [63] with permission of Elsevier. Copyright 2015. (E) N$_2$ adsorption/desorption isotherms at -196 °C of porous carbons obtained by sol-gel polycondensation of melamine/resorcinol/formaldehyde mixtures, showing the evolution of the micro- and mesoporosity upon the synthesis conditions. Reprinted from [61] with permission from Elsevier. Copyright 2015. (F) N$_2$ adsorption/desorption isotherms at -196 °C for a series of activated carbons obtained upon self-activation of seaweeds (biomass).

sawdust). In general, the yields of activation of hydrochars are similar regardless of the activating agent, although KOH is the preferred reactant to generate microporous materials [31,32]. The use of $ZnCl_2$ and others favours micro-mesoporous carbons [33–35].

The temperature of the hydrothermal carbonization step and the structural order of the precursor also influence the porosity of the activated carbons obtained after post-synthesis activation of the hydrochar. In this regard, hydrothermal carbonizations between 180–240 °C favour further development of the microporosity upon KOH activation. On the other hand, as also observed for the KOH activation of coals, a low porous development is obtained upon activation of hydrochars with a high degree of aromatization and structural order [36].

4.3.2 NANOCASTING

The synthesis of nanoporous carbons by nanocasting methods involves the use of a sacrificial template, which can be generated in-situ or ex-situ: endo- or exo-templating routes, also known as soft- or hard-templating.

Endo-templating (or soft-templating) methods refer to the use of an organic template that is self-assembled in a solvent in the presence of a polymerizable carbon precursor. The carbon material is obtained after a carbonization process, where the template is destroyed [37,38]. This procedure has been mainly used for the preparation of carbon materials with hierarchical meso- and macroporosity [38], although it can also be extended to microporous carbons and molecular sieves. In this regard, the preparation of carbon materials with pores of different sizes is strongly related to the nature of the template. For instance, polyethylene oxide (PEO) blocks in block copolymeric surfactants tend to produce microporosity; ionic surfactants produce micropores and small mesopores (2–4 nm); and non-ionic surfactants (e.g., Brij and Pluronics) generate mesopores of up to 10 nm [39]. Typical precursors in soft-templating methods are hydroxylated aromatic molecules (e.g., phenol, phloroglucinol, resorcinol) and functionalized compounds (e.g., melamine, pyrrol) when doped carbons are targeted. For example, carbon materials with surface areas between 700–800 m^2/g and high pore volumes have been reported using phloroglucinol/glyoxylic acid as precursors and triblock copolymers as templates [40,41]. It has been reported that the chemistry of the organic precursor allows controlling the degree of cross-linking, and thus the porosity of the final carbons. As an example, the photoinduced assembly of linear dihydroxylated precursors forms dense porous carbon matrices mainly composed of narrow micropores, whereas trihydroxylated molecules render micro- and mesoporous carbons [41]. Some other soft-templating strategies based on the use of inorganic salts and oxides, high internal phase emulsions, or mechanochemical soft-template procedures have been reported in the literature [42].

On the other hand, exo-templating (or hard-templating) approaches use a porous solid as a sacrificial scaffold, and the porous carbon obtained is an inverse replica of the former. Typical sacrificial scaffolds are zeolites, mesoporous silica/oxides, opals, and colloidal silica suspensions, rendering micro-, meso- and macroporous materials, respectively [43]. Although hard-templating allows controlling the porosity of the resulting carbon (by the choice of the scaffold), the process is still complex and resources-consuming (the template needs to be removed). The hard-templating process requires several steps: (i) deposition of the carbon precursor (i.e., sucrose, phenol resin, polyacrylonitrile, pitch) inside the template, (ii) carbonization, and (iii) removal of the scaffold. Among them, the most crucial steps are the deposition of the carbon precursor and the carbonization.

Pioneering research on preparing microporous carbons via hard-nanocasting was performed by Kyotani and co-workers using zeolites and layered clay minerals as templates, reporting the synthesis of zeolite-templated carbons with a periodic arrangement and uniform microporosity in the range of 1–2 nm [44–47]. The choice of the scaffold is important, as it may affect the formation of the carbon layer upon the infiltration (e.g., acidic zeolites catalyze the decomposition of the carbon source by chemical vapour deposition (CVD), rendering higher porosity in the carbon replica). The carbon source is also important: small hydrocarbon gases (acetylene, propylene, benzene) are preferred for chemical vapour deposition, and polymeric monomers (e.g., furfuryl alcohol, acrylonitrile) dissolved in solvents for wet/vacuum impregnation. Better results in terms of higher surface area and micropore volumes have been reported for two-step methods combining CVD/wet impregnation protocols, allowing the uniform carbon deposition of the polymer carbonization and the formation of a robust carbon framework. As a result, microporous carbons with surface areas between 3600–4100 m^2/g, and micropore volumes of up to 1.5 cm^3/g have been reported. Unlike carbons obtained from KOH activation, such zeolite-templated carbons have a very narrow micropore-sized distribution without mesoporosity, as well as high mechanical stability and electrical conductivity (when annealed above 1000 °C). Besides zeolites, other scaffolds have been used such as metallic oxides (e.g., MgO, Fe$_2$O$_3$, MnO$_2$), clays, colloidal crystals, mesoporous silica and MOFs (Figure 4.2 B) [48–50]. As an example, highly microporous carbons with large pore volumes (*ca.* 2.06 cm^3/g) and surface areas of up to 2872 m^2/g have been obtained using various MOFs as templates [51]. Furthermore, porous carbons with a micro- and mesopore structure have been obtained using ZnO nanoparticles as a scaffold, due to the dual role of Zn as an activating agent and graphitization catalyst (Figure 4.2C) [52,53].

4.3.3 SOL-GEL

Sol-gel approaches are based on the polycondensation of organic precursors and aldehydes in the presence of a catalyst. The resulting polymeric organic gels or resins can be converted to a rich-carbon material upon carbonization. In 1989, Pekala et al. [54] reported for the first time the sol-gel polymerization and condensation of resorcinol (R) and formaldehyde (F) in the presence of a catalyst (C) in water. The sol-gel polycondensation of R-F mixtures is endothermic and is typically catalyzed at *ca.* 80 °C in aqueous medium. A curing step usually at 80–90 °C allows the polymeric clusters to crosslink and form a wet organic gel. This is usually carried out by conventional heating furnaces, although microwave heating [55], photo-assisted [41] and laser-induced [56] have also been reported to reduce the reaction time and energy consumption.

Regarding synthesis, a large number of parameters may affect the properties of the organic gels and the carbons (pH of the solution, molar ratios of the components, and curing temperature, among others) (Figure 4.2D) [57–60]. The nature and concentration of the catalyst affect the porosity of the carbon gel. The most widely used catalyst is Na_2CO_3, and as a general trend, a low R/C molar ratio favours the formation of predominantly microporous carbons with low surface areas. In contrast, high R/C ratios render carbon gels with larger surface areas and a well-developed mesoporous structure [59,61,62]. Generally, a gelation temperature increase does not affect the microporosity but raises the mesoporosity [63]. On the other hand, the temperature of carbonization, time, and heating rate become significant parameters in the final porous network. Furthermore, carbonized gels can be activated to improve porosity, similarly to any other carbon precursor [20].

Other substituted phenolic compounds (e.g., catechol, phenol, hydroquinone), and heteroatom-containing molecules (e.g., urea, melamine, 3-hydroxypyridine) have been used as monomers for the preparation of functionalized carbon gels. The resulting porosity depends on the reactivity of the reactant's mixture, rendering predominantly microporous or micro-/mesoporous materials upon the operating conditions (Figure 4.2E). Furthermore, the incorporation of conductive additives (e.g., carbon black, graphene-derivatives) or metallic compounds opens up the possibility of preparing highly porous carbons with improved electrical conductivity carbons, in terms of efficiency or energetic yield [13,64,65]. Indeed, it was reported that the amount of carbonaceous additive and its nature (i.e., particle size, hydrophobic/hydrophilic character, structure, surface chemistry) strongly affect the final physicochemical properties of the carbon gel [66]. Another advantage of sol-gel routes is that they enable production of monolithic carbons, which offers important advantages compared to powder or pellets (e.g., better diffusion and mass transfer rates of fluids, resistance to compression) [67,68].

4.3.4 SELF-ACTIVATION

Self-activation consists of the single-step thermal treatment of a carbon precursor, rendering a porous carbon material in the absence of an added activating agent. Indeed, the composition of the precursor enables self-activation (chemical or physical) during the thermal degradation of the matrix. For example, highly porous carbons have been obtained by carbonization of organic polymers, biopolymers and their salts (i.e., gluconates, alginates, citrates, sulfonates, maleic acid), and biomass containing alkaline or alkali earth elements as precursors [69–83]. These compounds combine a carbon precursor and certain elements that during the heat treatment generate species capable of acting as activating agents (releasing CO_2, K, Na, etc.).

The pyrolysis yield and porous features of the carbon materials strongly depend on the composition of the feedstock and for a given precursor on the operating parameter (temperature and dwelling time). For instance, alkali (K and Na) and nickel salts render essentially microporous carbons, whereas calcium, cobalt and iron favour the formation of mesopores [69–72]. Following these procedures, nanoporous carbons with surface areas ranging between 650–1960 m^2/g and pore volumes of up to 2.6 cm^3/g have been reported. Another example is the one-step self-activation of biomass (e.g., seaweeds) rendering carbons with surface areas over 1000 m^2/g (Figure 4.2F) and a slightly developed mesoporosity [73].

4.4　CONCLUSIONS

Owing to their inherent attributes in terms of high surface areas and pore volumes, tuneable porosity and versatile surface chemistry, porous carbons are key materials in many application fields such as water treatment, gas adsorption/separation, environmental remediation, and energy conversion and storage. Activated carbons are manufactured at an industrial scale by the thermal or chemical gasification of a carbon precursor. While much knowledge has been accumulated over the years regarding the optimal activation conditions depending on the carbon precursor, the control of the porous features with these processes is still a challenge; particularly when high micropores volumes of adequate pore sizes are required. For this reason, much research has been directed towards the development of strategies to prepare nanoporous carbons with controlled textural features and increased carbon yield using renewable and abundant sources as precursors and sustainable and less energy-demanding processes. Among those strategies, solvothermal carbonization, nanocasting, sol-gel methods and self-activating processes have been widely explored. Matching the precursor to the synthetic procedure, using earth-abundant activating agents, or applying post-synthetic activation methodologies have appeared as interesting approaches to obtain carbons with controlled porous features in the full nanometric scale.

The ever-increasing interest in these materials in different application fields triggers the continuous research on new synthetic methods to obtain carbon materials with controlled porosity without sacrificing other key performance parameters such as composition, wettability, electrical conductivity, or structural order (among others). Yet, the critical issue of all these procedures is to demonstrate the scalability of the synthetic routes and to evaluate the final cost to verify whether the performance of such carbons with controlled pore architectures is compatible with the overall cost at large scale for near-market implementation.

ACKNOWLEDGEMENTS

A.C. thanks European Union NextGenerationEU (Margarita Salas) for the funding. This work has received funding from the European Union's Horizon 2020 research and innovation program under the grant agreement No 776816. ER-P and COA thank the Région Centre Val de Loire (Project APR-IA PRESERVE convention no 00134933; MATHYFON convention no 240602) for financial support. AG-B thanks European Union NextGenerationEU (ZAMBRANO21-10) for the funding.

REFERENCES

[1] V. Meunier, C. Ania, A. Bianco, Y. Chen, G.B. Choi and Y.A. Kim et al., Carbon science perspective in 2022: Current research and future challenges, *Carbon* 195 (2022), pp. 272–291.

[2] E. Raymundo-Pinero and F. Béguin, Application of nanotextured carbons for supercapacitors and hydrogen storage, in *Interface Science and Technology*, T.J. Bandosz, ed., Elsevier, (2006), pp. 293–343.

[3] M. Sevilla and R. Mokaya, Energy storage applications of activated carbons: Supercapacitors and hydrogen storage, *Energy Environ Sci* 7 (2014), pp. 1250–1280.

[4] C. Hu, Q. Dai and L. Dai, Multifunctional carbon-based metal-free catalysts for advanced energy conversion and storage, *Cell Rep Phys Sci* 2 (2021), pp. 100328.

[5] C. Hu, X. Chen, Q. Dai, M. Wang, L. Qu and L. Dai, Earth-abundant carbon catalysts for renewable generation of clean energy from sunlight and water, *Nano Energy* 41 (2017), pp. 367–376.

[6] Y. Peng, A. Rendón-Patiño, A. Franconetti, J. Albero, A. Primo and H. García, Photocatalytic overall water splitting activity of templateless structured graphitic nanoparticles obtained from cyclodextrins, *ACS Appl Energy Mater* 3 (2020), pp. 6623–6632.

[7] R. Ma, G. Lin, Y. Zhou, Q. Liu, T. Zhang and G. Shan et al., A review of oxygen reduction mechanisms for metal-free carbon-based electrocatalysts, *Comput Mater* 5 (2019), pp. 1–15.

[8] W. Li, M. Seredych, E. Rodríguez-Castellón and T.J. Bandosz, Metal-free nanoporous carbon as a catalyst for electrochemical reduction of CO_2 to CO and CH_4, *ChemSusChem* 9 (2016), pp. 606–616.

[9] H. Wang, J. Jia, P. Song, Q. Wang, D. Li and S. Min et al., Efficient electrocatalytic reduction of CO_2 by Nitrogen-doped nanoporous carbon/carbon nanotube membranes: A step towards the electrochemical CO_2 refinery, *Angew Chem Int Ed* 56 (2017), pp. 7847–7852.

[10] I. Harmanli, N.V. Tarakina, M. Antonietti and M. Oschatz, "Giant" nitrogen uptake in ionic liquids confined in carbon pores, *J Am Chem Soc* 143 (2021), pp. 9377–9384.

[11] F. Béguin and E. Frackowiak, *Carbons for Electrochemical Energy Storage and Conversion Systems*, 1st ed. CRC Press, Boca Raton, (2009).

[12] J. Iniesta, L. García-Cruz, A. Gomis-Berenguer and C.O. Ania, Carbon materials based on screen-printing electrochemical platforms in biosensing applications, *SPR Electrochem* 13 (2015), pp. 133–169.

[13] A. Casanova, A. Gomis-Berenguer, A. Canizares, P. Simon, M.D. Calzada and C.O. Ania, Carbon black as conductive additive and structural director of porous carbon gels, *Materials* 13 (2020).

[14] H. Marsh and F. Rodríguez-Reinoso, *Activated Carbon*, 1st ed. Elsevier Science, (2006).

[15] J.B. Parra, J.J. Pis, J.C. de Sousa, J.A. Pajares and R.C. Bansal, Effect of coal preoxidation on the development of microporosity in activated carbons, *Carbon* 34 (1996), pp. 783–787.

[16] J.J. Pis, M. Mahamud, J.A. Pajares, J.B. Parra and R.C. Bansal, Preparation of active carbons from coal: Part III: Activation of char, *Fuel Process Technol* 57 (1998), pp. 149–161.

[17] F. Rodriguez-Reinoso, Controlled gasification of carbon and pore structure development, in *Fundamental Issues in Control of Carbon Gasification Reactivity*, J. Lahaye and P. Ehrburger, eds., Springer, Dordrecht, Dordrecht, (1991), pp. 533–571.

[18] F. Rodríguez-Reinoso, M. Molina-Sabio and M.T. González, The use of steam and CO_2 as activating agents in the preparation of activated carbons, *Carbon* 33 (1995), pp. 15–23.

[19] M. Molina-Sabio and F. Rodríguez-Reinoso, Role of chemical activation in the development of carbon porosity, *Colloids Surf A Physicochem Eng Asp* 241 (2004), pp. 15–25.

[20] A. Gomis-Berenguer, R. García-González, A.S. Mestre and C.O. Ania, Designing micro- and mesoporous carbon networks by chemical activation of organic resins, *Adsorption* 23 (2017), pp. 303–312.

[21] D. Lozano-Castelló, M.A. Lillo-Ródenas, D. Cazorla-Amorós and A. Linares-Solano, Preparation of activated carbons from Spanish anthracite: I. Activation by KOH, *Carbon* 39 (2001), pp. 741–749.

[22] Y. Nakagawa, M. Molina-Sabio and F. Rodríguez-Reinoso, Modification of the porous structure along the preparation of activated carbon monoliths with H3PO4 and ZnCl2, *Micropor Mesopor Mater* 103 (2007), pp. 29–34.

[23] T. Fujino, J.M. Calderon-Moreno, S. Swamy, T. Hirose and M. Yoshimura, Phase and structural change of carbonized wood materials by hydrothermal treatment, *Solid State Ion* 151 (2002), pp. 197–203.

[24] M. Sevilla and A. B. Fuertes, Chemical and structural properties of carbonaceous products obtained by hydrothermal carbonization of saccharides, *Chem Eur J* 15 (2009), pp. 4195–4203.

[25] E. Dinjus, A. Kruse and N. Tröger, Hydrothermale Karbonisierung: 1. Einfluss des Lignins in Lignocellulosen, *Chem Ing Tech* 83 (2011), pp. 1734–1741.

[26] M.M. Titirici and M. Antonietti, Chemistry and materials options of sustainable carbon materials made by hydrothermal carbonization, *Chem Soc Rev* 39 (2009), pp. 103–116.

[27] A. Marinovic, F.D. Pileidis and M.M. Titirici, Hydrothermal carbonisation (HTC): History, state-of-the-art and chemistry, in *Porous Carbon Materials from Sustainable Precursors*, R.J. White, ed., Royal Society of Chemistry, London, (2015), pp. 129–155.

[28] A. Jain, R. Balasubramanian and M.P. Srinivasan, Hydrothermal conversion of biomass waste to activated carbon with high porosity: A review, *Chem Eng J* 283 (2016), pp. 789–805.

[29] N. Fechler, S.A. Wohlgemuth, P. Jäker and M. Antonietti, Salt and sugar: Direct synthesis of high surface area carbon materials at low temperatures via hydrothermal carbonization of glucose under hypersaline conditions, *J Mater Chem A Mater* 1 (2013), pp. 9418–9421.

[30] X. Cui, M. Antonietti and S.H. Yu, Structural effects of iron oxide nanoparticles and iron ions on the hydrothermal carbonization of starch and rice carbohydrates, *Small* 2 (2006), pp. 756–759.

[31] Z. Liu and F.S. Zhang, Removal of copper (II) and phenol from aqueous solution using porous carbons derived from hydrothermal chars, *Desalination* 267 (2011), pp. 101–106.

[32] A.J. Romero-Anaya, M. Ouzzine, M.A. Lillo-Ródenas and A. Linares-Solano, Spherical carbons: Synthesis, characterization and activation processes, *Carbon* 68 (2014), pp. 296–307.

[33] A.B. Fuertes and M. Sevilla, Superior capacitive performance of hydrochar-based porous carbons in aqueous electrolytes, *ChemSusChem* 8 (2015), pp. 1049–1057.

[34] T. Liu, Y. Li, N. Peng, Q. Lang, Y. Xia and C. Gai et al., Heteroatoms doped porous carbon derived from hydrothermally treated sewage sludge: Structural characterization and environmental application, *J Environ Manage* 197 (2017), pp. 151–158.

[35] S. Masoumi and A.K. Dalai, Optimized production and characterization of highly porous activated carbon from algal-derived hydrochar, *J Clean Prod* 263 (2020), pp. 121427.

[36] C. Falco, J.P. Marco-Lozar, D. Salinas-Torres, E. Morallón, D. Cazorla-Amorós and M.M. Titirici et al., Tailoring the porosity of chemically activated hydrothermal carbons: Influence of the precursor and hydrothermal carbonization temperature, *Carbon* 62 (2013), pp. 346–355.

[37] A.H. Lu and F. Schüth, Nanocasting: A versatile strategy for creating nanostructured porous materials, *Adv Mater* 18 (2006), pp. 1793–1805.

[38] L. Chuenchom, R. Kraehnert and B.M. Smarsly, Recent progress in soft-templating of porous carbon materials, *Soft Matter* 8 (2012), pp. 10801–10812.

[39] N.D. Petkovich and A. Stein, Controlling macro- and mesostructures with hierarchical porosity through combined hard and soft templating, *Chem Soc Rev* 42 (2013), pp. 3721–3739.

[40] C. Matei Ghimbeu, L. Vidal, L. Delmotte, J.M. le Meins and C. Vix-Guterl, Catalyst-free soft-template synthesis of ordered mesoporous carbon tailored using phloroglucinol/glyoxylic acid environmentally friendly precursors, *Green Chem* 16 (2014), pp. 3079–3088.

[41] L. Balan, M.C. Fernández de Córdoba, M. Zaier and C.O. Ania, A green and fast approach to nanoporous carbons with tuned porosity: UV-assisted condensation of organic compounds at room temperature, *Carbon* 116 (2017), pp. 264–274.

[42] N. Díez, M. Sevilla and A.B. Fuertes, Synthesis strategies of templated porous lcarbons beyond the silica nanocasting technique, *Carbon* 178 (2021), pp. 451–476.

[43] R. Ryoo, S.H. Joo and S. Jun, Synthesis of highly ordered carbon molecular sieves via template-mediated structural transformation, *J Phys Chem B* 103 (1999), pp. 7743–7746.

[44] T. Kyotani, N. Sonobe and A. Tomita, Formation of highly orientated graphite from polyacrylonitrile by using a two-dimensional space between montmorillonite lamellae, *Nature* 331 (1988), pp. 331–333.

[45] T. Kyotani, Z. Ma and A. Tomita, Template synthesis of novel porous carbons using various types of zeolites, *Carbon* 41 (2003), pp. 1451–1459.

[46] Z. Ma, T. Kyotani and A. Tomita, Preparation of a high surface area microporous carbon having the structural regularity of Y zeolite, *Chem Commun* 11 (2000), pp. 2365–2366.

[47] T. Kyotani, T. Nagai, S. Inoue and A. Tomita, Formation of new type of porous carbon by carbonization in zeolite nanochannels, *Chem Mater* 9 (1997), pp. 609–615.

[48] Y. Xia, Z. Yang and R. Mokaya, Templ+/a.cwted nanoscale porous carbons, *Nanoscale* 2 (2010), pp. 639–659.

[49] C. Zhu, M. Takata, Y. Aoki and H. Habazaki, Nitrogen-doped porous carbon as-mediated by a facile solution combustion synthesis for supercapacitor and oxygen reduction electrocatalyst, *Chem Eng J* 350 (2018), pp. 278–289.

[50] T. Morishita, T. Tsumura, M. Toyoda, J. Przepiórski, A.W. Morawski and H. Konno et al., A review of the control of pore structure in MgO-templated nanoporous carbons, *Carbon* 48 (2010), pp. 2690–2707.

[51] B. Liu, H. Shioyama, T. Akita and Q. Xu, Metal-organic framework as a template for porous carbon synthesis, *J Am Chem Soc* 130 (2008), pp. 5390–5391.

[52] J. Yin, W. Zhang, N.A. Alhebshi, N. Salah and H.N. Alshareef, Synthesis strategies of porous carbon for supercapacitor applications, *Small Methods* 4 (2020), pp. 1900853.

[53] B. Yan, J. Zheng, F. Wang, L. Zhao, Q. Zhang and W. Xu et al., Review on porous carbon materials engineered by ZnO templates: Design, synthesis and capacitance performance, *Mater Des* 201 (2021), pp. 109518.

[54] R.W. Pekala, Organic aerogels from the polycondensation of resorcinol with formaldehyde, *J Mater Sci* 24 (1989), pp. 3221–3227.

[55] M.L. Rojas-Cervantes, Some strategies to lower the production cost of carbon gels, *J Mater Sci* 50 (2015), pp. 1017–1040.

[56] W. Ma, J. Zhu, Z. Wang, W. Song and G. Cao, Recent advances in preparation and application of laser-induced graphene in energy storage devices, *Mater Today Energy* 18 (2020), pp. 100569.

[57] S.A. Al-Muhtaseb and J.A. Ritter, Preparation and properties of resorcinol-formaldehyde organic and carbon gels, *Adv Mater* 15 (2003), pp. 101–114.

[58] N. Job, R. Pirard, J. Marien and J.P. Pirard, Porous carbon xerogels with texture tailored by pH control during sol-gel process, *Carbon* 42 (2004), pp. 619–628.

[59] G. Rasines, P. Lavela, C. Macías, M.C. Zafra, J.L. Tirado and C.O. Ania, Mesoporous carbon black- aerogel composites with optimized properties for the electro-assisted removal of sodium chloride from brackish water, *J Electroanal Chem* 741 (2015), pp. 42–50.

[60] A. Arenillas, J Angel Menéndez, G. Reichenauer, A. Celzard, V. Fierro and F. José et al., Advances in sol-gel derived materials and technologies. Series Editors: M.A. Aegerter and M. Prassas *Organic and Carbon Gels*, 1st ed. Springer Cham, New York, (2019).

[61] G. Rasines, P. Lavela, C. Macías, M.C. Zafra, J.L. Tirado and J.B. Parra et al., N-doped monolithic carbon aerogel electrodes with optimized features for the electrosorption of ions, *Carbon* 83 (2015), pp. 262–274.

[62] G. Rasines, P. Lavela, C. Macías, M.C. Zafra, J.L. Tirado and C.O. Ania, On the use of carbon black loaded nitrogen-doped carbon aerogel for the electrosorption of sodium chloride from saline water, *Electro Acta* 170 (2015), pp. 154–163.

[63] E. Isaacs Páez, M. Haro, E.J. Juárez-Pérez, R.J. Carmona, J.B. Parra and R. Leyva Ramos et al., Fast synthesis of micro/mesoporous xerogels: Textural and energetic assessment, *Micropor Mesopor Mater* 209 (2015), pp. 2–9.

[64] C. Macías, G. Rasines, P. Lavela, M.C. Zafra, J.L. Tirado and C. Ania, Mn-containing N-doped monolithic carbon aerogels with enhanced macroporosity as electrodes for capacitive deionization, *ACS Sustain Chem Eng* 4 (2016), pp. 2487–2494.

[65] M. Canal-Rodríguez, J.A. Menéndez, M.A. Montes-Morán and A. Arenillas, The relevance of conductive additive addition methodology for optimizing the performance of electrodes based on carbon xerogels in aqueous supercapacitors, *J Electroanal Chem* 836 (2019), pp. 45–49.

[66] A. Casanova, PhD Thesis 2020. http://hdl.handle.net/10396/20501

[67] M. Antonietti, N. Fechler and T.P. Fellinger, Carbon aerogels and monoliths: Control of porosity and nanoarchitecture via Sol-Gel routes, *Chem Mater* 26 (2014), pp. 196–210.

[68] M. Salihovic, P. e, S. Herou, M.M. Titirici, N. Hüsing and M.S. Elsaesser, Monolithic carbon spherogels as freestanding electrodes for supercapacitors, *ACS Appl Energy Mater* 4 (2021), pp. 11183–11193.

[69] D. Hines, A. Bagreev and T.J. Bandosz, Surface properties of porous carbon obtained from polystyrene sulfonic acid-based organic salts, *Langmuir* 20 (2004), pp. 3388–3397.

[70] C.O. Ania and T.J. Bandosz, Metal-loaded polystyrene-based activated carbons as dibenzothiophene removal media via reactive adsorption, *Carbon* 44 (2006), pp. 2404–2412.

[71] M. Sevilla and A.B. Fuertes, A general and facile synthesis strategy towards highly porous carbons: Carbonization of organic salts, *J Mater Chem A Mater* 1 (2013), pp. 13738–13741.

[72] G.A. Ferrero, M. Sevilla and A.B. Fuertes, Mesoporous carbons synthesized by direct carbonization of citrate salts for use as high-performance capacitors, *Carbon* 88 (2015), pp. 239–251.

[73] E. Raymundo-Piñero, M. Cadek and F. Béguin, Tuning carbon materials for supercapacitors by direct pyrolysis of seaweeds, *Adv Funct Mater* 19 (2009), pp. 1032–1039.

[74] J. He, D. Zhang, M. Han, X. Liu, Y. Wang and Y. Li et al., One-step large-scale fabrication of nitrogen doped microporous carbon by self-activation of biomass for supercapacitors application, *J Energy Storage* 21 (2019), pp. 94–104.

[75] Y. Yao, Q. Zhang, P. Liu, L. Yu, L. Huang, S.Z. Zeng et al., Facile synthesis of high-surface-area nanoporous carbon from biomass resources and its application in supercapacitors, *RSC Adv* 8 (2018), pp. 1857–1865.

[76] P. Kleszyk, P. Ratajczak, P. Skowron, J. Jagiello, Q. Abbas and E. Frąckowiak et al., Carbons with narrow pore size distribution prepared by simultaneous carbonization and self-activation of tobacco stems and their application to supercapacitors, *Carbon* 81 (2015), pp. 148–157.

[77] Z. Li, W. Lv, C. Zhang, B. Li, F. Kang and Q.H. Yang, A sheet-like porous carbon for high-rate supercapacitors produced by the carbonization of an eggplant, *Carbon* 92 (2015), pp. 11–14.

[78] K. Sun, C.Y. Leng, J.C. Jiang, Q. Bu, G.F. Lin and X.C. Lu et al., Microporous activated carbons from coconut shells produced by self-activation using the pyrolysis gases produced from them, that have an excellent electric double layer performance, *New Carbon Mater* 32 (2017), pp. 451–459.

[79] A. Wang, K. Sun, R. Xu, Y. Sun and J. Jiang, Cleanly synthesizing rotten potato-based activated carbon for supercapacitor by self-catalytic activation, *J Clean Prod* 283 (2021), pp. 125385.

[80] C. Xia and S.Q. Shi, Self-activation for activated carbon from biomass: theory and parameters, *Green Chem* 18 (2016), pp. 2063–2071.

[81] C. Bommier, R. Xu, W. Wang, X. Wang, D. Wen and J. Lu et al., Self-activation of cellulose: A new preparation methodology for activated carbon electrodes in electrochemical capacitors, *Nano Energy* 13 (2015), pp. 709–717.

[82] C. Xia, C. Kang, M.D. Patel, L. Cai, B. Gwalani and R. Banerjee et al., Pine wood extracted activated carbon through self-activation process for high-performance lithium-ion battery, *ChemistrySelect* 1 (2016), pp. 4000–4007.

[83] Z. Zhang, J. He, X. Tang, Y. Wang, B. Yang and K. Wang et al., Supercapacitors based on a nitrogen doped hierarchical porous carbon fabricated by self-activation of biomass: Excellent rate capability and cycle stability, *Carbon Lett* 29 (2019), pp. 585–594.

5 Perspectives on Natural Clay Composites for Energy Generation, Storage, and Conversion

Revathy K.P and Vinod V.T Padil

5.1 INTRODUCTION

The search for green and sustainable energy sources is an important credential of the growing population in the world. Severe environmental pollution, water scarcity, and uncontrollable population explosion are the major causes of a world population with growing economic needs and high demand for energy for its uses in daily life. Scientists, engineers, and other professionals are looking for clean, renewable, and sustainable energy of the future such as solar energy, wind, and tidal, which will help reduce dependence on fossil fuels significantly. However, these renewable energies have alternating and recurrent issues, thus making them difficult to contribute uninterruptedly. In this context, high preferences are required for the development of competent energy storage and conversion systems. Today most energy storage depends on batteries and supercapacitors with their efficiency to provide high energy density and power density. Various components of the batteries such as electrodes, electrolytes, and separates will play a key role in energy storage systems. The major energy conversion procedures involve fuel cells and solar panels, and they could be integrated with energy storage devices. Developing green, cheap, abundant, high-performance, and sustainable materials for energy storage and its conversion is the need of the hour.

Many energy conversion and storage devices, such as solar cells, flywheels, compressed air, fuel cells, supercapacitors, and batteries, also have been developed to date from natural renewable sources such as solar, wind, geothermal, tidal, or biomass energy(1) The current chapter explores the recent developments in clay-based energy materials, including their physical, chemical, and structural modifications. Further, the chapter envisages the porous clay materials and their electrochemical properties, which are reflected in lithium-ion batteries (LIB), lithium-sulfur (Li–S) batteries, zinc-ion batteries (ZIBs), chloride-ion batteries (CIBs), supercapacitors, solar cells, and fuel cells(2).

DOI: 10.1201/9781003318859-5

5.2 CLASSIFICATION AND STRUCTURAL ASSIGNMENTS OF NATURAL CLAY

In this current scenario, natural clay materials and their composites are the emblematic natural cradles for the development of energy storage and conversion medium. The major availability of natural clay comes from South America, Central Africa, India, and East Australia(3). A brief description of natural clay and its predominant subgroups and their physicochemical properties are presented in Figures 5.1, 5.2 and Table 5.1, respectively.

5.3 ENERGY STORAGE APPLICATIONS OF CLAY COMPOSITES

The major advantages of clay and their nanocomposite materials in the areas of energy storage and conversion applications rely on their unique physical, chemical, thermal, mechanical, high stability, porosity, high surface area, active functional sites, and resistance to fire apart from their cost effectiveness and non-toxicity. Furthermore, the ionic conductivity and hydrophilicity have added benefits for their application in battery components as solid-state electrolytes(12). Various structural architectures such as 1D, 2D and other types of clay prevalently exhibited in various energy storage devices are presented schematically in Figure 5.3.

Natural clay materials have interacted with many polymers or additives and their composite polymer composite materials have been developed as separators., solid polymer electrolytes, solid-state electrolytes, and gel polymer electrolytes for energy storage applications. Cite the examples of polymer materials/additives such as PVDF (poly(vinylpyrrolidone), bacterial cellulose, nonwoven fabrics, methoxy poly (ethylene glycol) acrylate, dimethyl sulfoxide, PEO and LiTFSI, and cellulose acetate/poly-L-lactic acid with natural clay-based energy storage materials(14–16). These composites showed excellent ionic conductivity and porosity which shows their application in the merits for developing clay-based energy materials.

Further to improve the electrochemical properties of clay-based materials, the functionalization or modification of these materials is imperative for various energy storage and conversion fields. There are different processes which are generally summarized as acid leaching(17,18), expansion of interlayer spacing(18–20), cationic exchange(19,20), wettability modification (20), and calcination treatment(21,22).

5.4 APPLICATIONS OF CLAY COMPOSITES FOR ENERGY STORAGE

Two major devices in electrochemical energy storage include supercapacitors and batteries. To improve the much more environmentally friendly components such as electrode materials in these devices, scientists and researchers look for biomass-derived materials as well as natural clay-based materials instead of graphite and other rare earth metals-based inorganic compounds. Various clay composites have been used for the fabrication of cathodes, anodes, separators, and electrolytes in rechargeable batteries. Table 5.1 represents the various clay-based materials composite that

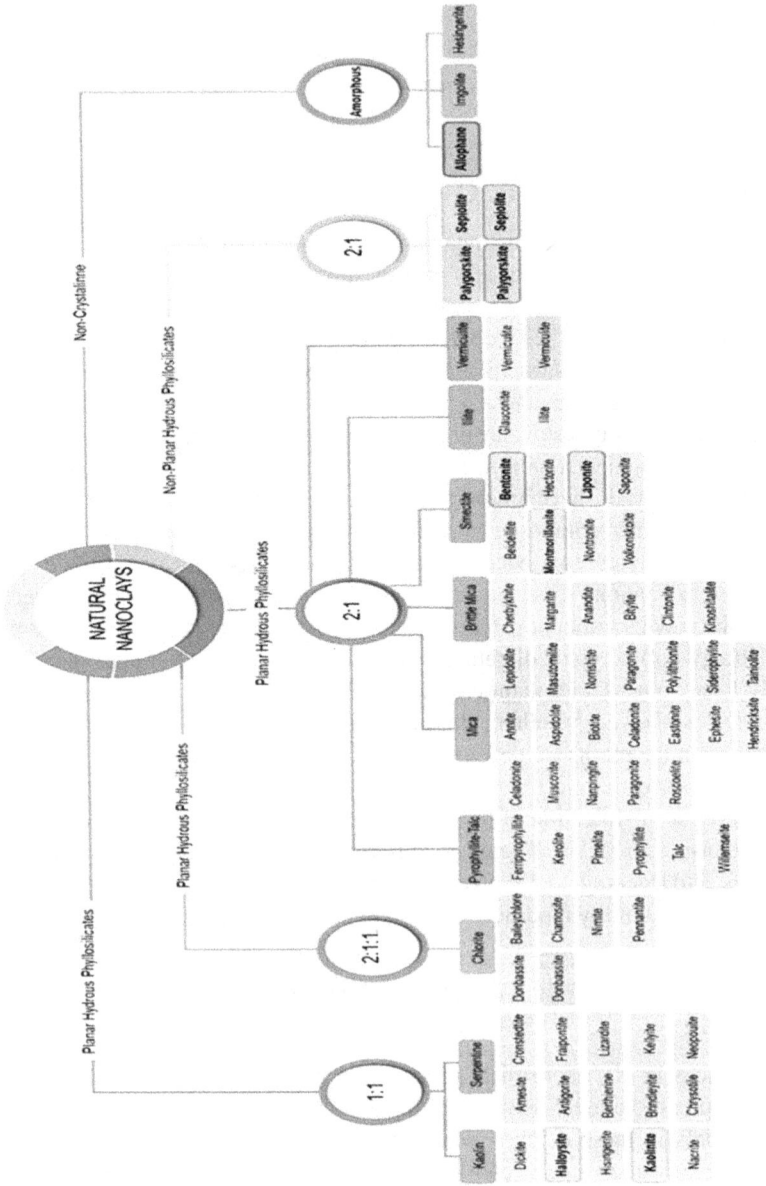

FIGURE 5.1 Crystalline and non-crystalline classification of natural clay. Reprinted with permission from Ref. (4).

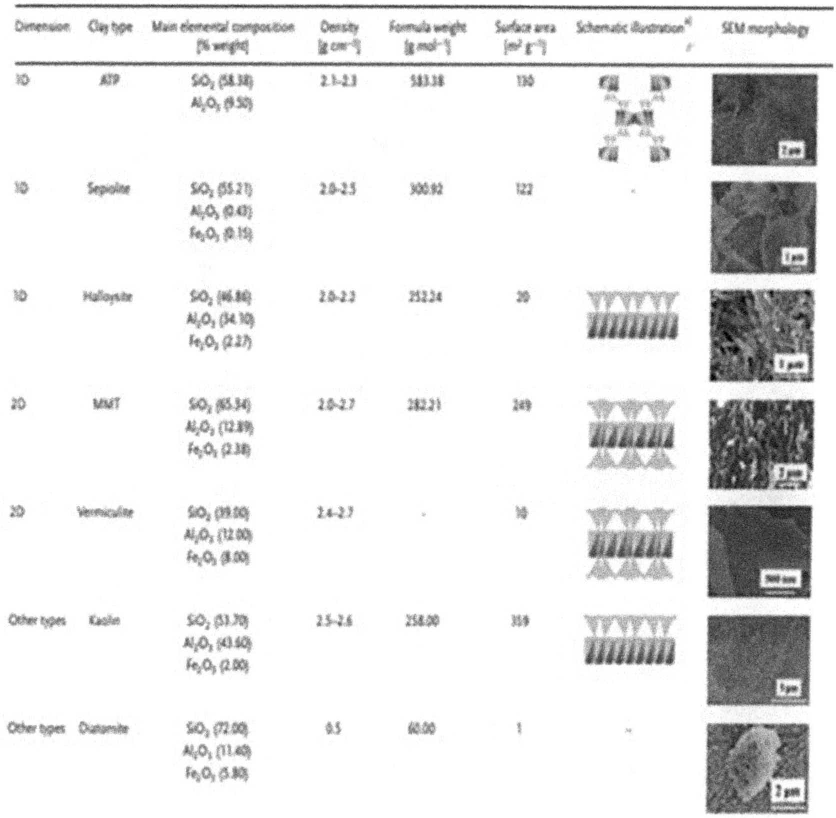

Dimension	Clay type	Main elemental composition [% weight]	Density [g cm⁻³]	Formula weight [g mol⁻¹]	Surface area [m² g⁻¹]	Schematic illustration	SEM morphology
1D	ATP	SiO_2 (58.38), Al_2O_3 (9.50)	2.1–2.3	583.38	130		
1D	Sepiolite	SiO_2 (55.21), Al_2O_3 (0.43), Fe_2O_3 (0.15)	2.0–2.5	300.92	122		
1D	Halloysite	SiO_2 (46.86), Al_2O_3 (34.10), Fe_2O_3 (2.27)	2.0–2.3	252.24	20		
2D	MMT	SiO_2 (65.34), Al_2O_3 (12.89), Fe_2O_3 (2.38)	2.0–2.7	282.21	249		
2D	Vermiculite	SiO_2 (39.00), Al_2O_3 (12.00), Fe_2O_3 (8.00)	2.4–2.7	-	10		
Other types	Kaolin	SiO_2 (53.70), Al_2O_3 (43.60), Fe_2O_3 (2.00)	2.5–2.6	258.00	359		
Other types	Diatomite	SiO_2 (72.00), Al_2O_3 (11.40), Fe_2O_3 (5.80)	0.5	60.00	1		

FIGURE 5.2 Various natural clay types and their physicochemical, morphological, and structural assignments. Reproduced with permission from Refs. (5–11).

have been used for various battery components, and their properties and merits are illustrated.

5.5 CLAY-BASED MATERIALS FOR SUPERCAPACITORS

Supercapacitors play an important role in electrochemical storage devices for practical applications. The charge is stored in a supercapacitor as a result of electrode/electrolyte interference via an electrical double layer or reversible faradic reactions(38). Due to their high capacitance, high power density, quick charging/discharging processes, and extended cycle life, supercapacitors have emerged as the supreme leaders in energy storage and are employed in biomedical, antimicrobial fabrics, flexible displays, wearable electronic devices, and mobile phones(39).

An ex-situ and in-situ ternary graphene/polypyrrole (PPy) nanocomposite was made using chemical reactive polymerization techniques using nanoclay(26). SEM and field emission microscopy surface morphology analysis revealed that both nanoclay

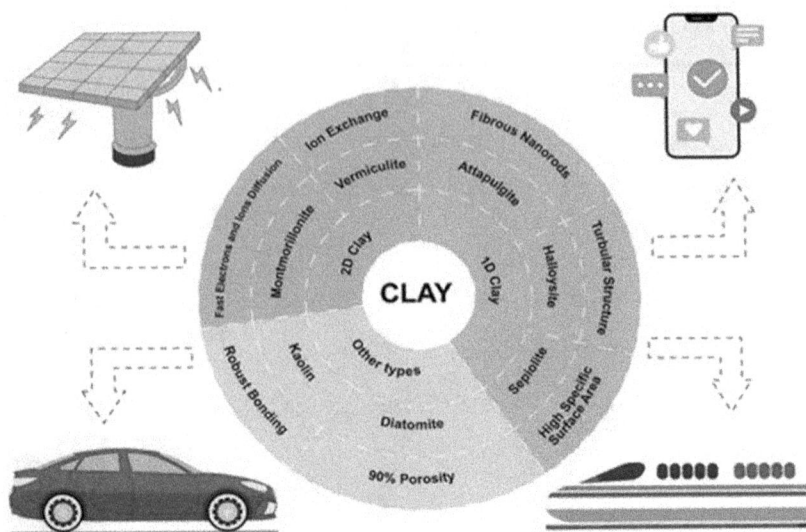

FIGURE 5.3 Various structural features of clay materials and their application in energy storage devices. Permission with Ref. from (13) here.

and graphene were coated with PPy. Utilizing cyclic voltammetry, potentiostatic charging-discharging, and electrochemical impedance spectroscopy, the electrochemical capabilities of each nanocomposite were assessed. The in-situ nanocomposite had a greater specific capacitance (347 F/g) than the ex-situ nanocomposite at a scan rate of 10 mV/s in 1 MKCl as a solution. The effect of sequential nanoclay addition on specific capacitance was investigated. This research highlighted the importance of clay-based nanocomposites and their appliances as energy storage devices.

A number of hierarchical open-linked mesoporous electrode materials for supercapacitors (SCs) based on nanoclay (Closite 30B) in combination with CNTs (carbon nanotube) and PPy (polypyrrole) via in-situ and ex-situ methods(40) The role of nanoclay as a dopant was investigated, as well as the significant doping effect it had on electrochemical performance for energy storage. A hierarchically interconnected mesoporous framework supported by nanoclay was visualized by FESEM analysis, which validated the coating of PPy over CNTs and nanoclay. A PPy-coated CNT array was also exposed to be extremely porous, have a high specific surface area, and show no obvious deterioration in the presence of nanoclay. The highest specific capacitance of in situ composite (CPN) and CP (i.e., PPy-coated CNTs) at a scan rate of 10 mV s^{-1}, respectively, was 425 Fg^{-1} and 317 Fg^{-1}.

5.6 CONCLUSION AND FUTURE PROSPECTIVES

The effects of adding various clay minerals to a polymer matrix on ion conductivity, thermal and mechanical stability, crystallinity, porosity, electrolyte absorption, and maximum capacity have been studied in previous studies. Clay minerals

TABLE 5.1
Clay Materials and their Reinforced Electrode Materials and Properties

Clay material/composites	Battery components	Specific capacity (mAh g⁻¹)	Excellence in properties	Ref.
diatomite	3D hierarchical PEO-DLSL composite anode	65	Solid-state Li metal battery, safety and high energy/power density and excellent cycling stability.	(23)
saponite	NiFe saponite with Na+ pillaring for LIBs	815	Hierarchical hybrid anodes for high-performance batteries, low cost and environmentally friendly material.	(24)
halloysite	Si anodes for LIBs- Acid etching and magnesiothermic reduction of Si nanoparticles with halloysite	800	Nano size and porous structures of Si anodes contributed to their high specific capacity.	(3)
MMT (Montmorillonite) clays	carbon cogenerated Si nanosheet anode for LIBs	865	Porous 2D Si nanosheets fashioned through molten salt exfoliation and a chemical redox process.	(25)
Tapulgite (ATP)	porous Si/graphite@carbon (Si/G@C) anode materials	799	Carbonization – aluminothermic reduction reaction at low temperatures – in situ synthesis of Si flakes.	(26)
Kaolin clay	anodes made of N-doped ordered mesoporous carbon	672	Melamine infiltration and carbonization in a mesoporous silica-sucrose framework. Excellent electrochemical properties, including a long life cycle and a high rate of performance.	(27)
ATP/polyacrylonitrile precursor aerogel	anode materials made of amorphous carbon coated silica	1628.4	- The porous structure of ATP could absorb more electrolytes and shorten lithium-ion diffusion pathways. -The cost effective and sustainable materials for the development anode for LIBs.	(28)

(Continued)

TABLE 5.1 (Continued)
Clay Materials and their Reinforced Electrode Materials and Properties

Clay material/composites	Battery components	Specific capacity (mAh g⁻¹)	Excellence in properties	Ref.
Kaolin clay	Kaolin clay was applied to the surface of Zn anodes	190	Cheap materials (precursors), large charge-transfer resistance, good cycling stability and	(29)
halloysite	in a Li-S battery, carbon skin wrapped halloysite/sulfur cathodes are used	657	solution impregnation. This process incorporates sulfur nanoparticles. Halloysite nanotubes (HNT), which increase the specific surface area (44.8 m^2 g^{-1}), resulted in higher volumetric expansion via g lithiation/delithiation process.	(30)
MMT	Li-MMT/S cathode for Li-S batteries	700 mAh g-1 over 600 cycles at a large current density of 5 mA cm⁻² a	Faster lithium-ion diffusion around lithium MMT structure. Ultra-wide interlayer distance and low Li-MMT/S cathode energy barrier promotes lithium ions resulting in diffusion rate, cycling stability and rate performance are both excellent.	(31)
vermiculite	porous vermiculite as the cathode for Li-S batteries	800	Surface cations on vermiculite (Mg^{2+}, Ca^{2+}, K^+, Li^+, Na^+, and so on) help absorb polysulfide anions (Sn^{2-}) and prevent polysulfide dissolution. The absorption effect aided in the formation of the double electric layer, resulting in exceptional rate capability and cycling stability.	(32)

Clay	Material	Performance	Notes	Ref.
halloysite	acid-treated halloysite, carbon nanotubes, and V_3S_4 material for aqueous zinc-ion batteries (ZIBs) as the cathode	148 mA h g⁻¹ under a current density of 0.5 A g⁻¹ (95% retention after 200 cycles)	Due to the low-cost design of flexible substrates and the impressive electrochemical performance of ZIBs, the fabricated devices show excellent potential applications in wearables and wearable electronics.	(33)
layered double hydroxide (LDH)	NiFe-based layered double hydroxide (LDH) with chloride ions intercalated as a cathode material for CIB	350.6 mAh g⁻¹ and a long lifetime of over 800 cycles at a current density of 100 mA g⁻¹	LDH with 2D diffusion path had high anion exchange capacity due to reversible anion intercalation/deintercalation. Chloride ions reversibly shuttled between the electrodes during electrochemical cycling with Ni^{2+}/Ni^{3+} and Fe^{2+}/Fe^{3+} redox reactions in LDH.	(34)
MMT clay	porous poly(vinylidene fluoride-co-trifluoroethylene) (P(VDF-TrFE)) composite separators of LIBs	173 mAh g⁻¹ at a current density of 0.1C.	The high surface area porous structure of MMT positively improved ion diffusion and reduced interfacial resistance.	(35)
halloysite nanotubes (HNT)	bacterial cellulose/HNT composite nanofiber separators for LIBs	162 mAh g⁻¹ at a current density of 0.2C and cycling stability (95% of origin capacity after 100 cycles)	-Enhanced electrolyte compatibility, thermal stability. -Improved ionic conductivity, electrolyte uptake and mechanical stability and reduces the interfacial resistance of composite separators.	(36)
Dickite	Dickey Clay on Crosslinked Nonwovens as a Separator for LIB	152 mAh g⁻¹ at 0.5C	A composite separator coated with expanded dickite showed high electrolyte absorption (0.861 g cm⁻³), ionic conductivity (3.157 mS cm⁻¹, porosity, electrolyte uptake ability, mechanical stability, and composite separator porosity (61.6%).	(37)

(Continued)

TABLE 5.1 (Continued)
Clay Materials and their Reinforced Electrode Materials and Properties

Clay material/ composites	Battery components	Specific capacity (mAh g^{-1})	Excellence in properties	Ref.
2D vermiculite	A 2D vermiculite layer with an interlayer structure enables selective transport of lithium ions, avoiding the shuttle effect in Li-S batteries	1000 mAh g^{-1} and an average Coulombic efficiency of 90.3% after 50 cycles	An exfoliated vermiculite separator with a compact stack structure was negatively charged and suppressed the electrostatic repulsion shuttle effect of polysulfides in Li–S batteries.	(38)
Porous dickite clays	Produced by intercalation/ deintercalation of urea into dickite interlayers in the presence of KClO$_3$ and coated on the ethylene/propylene separators	132 mAh g^{-1} at 0.2C and 89 mAh g^{-1} at 4C	The pore size of the expanded dickite was 0.3-1.2 μm. After changing the organic separator of the rechargeable aqueous hybrid battery, the composite separator exhibited better porosity (67.1%), electrolyte absorption (1.378 g cm^{-3}), and ionic conductivity (13.12 mS cm^{-1}). showed.	(39)
halloysite	halloysite/poly(ethylene oxide) (PEO)-LiFePO$_4$ solid polymer electrolyte for LIBs	156 mAh g^{-1} after 100 cycles at 0.1C and 90.9% capacity retention	-The interaction between halloysite, lithium bis(trifluoromethanesulfonyl)imide (LiTFSI) and PEO resulted in a unique 3D structure for lithium-ion diffusion. -The solid polyelectrolyte filler used in halloysite has an ionic conductivity (9.23 × 10^{-5} S cm^{-1} at 25°C), a lithium-ion transference number (0.46), and an electrochemical voltage window (5.14 V). improved. -The addition of HNT and LiFePO$_4$ improved both the compatibility and charge transfer between the electrode and electrolyte, thus contributing to a stable interfacial resistance.	(26)

cellulose acetate/ poly-L-lactic acid/ halloysite nanotube (CA/PLLA/HNT) composite	gel polymer electrolytes (GPEs) for LIBs	125.2 mAh g^{-1} at 0.1C and a capacity retention of 93.1% after 50 cycles.	CA/PLLA/HNT composite nanofiber membrane can be used as a new green skeleton material in GPE for high-performance lithium-ion batteries, offering a perfect combination of high performance and environmental protection.	(40)
Boehmite	PAN/boehmite separator for lithium-ion batteries (LIBs)	162	Acquired higher porosity, greater electrolyte uptake, higher thermal stability, and better electrochemical performance with their three-dimensional (3D) interconnected structure.	20
Laponite	CNF/Laponite hierarchically porous, ultralight silica membranes as a separator for high-performance LIBs	161	Improving thermal stability, achieving robust high-temperature mechanical strength, superior electrolyte wettability, and enhanced ionic conductivity	21

are an excellent choice for use as a filler in polymeric separators and Li-ion battery electrolytes due to their significant impact on the polymer matrix when incorporated into polymer-based nanocomposites, as well as their low cost, availability, eco-friendliness, and high specific surface area.

It is possible to explore the compatibility of different polymers with different clay minerals and in varied ratios for usage as separators or solid electrolytes. Although it has not been investigated yet, clay minerals having a 3D/2D structure may be taken into account in future research. There is much research to be done on 1D clay minerals, specifically the impact of orientation behaviours on Li-ion batteries. Although 1D clay minerals are prone to aggregation, workable methods can be looked into. The performance of clay minerals as an addition to a polymeric separator and electrolyte has been documented in the literature, but less attention has been paid to the specific mechanisms of clay minerals and their function as separators and electrolyte fillers. Furthermore, biodegradable polymer-based clay composites would be challenging materials for developing various battery components which can be a green and sustainable progression in this field.

ACKNOWLEDGMENTS

This work is supported by Amrita School for Sustainable Development (AST), Amrita University, India.

REFERENCES

1. Wang J, Nie P, Ding B, Dong S, Hao X, Dou H, et al. Biomass derived carbon for energy storage devices. *J. Mater. Chem. A.* 2017;5:2411–2428.
2. Du P, Liu D, Chen X, Xie H, Qu X, Wang D, et al. Research progress towards the corrosion and protection of electrodes in energy-storage batteries. *Energy Storage Mater.* 2023 Mar 1;57:371–99.
3. Shangguan W, Dai Y, Duan Q, Liu B, Yuan H. A global soil data set for earth system modeling. *J Adv Model Earth Syst.* 2014 Mar 1;6(1):249–63.
4. Peixoto D, Pereira I, Pereira-Silva M, Veiga F, Hamblin MR, Lvov Y, et al. Emerging role of nanoclays in cancer research, diagnosis, and therapy. *Coord Chem Rev.* 2021 Aug 1;440:213956.
5. Wu J, Wang Y, Wu Z, Gao Y, Li X. Adsorption properties and mechanism of sepiolite modified by anionic and cationic surfactants on oxytetracycline from aqueous solutions. *Sci. Total Environ.* 2020;708:134409.
6. Jee SC, Kim M, Shinde SK, Ghodake GS, Sung JS, Kadam AA. Assembling ZnO and Fe3O4 nanostructures on halloysite nanotubes for anti-bacterial assessments. *Appl Surf Sci.* 2020 Apr 15;509:145358.
7. Lan Y, Chen D. Fabrication of silver doped attapulgite aerogels as anode material for lithium ion batteries. *J. Mater. Sci.: Mater. Electron.* 2018;29(23):19873–19879.
8. Ramadass K, Sathish CI, Mariaruban S, Kothandam G, Joseph S, Singh G, et al. Carbon nanoflakes and nanotubes from halloysite nanoclays and their superior performance in CO2 capture and energy storage. *ACS Appl Mater Interfaces.* 2020;12(10):11922–11933.
9. Hu G, Guo D, Shang H, Sun Y, Zeng J, Li J. Microwave-Assisted rapid preparation of vermiculite-loaded nano-nickel oxide as a highly efficient catalyst for acetylene carbonylation to synthesize acrylic acid. *ChemistrySelect.* 2020;5(10):2940.

10. Oluwatuyi OE, Ojuri OO, Khoshghalb A. Cement-lime stabilization of crude oil contaminated kaolin clay. *J. Rock Mech. Geotech. Eng.* 2020;12(1):160–167.

11. Fu N, Zhang S, Ma Y, Yang Z, Liu W. Diatomite/Cu/Al layered double hydroxide hybrid composites for polyethylene degradation. *RSC Adv.* 2020;10(17):9808–9813.

12. Lan Y, Liu Y, Li J, Chen D, He G, Parkin IP. Natural clay-based materials for energy storage and conversion applications. Adv Sci. 2021;8:2004036.

13. Dyartanti ER, Purwanto A, Widiasa IN, Susanto H. Ionic conductivity and cycling stability improvement of PVDF/Nano-clay using PVP as polymer electrolyte membranes for liFePo4 batteries. *Membranes (Basel).* 2018;8(3):36.

14. Huang C, Ji H, Guo B, Luo L, Xu W, Li J, et al. Composite nanofiber membranes of bacterial cellulose/halloysite nanotubes as lithium ion battery separators. *Cellulose.* 2019;26(11):6669–6681.

15. Liu Y, Jiang Y, Li F, Xue B, Cao X. Pore structure control of expanded dickite and its application as a clay coating layer on cross-linked nonwoven fabrics for lithium-Ion batteries. *J Electrochem Soc.* 2019;166(6):135995.

16. Feng J, Ao X, Lei Z, Wang J, Deng Y, Wang C. Hollow nanotubular clay composited comb-like methoxy poly(ethylene glycol) acrylate polymer as solid polymer electrolyte for lithium metal batteries. *Electrochim Acta.* 2020;340.

17. Lee S, Hwang HS, Cho W, Jang D, Eom T, Martin DC, et al. Eco-Degradable and flexible solid-state ionic conductors by clay-nanoconfined DMSO composites. *Adv Sustain Syst.* 2020;4(5):1900134.

18. Lin Y, Wang X, Liu J, Miller JD. Natural halloysite nano-clay electrolyte for advanced all-solid-state lithium-sulfur batteries. *Nano Energy.* 2017;31:478–485.

19. Chen Q, Zhu R, He Q, Liu S, Wu D, Fu H, et al. In situ synthesis of a silicon flake/nitrogen-doped graphene-like carbon composite from organoclay for high-performance lithium-ion battery anodes. *Chem Comm.* 2019;55(18):2644–2647.

20. Steudel A, Friedrich F, Boháč P, Lieske W, Baille W, König D, et al. Equimolar cation exchange of polyacrylamide in smectite. *Appl Clay Sci.* 2020;188:105501.

21. Daud SM, Daud WRW, Bakar MHA, Kim BH, Somalu MR, Muchtar A, et al. Low-cost novel clay earthenware as separator in microbial electrochemical technology for power output improvement. *Bioprocess Biosyst Eng.* 2020;43(8):1369–1379.

22. Dong W, Ding J, Wang W, Zong L, Xu J, Wang A. Magnetic nano-hybrids adsorbents formulated from acidic leachates of clay minerals. *J Clean Prod.* 2020;256:120383.

23. Zhou F, Li Z, Lu YY, Shen B, Guan Y, Wang XX, et al. Diatomite derived hierarchical hybrid anode for high performance all-solid-state lithium metal batteries. *Nat Commun.* 2019;10(1):2482.

24. Zhang J, Yin Q, Luo J, Han J, Zheng L, Wei M. NiFe saponite as a new anode material for high-performance lithium-ion batteries. *J Mater Chem A Mater.* 2020;8(14):6539–6545.

25. Ryu J, Hong D, Choi S, Park S. Synthesis of Ultrathin Si Nanosheets from Natural Clays for Lithium-Ion Battery Anodes. *ACS Nano.* 2016;10(2):2843–2851.

26. Oraon R, De Adhikari A, Tiwari SK, Sahu TS, Nayak GC. Fabrication of nanoclay based graphene/polypyrrole nanocomposite: An efficient ternary electrode material for high performance supercapacitor. *Appl Clay Sci.* 2015;118:231–238.

27. Le HTT, Dang TD, Chu NTH, Park CJ. Synthesis of nitrogen-doped ordered mesoporous carbon with enhanced lithium storage performance from natural kaolin clay. *Electrochim Acta.* 2020;332:135399.

28. Loaiza LC, Monconduit L, Seznec V. Si and Ge-based anode materials for Li-, Na-, and K-ion batteries: A perspective from structure to electrochemical mechanism. *Small.* 2020;16:1905260.

29. Lan Y, Chen D. Fabrication of nano-sized attapulgite-based aerogels as anode material for lithium ion batteries. *J Mater Sci.* 2018;53(3).

30. Chen W, Lei T, Lv W, Hu Y, Yan Y, Jiao Y, et al. Atomic interlamellar Ion path in high sulfur content lithium-montmorillonite host enables high-rate and stable lithium–sulfur battery. *Adv. Mater.* 2018;30(40):1804084.

31. Wu F, Lv H, Chen S, Lorger S, Srot V, Oschatz M, et al. Natural vermiculite enables high-performance in lithium–sulfur batteries via electrical double layer effects. *Adv Funct Mater.* 2019;29(27):1902820.

32. Liu S, Chen X, Zhang Q, Zhou J, Cai Z, Pan A. Fabrication of an inexpensive hydrophilic bridge on a carbon substrate and loading vanadium sulfides for flexible aqueous zinc-ion batteries. *ACS Appl Mater Interfaces.* 2019;11(40):36676–36684.

33. Yin Q, Zhang J, Luo J, Han J, Shao M, Wei M. A new family of rechargeable batteries based on halide ions shuttling. *J. Chem. Eng.* 2020;389:124376.

34. Nunes-Pereira J, Kundu M, Gören A, Silva MM, Costa CM, Liu L, et al. Optimization of filler type within poly(vinylidene fluoride-co-trifluoroethylene) composite separator membranes for improved lithium-ion battery performance. *Compos B Eng.* 2016;96:94–102.

35. Xu R, Sun Y, Wang Y, Huang J, Zhang Q. Two-dimensional vermiculite separator for lithium sulfur batteries. *Chin Chem Lett.* 2017;28(12):2235–2238.

36. Liu Y, Li D, Xu H, Jiang Y, Li F, Xue B. An expanded clay-coated separator with unique microporous structure for enhancing electrochemical performance of rechargeable hybrid aqueous batteries. *J Solid State Electrochem.* 2019;23(1):215–226.

37. Li X, Chen S, Xia Z, Li L, Yuan W. High performance of boehmite/polyacrylonitrile composite nanofiber membrane for polymer lithium-ion battery. *RSC Adv.* 2020;10(46):27492–27501.

38. Wang J, Liu Y, Cai Q, Dong A, Yang D, Zhao D. Hierarchically porous silica membrane as separator for high-performance lithium-ion batteries. *Adv Mater.* 2022;34(3):2107957.

39. Yu D, Qian Q, Wei L, Jiang W, Goh K, Wei J, et al. Emergence of fiber supercapacitors. *Chem Soc Rev.* 2015;44:647–662.

40. Oraon R, De Adhikari A, Tiwari SK, Nayak GC. Nanoclay-based hierarchical interconnected mesoporous CNT/PPy electrode with improved specific capacitance for high performance supercapacitors. *Dalton Trans.* 2016;45(22):9113–9126.

6 Graphene-based Materials for Energy Production, Conversion, and Storage

Nur Ezyanie Safie, Mohd Asyadi Azam, Raja Noor Amalina Raja Seman, and Akito Takasaki

6.1 INTRODUCTION

A worldwide energy crisis has resulted from increased energy demand and the dwindling of existing energy sources. Fossil fuels are the primary energy sources such as coal, oil, and natural gas that generate electricity. Fossil fuel combustion alters the climate through the presence of carbon dioxide and harms food chains together with the ocean. Supplies of fossil fuels are running low, and no new sources are being recognized or generated at the same pace. Renewable and sustainable energy production, conversion, and storage are critical issues. As the world's energy consumption rises at an alarming pace, developing clean and renewable energy storage and conversion technologies is more critical now. The structure and quality of the component materials significantly affect the overall performance of energy storage and conversion devices. Nanotechnology has revealed new vistas in materials science and engineering by producing novel materials, notably graphene and its derivatives, for efficient energy conversion, harvesting, and storage.

Graphene's one-of-a-kind mix of remarkable qualities has helped propel it to the forefront of the race to be one of the rising star nanomaterials. Graphene is a sheet of sp^2-hybridized carbon composed in a two-dimensional (2D) monolayer structure. As a high-performance electrode in the application of energy devices, graphene, a material with a fundamentally layered structure that is effectively 2D, has drawn much scientific interest. Graphene is a fascinating 2D nanomaterial in this century due to its exceptional qualities, including its surface area, electrical conductivity, thermal conductivity, mechanical strength, and optical transmittance in the visible-infrared range. Modifying graphene's surface and structure has received significant focus in recent years to enhance the material's performance in energy-related devices [1]–[3]. Graphene preparation technologies have evolved over a short period. Graphene synthesis relies on two primary techniques: top-down (destruction) and bottom-up (building). In most cases, graphene is produced in one of two ways. The chemical vapor deposition (CVD) [4] and the Hummer methods [5] are two of the most popular methods. In the presence of a CVD catalyst, precursors decompose and transform

DOI: 10.1201/9781003318859-6

into graphene. Many examples of materials may be derived from graphite and bio-mass, which are both plentiful and naturally occurring. The graphite is oxidized in a high acid environment for the latter process. Single-sheet graphene oxide (GO) is formed due to the exfoliation process. GO is decreased when it is reduced at high temperatures or by decreasing a chemical agent (rGO). GO and rGO, two graphene derivatives with many surfaces' functional groups, also exhibit many remarkable behaviors comparable to pristine graphene [6].

The role of graphene in energy conversion technology is researched using graphene as a Schottky junction, transparent electrode, active layer material, and catalytic counter electrode for different solar systems [7], [8]. Due to lower sheet resistance and high transmittance, graphene is utilized as a transparent electrode in organic, inorganic, and hybrid electronics. The inclusion of graphene considerably enhances the power conversion efficiency of solar devices. Besides, graphene plays a crucial role in flexible and long-lasting solar systems. Increasing solar cell effi-ciency using graphene might be helpful for future commercial applications of highly efficient photovoltaic systems. In addition, graphene's employment in fuel cells is not restricted to its position as catalyst support. Because of its remarkable mechan-ical strength and electrical conductivity, graphene may be used as a filler in polymer electrolytes. Furthermore, graphene may be utilized as a metal-free catalyst in substi-tuting the metals that have been conventionally used. Furthermore, batteries [9] and supercapacitors [10] are two primary energy storage devices, and significant research is being done to improve the amount of energy stored in each device. The research is being done to handle the fast development of renewable energy sources. Additionally, graphene is a good material for manufacturing energy storage applications which is feasible, mainly when the metal oxides have been treated, which displays restricted sheet restacking. The high conductivity of the linked graphene networks is another factor that adds to why these materials are of interest for energy storage applications. Regarding the usage of graphene in energy storage devices, some benefits include the material's porous microstructure, its electrochemical and mechanical stability, and its electrochemical and mechanical stability.

The unique characteristics and requirements of graphene and its derivatives will prolong graphene's production for decades and accelerate its usage in specialized industries, especially for energy conversion and storage. Its remarkable characteristics have made it desirable for industry and academic study, increasing its market demand. Graphene has raised the bar for similar materials, resulting in breakthroughs in the production of graphene and its derivatives. This chapter provides a significant over-view of graphene employment in such energy technologies that positively enhance the functional system of energy devices. Figure 6.1 shows the structure of graphene in different conditions.

6.2 GRAPHENE-BASED FOR ENERGY CONVERSION AND HARVESTING

6.2.1 PHOTOVOLTAIC TECHNOLOGY

Photovoltaics is the system used to generate electricity from solar radiation. Solar panels convert sunlight into electrical energy by splitting the incoming light into

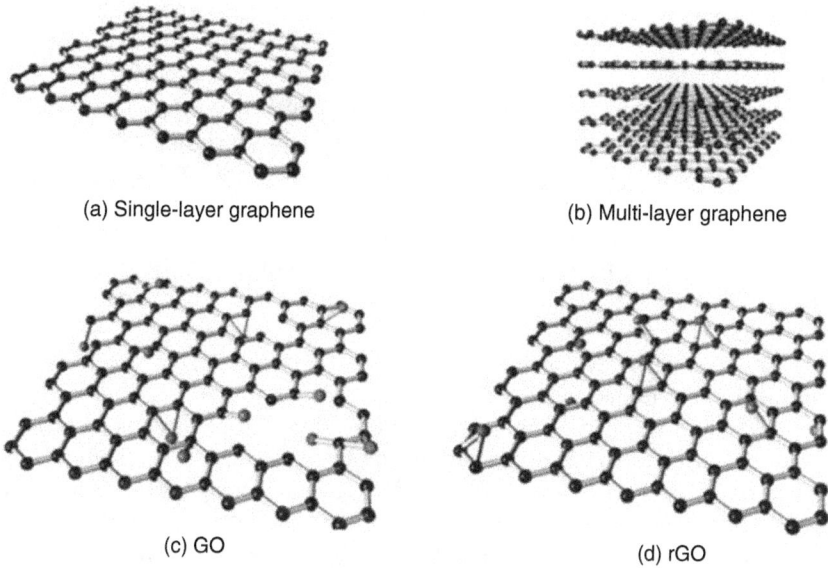

(a) Single-layer graphene

(b) Multi-layer graphene

(c) GO

(d) rGO

FIGURE 6.1 Structure of Graphene and its derivatives. Reprint with permission [11].

individual colors using a semiconductor material. The most usual form of the photovoltaic cell consists of silicon wafers coated with silicon dioxide [12]. Monocrystalline cells are the most prevalent form of cell used in solar panels. A solar panel module is a cell composed of thin sheets of silicon arranged in a grid-like pattern known as a polycrystalline cell, which produces more electricity than a monocrystalline cell. Glass, plastic, or metal solar panels may also be utilized to capture energy from the sun. These panels are referred to as thin-film modules and are less prevalent than the other two varieties.

Inverted architecture, solvent modification, and deposition techniques of stack devices are all applied to improve efficiency and stability. Organo- and perovskite-semiconductors have a more straightforward band gap tuning process and may be used with other materials with appropriate band gaps to create highly efficient solar cells [13]. Curtain walls or glass windows adorned solar cells have arisen as an unusually demanding solar cell technology to address concerns about large-area demands in the installation of rooftop solar cells. Building integrated photovoltaic is the name given to this kind of solar power. As with the building's curtain walls, the integrated photovoltaics should be able to let the weather leave while enabling the building's occupants and natural light to enter as required. Since organic and perovskite solar cells do not meet all these requirements, researchers have turned to semi-transparent solar cells as a possible solution. Third-generation solar cells offer the benefits of cheap manufacturing cost and excellent mechanical flexibility over first- and second-generation solar cells. In organic solar cells (OSCs) [14], dye-sensitized solar cells (DSSCs) [15], and perovskite solar cells (PSCs) [16], graphene has been explored.

Graphene's extreme transparency, ultrahigh carrier mobilities, and chemical and mechanical durability [17] make it an attractive contender for flexible transparent electrodes. Du et al. constructed OCSc with graphene-based anodes. Three layers-doped CVD graphene anode cells achieved the best performance. A 0.2 cm^2 cell with a benzimidazole (BI)-doped graphene anode achieved a 6.85% PCE, 96% of the highest efficiency for a similar device with an ITO on a glass anode [18]. Meanwhile, by directly integrating polyimide (PI) onto CVD graphene, Koo et al. present a highly flexible and resilient electrode with excellent thermal stability. PI's direct integration enhanced the graphene electrode's durability by minimizing graphene delamination under mechanical stress. A flexible OSC with a PCE of 15.2% and excellent mechanical durability was manufactured [19]. The obtained data show the significance of graphene TCE for next-generation wearable and flexible optoelectronics.

Besides, Liu et al. developed ultra-flexible PSCs using CVD graphene as transparent bottom electrodes and reported a device efficiency of 11.5% with an output power per unit weight of 5.07 W/g [20]. Because of the strong bending resilience of graphene electrodes, flexible electronics may function at varying bending radii with slight performance deterioration during bending testing. Doping CVD graphene (GR) with bis (tri-fluoromethanesulfonyl)-amide (TFSA) and triethylenetetramine (TETA) produced semi-transparent and significant-aperture flexible PSCs with a PCE achieved of 11.6%, as reported by Jang et al. [21]. Due to the remarkable-transparency component materials, such as the doped GR TCEs, the flexible PSCs also demonstrated good transmittance in the visible NIR region. Moreover, after 1000 bending cycles at R=8 mm, the PCE values of the FPSCs remained more significant than 70% of their original values. Furthermore, Villarreal et al. revealed that significant-quality 4-layer CVD graphene could cope with FTO in DSSC photoanodes. Graphene's enhanced optical transparency increased short circuit current density, as well as open circuit voltage. Graphene's higher sheet resistance reduced the fill factor of graphene-DSSCs. Despite being 250 times wafer-thin compared to FTO, graphene- and FTO-DSSCs had an equal power conversion efficiency of 0.4% [22]. The results have substantial financial and environmental insinuations for the wide-ranging manufacturing of DSSCs, perovskites, and other photo-electrochemical devices. Replacing standard tin oxides with graphene in TCs will lower PV devices' weight, cost, and social and environmental impact while satisfying industry criteria for sustainable, robust, adaptable, and portable technology.

6.2.2 FUEL CELLS

Electrochemical devices used to convert hydrogen and oxygen into water and produce electricity are known as fuel cells. They are pollution-free energy sources because they do not produce greenhouse gases or other pollutants. Fuel cells also have a minimal environmental impact because they require little energy to operate and generate no waste products. Water vapor is the only end product of a fuel cell that is much less harmful than air pollution. The primary type of fuel cell is the proton exchange membrane (PEM) cell [23]. Hydrogen gas is passed through the anode and forms positively charged hydrogen ions. The negatively charged electrons are driven to the cathode, where they merge with oxygen from the air to form water vapor and electricity.

Bipolar plates, which carry electric current between cells, are an essential component of PEM fuel cells. Bipolar plates are often made of graphite or other carbon-based materials. Carbon-based bipolar plates, on the other hand, often have significant material and production costs that would result in vast and heavy fuel cell stacks, which would be particularly unsuitable for transportation or portable devices. Chen et al. proposed a promising method to protect aluminium bipolar plates by layered graphene over aluminium [24]. The graphene synthesis method uses affordable graphite powder as its raw material and does not need hazardous organic solvents, complex equipment, or severe temperatures. The interfacial contact resistance (ICR) of graphene sheet over aluminium is 5 mΩ cm^2, which is much less and more stable. The graphene-coated aluminium sheet fulfills corrosion and electrical resistance specifications for bipolar plates established by the U.S. DOE in 2020. In contrast to the low-temperature PEM fuel cells, the high-temperature (HT)-PEM Fuel cells based on phosphoric acid (PA) [25] and polybenzimidazole (PBI) [26] conduct protons through the Grotthuss mechanism. In this process, charge carriers hop over an extended hydrogen bond network formed by PA and PBI. The relatively high operating temperature of an HT-PEMFC system makes it easier for the system to have a high proton conductivity. Compared to pure polybenzimidazole membranes with a significant doping level of phosphoric acid, membrane electrode assembly (MEA) assembled with single-layer graphene (SLG) at divergent locations displays a larger electrochemical active surface area, lower electrode resistances, and a higher peak power density [27]. With the power density, zero-carbon emissions, and significant energy efficiency, fuel cells are one of the promising stars in renewable energy conversion technology. The hydrogen fuel cell is driven by the hydrogen oxidation process (HOR) and the oxygen reduction reaction (ORR). Platinum (Pt) and Pt alloy catalysts have enhanced ORR activity while lowering costs. However, excessive Pt catalyst loading and hydrogen fuel cell durability impede practical operation and commercial acceptability. Ji et al. show that using p-phenyl groups and graphitic nitrogen (N) as bridges between platinum and graphene/carbon black hybrid support materials in platinum nanoparticles shows an average particle size of (4.88 ± 1.79) nm [28]. Compared to a single carbon material, -NH$_2$ groups and graphitic N-doped electrochemical exfoliation graphene oxide and carbon black (N-C) were sturdier, boosting corrosion resistance.

6.2.3 NANOGENERATORS

Triboelectric nanogenerators (TENGs) [29] have applications aligning from self-powered wearable sensors to sea wave and wind energy collecting. The rapidly increasing urge for new sustainable energy sources has increased intrigue in energy harvesters that can transform mechanical energy into electrical power. TENGs are capable of supplying such green energy. TENGs power density varies because of the usage of different materials and device structures and the various ambient and mechanical energies that TENGs may gather. TENGs, or flexible triboelectric nanogenerators, can absorb mechanical energy from various motion-related energy capture mechanisms, such as random wind drifts. Much study has been devoted to pushing the maximum output power of TENGs. Incorporating low-dimension carbon compounds, such as graphene, can provide various synergistic benefits, including output augmentation and multi-functionality.

Owing to their extraordinary mechanical, thermal, and electrical properties, graphene nanoplatelets (GNPs) are superb nanofillers capable of producing stiff materials with minimal interface resistance. GNPs improve the material's relative permittivity in which they are embedded, enhancing triboelectric output. According to Shabbir et al., a Graphene–nanoparticle (GNP)-filled Polydimethylsiloxane (PDMS) composite might be utilized to generate a high triboelectric output in a rolling-mode TENG while minimizing surface abrasion (RL-TENG). This investigation produced a significant voltage of 75.2 V and a current of 7.36 µA. By achieving such remarkable results, the RL-TENG maintained its stability for up to 16,000 cycles of repetitive motion and could power 40 red LEDs utilizing energy gathered from motion. This study demonstrates that these devices may be employed in harsh working conditions for applications requiring remarkable voltage but low crest [30]. Furthermore, Graphene/Copper heterostructures developed by electrodeposition and spin-coating are used in energy harvesting for triboelectric nanogenerators (TENGs). Li et al. confirm graphene's role in preventing copper nanostructure oxidation and increasing graphene/Cu heterostructure durability [31]. Copper's sensitivity to oxidation by copper oxides is one of its significant drawbacks. The enthalpy and Gibbs free energy changes at 298.15 K are negative, suggesting that the reaction might happen suddenly at ambient temperature and that copper's characteristics may be damaged after oxidation. Copper is a metal that is used in wide applications. Physical exfoliation of graphene and electrodeposition of a Cu nanostructure to expose the working nature of graphene/Cu heterostructures, as shown in Figure 6.2, achieve the primary objective of inhibiting copper oxidation and utilizing its role in TENG via internal stress and electron transport.

The few layers of graphene (FLG) electrodes studied by Pace et al. outperform traditional metal electrodes in TENGs. They are fully adapted into flexible devices by solution processing under ambient settings making FLG electrodes an excellent charge collector in TENGs [32]. Replacing the gold electrodes with flexible few-layer graphene-based electrodes in TENGs that operate in vertical contact resulted in a 26-fold boost in power density mode (28.6 W/cm^2 at 100 MΩ). Controlling electrodes' work function is crucial for developing an outstanding built-in potential in TENGs. It influences electrical performance by raising the charge density between the electrode/tribomaterials interfaces. To tune the graphene sheet's work function (positive triboelectric material), Zhou et al. developed a simple defect-mediated approach (DMS) [33]. rGO was annealed at 2000°C and had the minimum work function to best promote the deficit of electrons in positive triboelectric material, hence increasing the output voltage and current of TENGs to 190 V and 14 µA at a power density of 5.04 W/m^2.

6.3   GRAPHENE-BASED FOR ENERGY STORAGE DEVICES

Energy storage devices play a crucial role, especially for mobile applications, in addition to the energy harvesting/conversion technologies (such as solar cells and fuel cells) outlined above. The major electrochemical energy storage devices that have

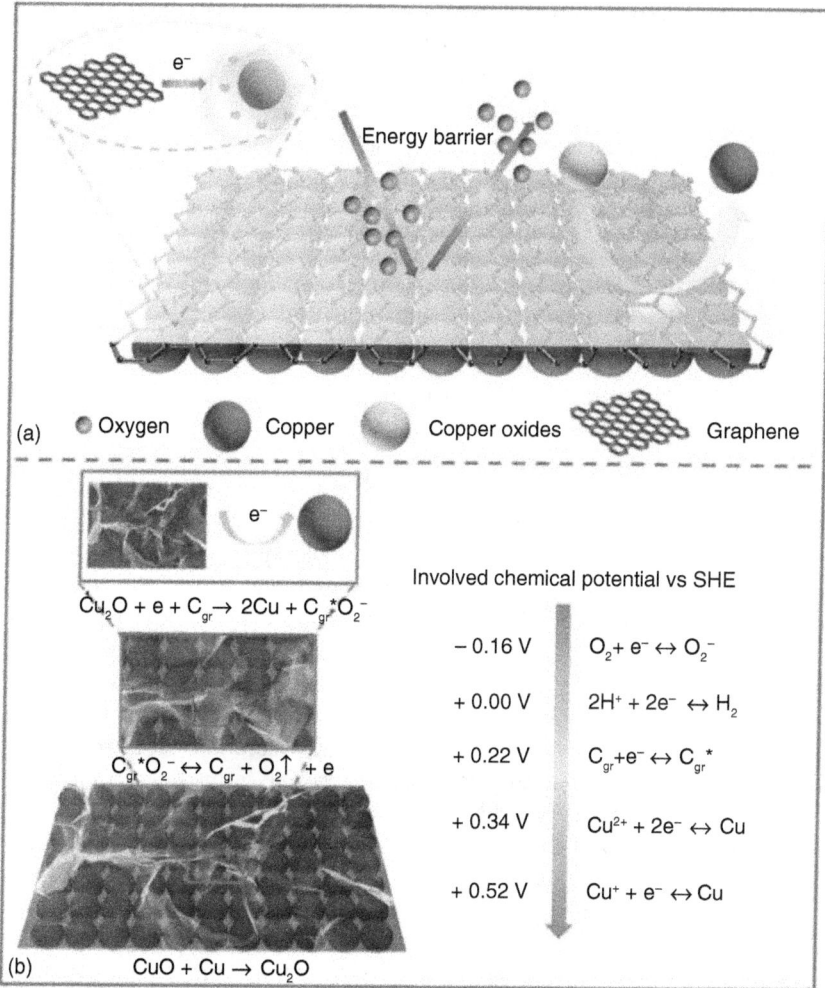

FIGURE 6.2 (a) The mechanism in graphene/Cu heterostructure. (b) Electron transfer and chemical reactions. Reprint with permission [31].

captivated vast interest are supercapacitors [34], micro-supercapacitors (MSCs) [35], and lithium-ion batteries (LIBs) [36]. Many studies have been done on supercapacitors, MSCs, and LIBs because of their performance in terms of power/energy density, safety, lifespan, and rate capability. Researchers are working to create lightweight, affordable, easily manufactured, and flexible substrate-compatible remarkable-performance supercapacitors, MSCs, and LIBs. Graphene-based materials have attracted researchers' curiosity to investigate their uses in energy storage devices because of their large surface area, as well as good electrical conductivity [37]. These features are crucial to the effectiveness of higher energy storage devices. With this, we primarily concentrate on the most recent development in the use of graphene in supercapacitors, MSCs and LIBs. Recently, graphene and its composite materials have been used in remarkable-performance supercapacitors, MSCs and LIBs. Supercapacitors are electrochemical devices that can store energy and release it in a short amount of time with significant power and current density. Because of their flexibility, lightweight, and low cost, supercapacitors are a highly desirable alternative to conventional inorganic solar cells and a possible source of renewable energy [38].

6.3.1 SUPERCAPACITORS

Recently, Azam et al. developed remarkable-performance supercapacitors using graphene and molybdenum disulfide as electrodes [39]. The G/MoS$_2$ hybrid device demonstrated the best performance with a significant specific gravimetric capacitance of 48.58 F g^{-1} in 6 M KOH electrolyte with a composition of 60:40, comparable to previous G/MoS$_2$ supercapacitors. Additionally, the electrodes demonstrated a longer cell lifespan than the other G/MoS$_2$ electrodes. Su et al. reported flexible binder-free electrodes based on Fe-Co binary oxides nanoparticles/graphene-painted on carbon cloth with a distinctive leaf-like shape that can be employed in supercapacitors without additional processing or additives. The resultant electrochemical characterization reveals that the electrode has a significant specific capacitance of 458 F g^{-1} at 1 A g^{-1} under a broad operating potential window of 1.45 V [40]. The distinctive characteristics of widely dispersed Fe-Co binary oxides provide sufficient active sites and participate in the electrochemical reactions.

In addition to these beneficial qualities, PANI's deformed structure produces poor energy storage or cycle performance in its pristine formed. Due to its remarkable, thermal, chemical, and mechanical stability, as well as electrical conductivity, nature of two-dimensional for significant contact area, and energy storage potential between the electrode/electrolyte interface, graphene in PANI can be a good choice for a successful application as electrodes in supercapacitors. Pal et al. have successfully produced pure form PANI along with its nanocomposites of highly conducting graphene using a chemical process based on chemistry to produce remarkable-efficiency supercapacitors [41]. By adding 8 wt% of graphene nanosheets to PANI electrodes, the charge storage capacity of these electrodes was significantly increased to 1412 F g^{-1} with ~89% of capacitance retention after 10,000 charging-discharging cycles suggesting graphene-doped PANI devices have higher porosity and excellent cyclic stability than pristine PANI devices. Electrolyte ions can smoothly intercalate and de-intercalate through nanopores due to the significant porosity of doped PANI devices.

6.3.2 MICRO-SUPERCAPACITORS

The micro-supercapacitors (MSCs) features include remarkable power densities (103 W cm^3), a long cycle lifetime (106 cycles), and fast charge-discharge rates. Generally, the performance of MSCs is determined by the electrode material, an essential structural component to a large extent. More active areas for electrochemical processes because of their significant specific surface area, graphene, have been extensively exploited for MSCs, along with its derivatives, including nanosheets, vertical arrays, and hybrids [42]. For example, Kamboj et al. [43] prepared highly conducting laser-irradiated graphene (LIG) film by electrochemically depositing reduced graphene oxide (rGO). PPy was modified by electrochemical polymerization to create a hybrid electrode material after laser irradiation of rGO. The constructed LIG-PPy hybrid material produces significant specific capacitance of 124 mF cm^{-2} (Figure 6.3(a)), 99.2 Wh cm^{-2} of energy density, and 47.75 mW cm^{-2} of power density. For instance, Yuan et al. [44] fabricated oxygen-sulfur co-doped porous graphene electrodes, a three-dimensional (3D) superhydrophilic. It is important to note that by taking advantage of these benefits, the constructed MSCs achieve excellent capacitive performance, as evidenced by the optimized MSCs' delivery of an areal capacitance of 53.2 mF cm^{-2} at 0.08 mA cm^{-2} (Figure 6.3(b)), around 39 times greater than an undoped MSC. They also attained excellent modular integration capabilities, outstanding mechanical flexibility, and good cycling stability (81.3% capacitance retention after 8000 cycles) have all been attained.

Furthermore, Li et al. [45] utilized vanadium carbide/reduced graphene oxide (V_8C_7/rGO) MSCs, that had been made using an effective continuous centrifugal casting process via laser scribing on ammonium metavanadate/graphene oxide

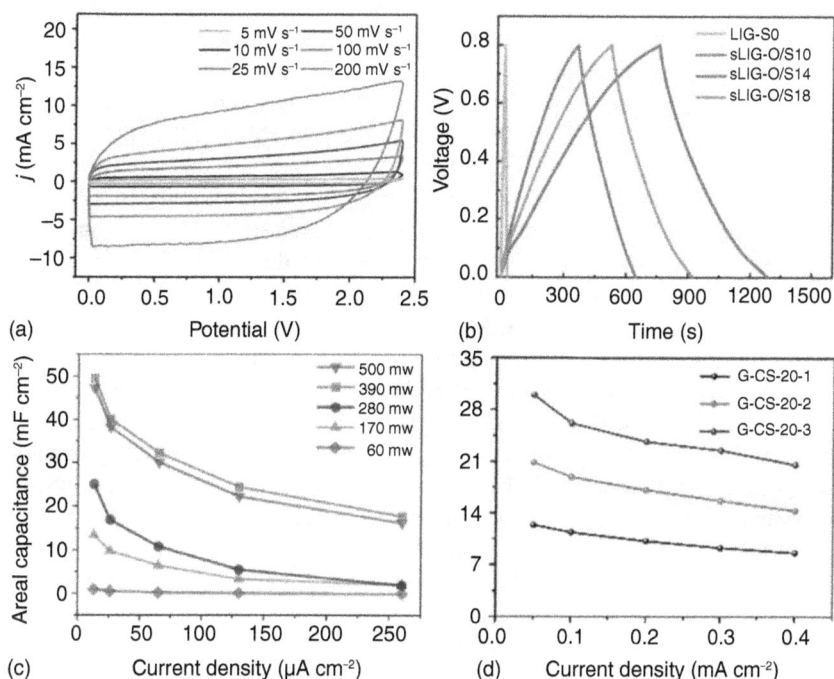

FIGURE 6.3 Electrochemical performances of the electrode. (a) LIG-PPy hybrid, (b) sLIG-O/S14, (c) V_8C_7/rGO, and (d) G-CS-20-3.

Source: [46].

(NH$_4$VO$_3$/GO) films. The built-in MSCs demonstrated a ground-breaking areal capacitance up to 49.5 mF cm^{-2} (Figure 6.3(c)), which is 11 times greater than rGO MSCs. The volumetric energy and power densities can be found up to 3.4 mWh cm^{-3} and 401 mW cm^{-3}, respectively. The graphene nanosheets are uniformly coated with V$_8$C$_7$ nanoparticles. The availability of charge carriers and active sites was made more accessible by the well-defined structure of the V$_8$C$_7$/rGO electrodes. Recently, Zhang et al. [46] reported a miniaturized graphene-carbon sphere (G-CS) that combined two approaches—adding carbon sphere (CS) spacers and using an ethyl cellulose (EC) template that would sacrifice itself—and developed a 3D printable G-CS composite ink to design the electrodes. The constructed MSCs displayed an ultra-areal capacitance of 30.00 mF cm^{-2} (0.22 mW cm^{-2}) (Figure 6.3(d)), good mechanical flexibility, a significant energy density (4.17 Wh cm^{-2}), and power density.

6.3.3 LITHIUM-ION BATTERIES

Currently, batteries have received significant attention for developing next-generation energy storage because of their low manufacturing costs, simple procedure approach, and excellent power conversion efficiency [47], [48]. The development of graphene and its usage in batteries opens up a new route for producing remarkable-quality and low-cost batteries. In one exciting example, Ou and colleagues [49] fabricated carbon-coated silicon, and the homogeneous amorphous carbon layer coating further increased the Si anode's electrical conductivity and mechanical stability. The as-prepared anode of Si@NC/Graphene for lithium-ion batteries (LIBs) exhibited a significant reversible capacity of 1870 mAh g^{-1} at 0.1 A g^{-1} with an increment Coulombic efficiency of 85.6% compared to the bare Si. In another interesting example, Lan et al. [50] evaluated SnO$_2$/graphene's electrochemical performance as an anode material in LIB. The authors employed a remarkable-quality SnO$_2$/graphene anode, resulting in a sizeable reversible capacity demonstrating that the graphene's significant specific surface area makes the effect on the lithium-ion storage more accessible. The first reversible specific capacity of the constructed SnO$_2$/G + CNT electrode was as high as 1227.2 mAh g^{-1} at a current density of 0.1 A g^{-1}, demonstrating the excellent reversible capacity of this electrode. The reversible capacity remained at 1630.1 mAh g^{-1} after 1000 cycles at a current density of 0.5 A g^{-1}.

Moreover, He et al. [51] also contributed to the explanation of graphene's fair use in lithium storage devices by linking the remarkable-rate discharge capability of LIBs to nanoscale holes found in the graphene sheets. In hierarchical SnS$_2$-rGO microspheres, the S-doped graphene promotes electron movement in electrodes, thus, turning into a conductive network. The constructed SnS$_2$-rGO retains its capacities of 1647.8 and 776.2 mAh g^{-1} at 0.1 A g^{-1}, respectively, after 100 cycles. It also provides capacities of 1177.2 mAh g^{-1} (at 1 A g^{-1}) for LIBs after 400 cycles. Xu et al. [35] reported reduced graphene oxide (rGO) composite materials based on thiophene-diketopyrrolopyrrole (TDPP) as organic anode materials for LIBs. The resulting P(C-TDPPAC)/rGO anode showed a significant discharge-specific capacity of 857 mAh g^{-1} at 1000 mA g^{-1} for 1000 cycles. Meanwhile, the cyclic stability is up to 83 %, its energy density is 643 Wh kg^{-1}, and its power density is 774 W kg^{-1}, respectively.

6.4 CONCLUSIONS

In this chapter, we have discussed current developments in the use of graphene-based materials for energy conversion/harvesting (solar cell, fuel cell, nanogenerator) and energy storage devices (supercapacitor, MSC, LIB). Due to their unique properties which include high electrical conductivity, outstanding mechanical stability, and large surface area, they have a tremendous potential to replace traditional materials in energy conversion/harvesting and storage device applications. Over the past few years, the activity in this research field has continued to grow rapidly. The findings of in-depth research have shown that graphene has demonstrated wide practical utilization and has significant potential in fabrication for particular applications (e.g., energy conversion and storage). Furthermore, graphene plays a vital role in flexible and long-lasting solar systems. Graphene can aid in the future development of lightweight, environmentally friendly, and economically viable high-performance solar cells because of its great mechanical flexibility, high environmental stability, low electrical resistivity, and minimal environmental impact.

The research into the incorporation of nanofillers such as graphene-based materials in polymers for better performance and to extend the long-term product lifecycle, particularly in fuel cells, was spurred by the increased need for optimal, dependable, and efficient energy supplies. Graphene-based materials added in specific amounts continue to be a suitable candidate for the evaluation of innovative membrane modifications in the future. Carbon materials have the potential to expand the scope of TENG applications and have a favourable effect on TENG technology-related problems that are frequently encountered. The future improvements to TENG technologies are the foundations of low-dimension carbon materials-based triboelectric production modes that should be thoroughly investigated. The interaction of graphene-based materials and their synergistic effects on the electrode materials in energy storage devices should receive much research attention. To acquire a deeper understanding of the deterioration effects at the novel interfaces produced with the inclusion of graphene materials in these devices, numerous in-depth interface investigations as well as additional stability studies are required.

REFERENCES

[1] K. Sakthi Velu *et al.*, "Photo-anode surface modification using novel graphene oxide integrated with methylammonium lead iodide in organic-inorganic perovskite solar cells," *J. Phys. Chem. Solids.*, vol. 154, no. September 2020, pp. 110036, 2021. doi: 10.1016/j.jpcs.2021.110036

[2] C. T. Tsai, Y. C. Wu, Y. T. Lin, and M. S. Wu, "Sulfur modification of worm-like exfoliated graphite with crumpled graphene sheets derived from mesocarbon microbeads for electrochemical supercapacitors," *J. Energy Storage.*, vol. 50, no. January, pp. 104250, 2022. doi: 10.1016/j.est.2022.104250

[3] M. A. Azam *et al.*, "Structural characterization and electrochemical performance of nitrogen doped graphene supercapacitor electrode fabricated by hydrothermal method," *Int. J. Nanoelectron. Mater.*, vol. 14, no. 2, pp. 127–136, 2021.

[4] X. Zhang *et al.*, "Evolution of copper step beams during graphene growth by CVD method," *Appl. Surf. Sci.*, vol. 610, Feb. 2023. doi: 10.1016/j.apsusc.2022.155518

[5] M. Helmi, A. Kudus, M. Razlan, H. Akil, and F. Ullah, "Oxidation of graphene via a simplified Hummers ' method for graphene- diamine colloid production," *J. King Saud. Univ. Sci.*, vol. 32, no. 1, pp. 910–913, 2019. doi: 10.1016/j.jksus.2019.05.002

[6] R. Tarcan *et al.*, "A new, fast and facile synthesis method for reduced graphene oxide in N,N-dimethylformamide," *Synth Met.*, vol. 269, no. October, pp. 116576, 2020. doi: 10.1016/j.synthmet.2020.116576

[7] R. Get, Sk. M. Islam, S. Singh, and P. Mahala, "Design and fabrication of graphene/cds schottky junction for photovoltaic solar cell applications," *Optik (Stuttg)*, vol. 266, no. January, pp. 169560, 2022. doi: 10.1016/j.ijleo.2022.169560

[8] N. E. Safie, M. A. Azam, M. F. Aziz, and M. Ismail, "Recent progress of graphene-based materials for efficient charge transfer and device performance stability in perovskite solar cells," *Int J Energy Res.*, vol. 45, no. 2, pp. 1347–1374, 2021. doi: 10.1002/er.5876

[9] Z. Chang, B. Ding, H. Dou, J. Wang, G. Xu, and X. Zhang, "Hierarchically porous multilayered carbon barriers for high-performance li–s batteries," *Chem. Eur. J.*, vol. 24, no. 15, pp. 3768–3775, 2018. doi: 10.1002/chem.201704757

[10] R. N. A. R. Seman, M. A. Azam, and M. H. Anib, "Graphene/transition metal dichalcogenides hybrid supercapacitor electrode: Status, challenges, and perspectives," *Nanotechnology*, vol. 29, no. 50, pp. 502001, 2018. doi: 10.1088/1361-6528/aae3da

[11] A. A. Iqbal, N. Sakib, A. K. M. P. Iqbal, and D. M. Nuruzzaman, "Graphene-based nanocomposites and their fabrication, mechanical properties and applications," *Materialia (Oxf)*, vol. 12, no. July, 2020. doi: 10.1016/j.mtla.2020.100815

[12] Z. Rui *et al.*, "On the passivation mechanism of poly-silicon and thin silicon oxide on crystal silicon wafers," *Solar Energy*, vol. 194, pp. 18–26, Dec. 2019. doi: 10.1016/j.solener.2019.10.064

[13] T. C. Yang, P. Fiala, Q. Jeangros, and C. Ballif, "High-bandgap perovskite materials for multijunction solar cells," *Joule*, pp. 1–16, 2018. doi: 10.1016/j.joule.2018.05.008

[14] D. Koo *et al.*, "Flexible organic solar cells over 15% efficiency with polyimide-integrated graphene electrodes," *Joule*, vol. 4, no. 5, pp. 1021–1034, May 2020. doi: 10.1016/j.joule.2020.02.012

[15] A. Vasanth, N. S. Powar, D. Krishnan, S. v. Nair, and M. Shanmugam, "Electrophoretic graphene oxide surface passivation on titanium dioxide for dye sensitized solar cell application," *J. Sci.: Adv. Mater. Devices.*, vol. 5, no. 3, pp. 316–321, 2020. doi: 10.1016/j.jsamd.2020.07.006

[16] R. Ishikawa, S. Yamazaki, S. Watanabe, and N. Tsuboi, "Layer dependency of graphene layers in perovskite/graphene solar cells," *Carbon N. Y.*, vol. 172, pp. 597–601, 2021. doi: 10.1016/j.carbon.2020.10.065

[17] F. Zhang *et al.*, "Recent advances on graphene: Synthesis, properties and applications," *Compos. Part A Appl. Sci. Manuf.*, vol. 160, Sep. 2022. doi: 10.1016/J.COMPOSITESA.2022.107051

[18] J. Du *et al.*, "Extremely efficient flexible organic solar cells with a graphene transparent anode: Dependence on number of layers and doping of graphene," *Carbon N Y.*, vol. 171, pp. 350–358, 2021. doi: 10.1016/j.carbon.2020.08.038

[19] D. Koo *et al.*, "Flexible organic solar cells over 15% efficiency with polyimide-integrated graphene electrodes," *Joule*, vol. 4, no. 5, pp. 1021–1034, 2020. doi: 10.1016/j.joule.2020.02.012

[20] Z. Liu, P. You, C. Xie, G. Tang, and F. Yan, "Ultrathin and flexible perovskite solar cells with graphene transparent electrodes," *Nano Energy*, vol. 28, pp. 151–157, 2016. doi: 10.1016/j.nanoen.2016.08.038

[21] C. W. Jang, J. M. Kim, and S. H. Choi, "Lamination-produced semi-transparent/flexible perovskite solar cells with doped-graphene anode and cathode," *J. Alloys Compd.*, vol. 775, pp. 905–911, 2019. doi: 10.1016/j.jallcom.2018.10.190

[22] C. C. Villarreal, J. I. Sandoval, P. Ramnani, T. Terse-Thakoor, D. Vi, and A. Mulchandani, "Graphene compared to fluorine-doped tin oxide as transparent conductor in ZnO dye-sensitized solar cells," *J. Environ. Chem. Eng.*, vol. 10, no. 3, pp. 107551, 2022. doi: 10.1016/j.jece.2022.107551

[23] S. Ahmad, T. Nawaz, A. Ali, M. F. Orhan, A. Samreen, and A. M. Kannan, "An overview of proton exchange membranes for fuel cells: Materials and manufacturing," *Int. J. Hydrogen Energy.*, vol. 47, no. 44, pp. 19086–19131, May 2022. doi: 10.1016/J.IJHYDENE.2022.04.099

[24] P. Chen, F. Fang, Z. Zhang, W. Zhang, and Y. Wang, "Self-assembled graphene film to enable highly conductive and corrosion resistant aluminum bipolar plates in fuel cells," *Int. J. Hydrogen Energy.*, vol. 42, no. 17, pp. 12593–12600, 2017. doi: 10.1016/j.ijhydene.2017.03.214

[25] Z. Guo, M. Perez-Page, J. Chen, Z. Ji, and S. M. Holmes, "Recent advances in phosphoric acid–based membranes for high–temperature proton exchange membrane fuel cells," *J. Energy Chem.*, vol. 63, pp. 393–429, Dec. 2021. doi: 10.1016/J.JECHEM.2021.06.024

[26] D. E. Hussin, Y. Budak, and Y. Devrim, "Development and performance analysis of polybenzimidazole/boron nitride composite membranes for high-temperature PEM fuel cells," *Int. J. Energy Res.*, vol. 46, no. 4, pp. 4174–4186, Mar. 2022. doi: 10.1002/ER.7418

[27] J. Chen *et al.*, "The performance and durability of high-temperature proton exchange membrane fuel cells enhanced by single-layer graphene," *Nano Energy*, vol. 93, no. September 2021, pp. 106829, 2022. doi: 10.1016/j.nanoen.2021.106829

[28] Z. Ji *et al.*, "Doped graphene/carbon black hybrid catalyst giving enhanced oxygen reduction reaction activity with high resistance to corrosion in proton exchange membrane fuel cells," *J. Energy Chem.*, vol. 68, pp. 143–153, 2022. doi: 10.1016/j.jechem.2021.09.031

[29] R. Walden, I. Aazem, A. Babu, and S. C. Pillai, "Textile-Triboelectric nanogenerators (T-TENGs) for wearable energy harvesting devices," *J. Chem. Eng.*, vol. 451, pp. 138741, Jan. 2023. doi: 10.1016/J.CEJ.2022.138741

[30] I. Shabbir *et al.*, "A graphene nanoplatelets-based high-performance, durable triboelectric nanogenerator for harvesting the energy of human motion," *Energy Rep.*, vol. 8, pp. 1026–1033, 2022. doi: 10.1016/j.egyr.2021.12.020

[31] Y. Li *et al.*, "Electron transfer mechanism of graphene/Cu heterostructure for improving the stability of triboelectric nanogenerators," *Nano Energy*, vol. 70, no. December 2019, pp. 104540, 2020. doi: 10.1016/j.nanoen.2020.104540

[32] G. Pace, A. Ansaldo, M. Serri, S. Lauciello, and F. Bonaccorso, "Electrode selection rules for enhancing the performance of triboelectric nanogenerators and the role of few-layers graphene," *Nano Energy*, vol. 76, no. June, pp. 104989, 2020. doi: 10.1016/j.nanoen.2020.104989

[33] J. Zhou *et al.*, "Defect-mediated work function regulation in graphene film for high-performing triboelectric nanogenerators," *Nano Energy*, vol. 99, no. March, pp. 107411, 2022. doi: 10.1016/j.nanoen.2022.107411

[34] M. A. Azam, N. S. N. Ramli, N. A. N. M. Nor, and T. I. T. Nawi, "Recent advances in biomass-derived carbon, mesoporous materials, and transition metal nitrides as new electrode materials for supercapacitor: A short review," *Int. J. Energy Res.*, vol. 45, no. 6, pp. 8335–8346, 2021. doi: 10.1002/er.6377.

[35] J. Qin *et al.*, "Recent advances and key opportunities on in-plane micro-supercapacitors: From functional microdevices to smart integrated microsystems," *J. Energy Chem.*, Feb. 2023. doi: 10.1016/J.JECHEM.2023.01.065

[36] L. Zeng, L. Qiu, and H. M. Cheng, "Towards the practical use of flexible lithium ion batteries," *Energy Storage Mater.*, vol. 23, pp. 434–438, Dec. 2019. doi: 10.1016/J.ENSM.2019.04.019

[37] S. Scaravonati *et al.*, "Combined capacitive and electrochemical charge storage mechanism in high-performance graphene-based lithium-ion batteries," *Mater. Today Energy*, vol. 24, pp. 100928, Mar. 2022. doi: 10.1016/J.MTENER.2021.100928

[38] H. Liu *et al.*, "Silicon doped graphene as high cycle performance anode for lithium-ion batteries," *Carbon N. Y.*, vol. 196, pp. 633–638, Aug. 2022. doi: 10.1016/J.CARBON.2022.05.018

[39] R. N. A. R. Seman and M. A. Azam, "Hybrid heterostructures of graphene and molybdenum disulfide: The structural characterization and its supercapacitive performance in 6M KOH electrolyte," *J. Sci.: Adv. Mater. Devices.*, vol. 5, no. 4, pp. 554–559, Dec. 2020. doi: 10.1016/J.JSAMD.2020.09.010

[40] S. Su, H. Dai, Y. Lin, X. Xiang, Y. Jiang, and X. Zhu, "Design and realization of binder-free electrodes with leaf-like structure based on Fe-Co binary oxides/graphene enabling efficient energy storage of flexible solid-state supercapacitors," *J. Energy Storage*, vol. 50, pp. 104699, Jun. 2022. doi: 10.1016/J.EST.2022.104699

[41] R. Pal, S. L. Goyal, I. Rawal, A. K. Gupta, and Ruchi, "Efficient energy storage performance of electrochemical supercapacitors based on polyaniline/graphene nanocomposite electrodes," *J. Phys. Chem. Solids*, vol. 154, pp. 110057, Jul. 2021. doi: 10.1016/J.JPCS.2021.110057

[42] X. Li, Y. Wang, Y. Zhao, J. Zhang, and L. Qu, "Graphene materials for miniaturized energy harvest and storage devices," *Small Struct.*, vol. 3, no. 1, pp. 2100124, Jan. 2022. doi: 10.1002/SSTR.202100124

[43] N. Kamboj and R. S. Dey, "Exploring the chemistry of organic/water-in-salt' electrolyte in graphene-polypyrrole based high-voltage (2.4V) microsupercapacitor," *Electrochim. Acta*, vol. 421, pp. 140499, Jul. 2022. doi: 10.1016/J.ELECTACTA.2022.140499

[44] M. Yuan *et al.*, "Laser synthesis of superhydrophilic O/S co-doped porous graphene derived from sodium lignosulfonate for enhanced microsupercapacitors," *J Power Sources.*, vol. 513, pp. 230558, Nov. 2021. doi: 10.1016/J.JPOWSOUR.2021.230558

[45] H. Li *et al.*, "Scalable fabrication of vanadium carbide/graphene electrodes for high-energy and flexible microsupercapacitors," *Carbon N. Y.*, vol. 183, pp. 840–849, Oct. 2021. doi: 10.1016/J.CARBON.2021.07.066

[46] Y. Zhang *et al.*, "High-Performance All-solid-state microsupercapacitors from 3D printing Structure-engineered graphene-carbon sphere electrodes," *Appl. Surf. Sci.*, vol. 597, pp. 153730, Sep. 2022. doi: 10.1016/J.APSUSC.2022.153730

[47] M. Ershadi, M. Javanbakht, Z. Kiaei, H. Torkzaban, S. A. Mozaffari, and F. B. Ajdari, "A patent landscape on Fe3O4/graphene-based nanocomposites in Lithium-Ion Batteries," *J Energy Storage.*, vol. 46, pp. 103924, Feb. 2022. doi: 10.1016/J.EST.2021.103924

[48] J. Ou, F. Jin, H. Wang, S. Wu, and H. Zhang, "Carbon coated Si nanoparticles anchored to graphene sheets with excellent cycle performance and rate capability for Lithium-ion battery anodes," *Surf. Coat Technol.*, vol. 418, pp. 127262, Jul. 2021. doi: 10.1016/J.SURFCOAT.2021.127262

[49] B. Lan, Y. Wang, X. Zhang, and G. Wen, "Interconnected SnO2/graphene+CNT network as high performance anode materials for lithium-ion batteries," *Ceram. Int.*, vol. 47, no. 17, pp. 24476–24484, Sep. 2021. doi: 10.1016/J.CERAMINT.2021.05.163

[50] C. J. He *et al.*, "Hierarchical microspheres constructed by SnS2 nanosheets and S-doped graphene for high performance lithium/sodium-ion batteries," *J Alloys Compd.*, vol. 889, pp. 161648, Dec. 2021. doi: 10.1016/J.JALLCOM.2021.161648

[51] Z. Xu *et al.*, "Thiophene-diketopyrrolopyrrole-based polymer derivatives/reduced graphene oxide composite materials as organic anode materials for lithium-ion batteries," *J. Chem. Eng.*, vol. 438, pp. 135540, Jun. 2022. doi: 10.1016/J.CEJ.2022.135540

7 Anodic Oxides-based Active Layers for Energy Production, Conversion, and Storage

Francisco Trivinho-Strixino, Mariana Sikora, and Janaina S. Santos

7.1 INTRODUCTION

The anodic oxidation of a metal is an electrochemical process known as anodization. It is a facile, scalable, and environmentally friendly surface treatment on metals, widely applied in research and industry for manufacturing nanostructured films on metal substrates with different morphologies [1–6], like nanoporous, nanotubular, nanowires, nanorods, and nanopetals, among others. This technique, employed for fabricating metal oxides with high surface area and unidirectional orientation; presents advantages like mild temperature conditions, non-toxic reagents, strict control of the morphology, microstructure, and composition by the anodizing conditions [7].

The synthetic routes employed to fabricate the anodic oxides generally consist of the following steps [1]: (*i*) pre-treatment of the metal substrate, (*ii*) anodization, (*iii*) heat treatment, and (*iv*) surface modification (if necessary). The pre-treatment (*i*) involves cleaning and polishing procedures to reduce the rugosity of the substrate surface. In the anodization (*ii*) step, the metal works as the anode of an electrochemical reactor being oxidized ($M \rightarrow M^{z+} + ze^-$) in a proper electrolyte. The composition and morphology of the nanostructured oxide film can be controlled by the metal substrate type, electrolyte composition, applied voltage/current, the rugosity of the substrate surface, temperature, anodizing duration, and the number of anodizing steps. Therefore, depending on the experimental conditions, the resulting anodic oxide film can form a rigid compact layer or a nanostructured layer over the metal surface with different geometries. The rigid compact layer is usually employed as a protective layer for the metal, not acting as the active layer. On the other hand, a nanostructured oxide is used in applications where the oxide is the active layer and a high surface area is required. After anodization, a heat treatment (*iii*) under a suitable gas atmosphere is performed to crystallize the oxide. To deposit another element over the nanostructured oxide or change the properties of its surface, additional modification steps (*iv*) might be necessary, depending on the application.

DOI: 10.1201/9781003318859-7

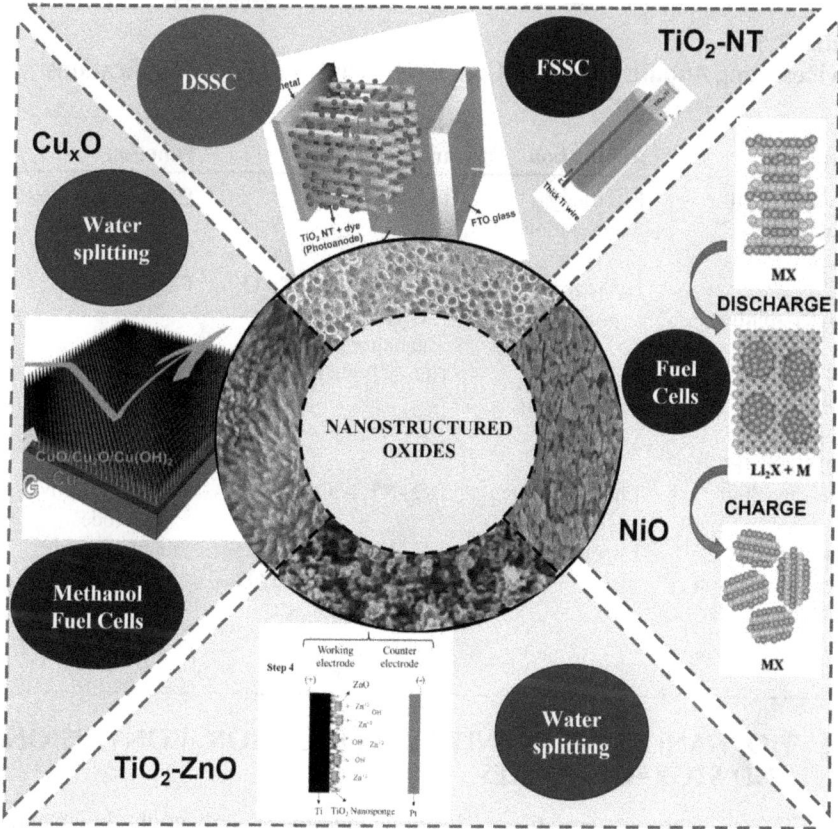

FIGURE 7.1 Micrographs of nanostructured anodic oxides produced via anodization technique and some related applications in energy field. Reproduced from Refs. [8–12] (Creative Commons License – CC BY 4.0).

Figure 7.1 depicts the micrographs of nanostructured oxides produced and some applications in energy devices. Table 7.1 depicts the systems that employed anodic oxides as part of their components and their function in the device.

In the following sections, we discuss the main uses of nanostructured anodic oxides in systems devoted to energy production, conversion, or storage. The discussions are based on the aspects and properties that motivated the utilization of these materials in those mentioned applications and the contribution that the anodic oxides can offer to these devices in terms of performance. A brief description of the synthetic route is presented, although we recommended consulting the original paper cited for specific details. Additionally, for more information on the fundamental aspects of the anodizing technique, we recommended some review articles on this topic [2,3,7,13].

TABLE 7.1
Devices using Anodic Oxides as Components in Energy-field Applications

Device	Application	Nanostructured anodic oxide	Function
Dye-sensitized fuel cell (DSSC)	Electricity generation	TiO_2-NT	Photoanode
PEC water-splitting	Hydrogen production	TiO_2-NT, Fe_2O_3, Cu_xO, ZnO, WO_3-NT, nanoporous WO_3	Photoanode
Fuel cells (PEMFC, DFAFC, DMFC, MFC)	Electricity generation	TiO_2-NT, NiO	Anode, cathode
Supercapacitor	Energy storage	TiO_2-NT, NiO, Cu_xO	Pseudocapacitive electrode
Lithium-ion (LIB) and sodium-ion batteries (SIB)	Energy storage	TiO_2-NT, Cu_xO, ZnO, Nb_2O_5, SnO_2, Ta_2O_5, WO_3 nanosheets	Anode, cathode

Source: [1].

7.2 TIO$_2$ NANOTUBES IN ENERGY PRODUCTION, CONVERSION, AND STORAGE DEVICES

The TiO_2-NT is the most versatile anodic oxide in terms of applications, being extensively employed in biomaterials, and explored as photocatalysts or electrodes in several systems in environmental and energy applications. Due to its chemical stability, low cost, high surface area, and unidirectional orientation, the TiO_2-NT film has been used in the energy field as a component of dye-sensitized solar cells (DSSC), photoelectrochemical (PEC) water-splitting systems for hydrogen generation, fuel cells, supercapacitors, and rechargeable batteries [1,8,14]. The most common procedure to fabricate TiO_2-NT films is the anodization of high-purity Ti foils at 10–60 V in organic media containing fluoride ions and a small amount of deionized water. NH_4F and ethylene glycol (EG) are preferred for composing the electrolyte. Under this condition, the inner nanotube diameter nanotube ranges from 40 to 160 nm and the thickness layer, usually in the µm range, is controlled by the anodizing time, which can vary from 20 min to 24 h [1]. A heat treatment at 450–600 °C for 1–3 h in air atmosphere is usually performed to form the anatase and rutile crystalline phases. Temperatures above 600 °C tend to destroy the nanotubular structure.

Figure 7.2 illustrates an experimental apparatus utilized in titanium anodizing, consisting of a glass reactor containing the electrolyte, a titanium specimen as anode placed at the center, and a stainless-steel foil cathode wrapped in the inner wall of the reactor. The electrodes are connected externally to a power supply and a multimeter. The inset shows a photo of a titanium specimen (3 cm × 1 cm) after the anodization, where the brown area corresponds to the anodized area, i. e. the region covered with

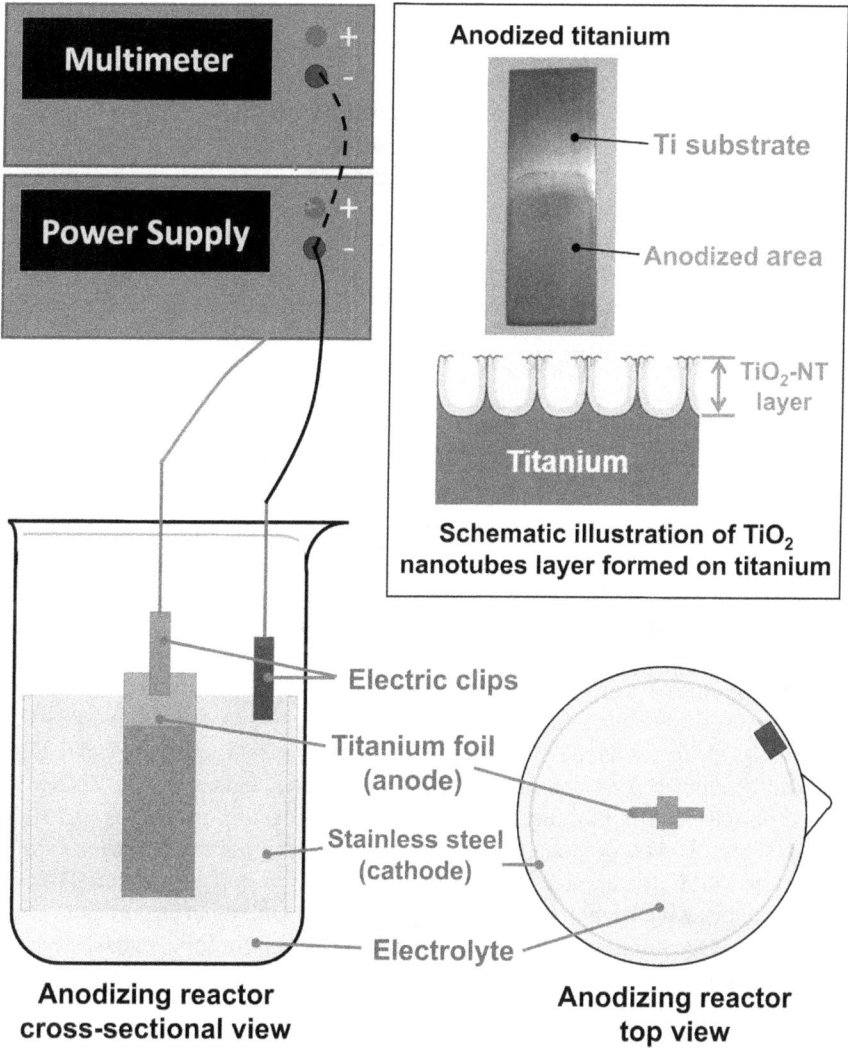

FIGURE 7.2 Illustration of experimental apparatus utilized in titanium anodizing. The inset shows a picture of a titanium specimen after the anodization and a schematic illustration of the oxide layer growth on the metal surface.

the anodic oxide layer. The figure also depicts a schematic illustration of the oxide layer (TiO$_2$-NT) formed on the titanium surface.

7.3 DYE-SENSITIZED SOLAR CELLS

Solar cells convert solar energy into usable electric energy via photovoltaic (PV) or PEC systems. These devices are classified according to the operational functions, composition, fabrication methods, and component materials: silicon solar cell, organic photovoltaic solar cell, perovskite solar cell, etc. The current demand for low-cost and non-toxic materials and simple manufacturing processes led to the development of DSSCs. Due to its electronic properties and high corrosion resistance, TiO_2 is the *n*-type semiconductor most explored as the photoanode in dye-sensitized solar cells (DSSC) [1]. Despite its low efficiency compared with the high-efficient Si-solar cells, the low cost and simplicity turn this device into a cost-effective and promising alternative for clean electricity production. A DSSC operates similarly to an electrochemical system, consisting of an anode, cathode, and electrolyte. In most DSSC devices, the photoanode comprises a transparent conductive oxide substrate and a nanocrystalline metal oxide layer with a high dye-sensitive surface area. The electrons generated in the dye interphase are injected into the metal oxide and collected in the conductive substrate. The two electrodes are connected externally so the current can flow. Between the electrodes, a non-aqueous redox electrolyte or an aqueous iodine electrolyte completes the circuit. The light source can be irradiated from the photoanode to the cathode or vice versa. The DSSC's overall efficiency depends on the optimization and operation of each component, particularly the processes occurring in the semiconductor film.

Unlike the random orientation of TiO_2 nanoparticles (TiO_2-NP), the nanotubular TiO_2 fabricated by anodization offers better electron transport along the channels. Its 3D nanostructure provides a large number of active sites for diffusion and adsorption/desorption of species. In the DSSC, the TiO_2-NT films are usually detached from the Ti substrate and binding into a transparent conductive substrate like FTO to improve performance as the photoanode. The photocatalytic properties of TiO_2-NT can also be enhanced by combining it with other semiconductors like ZnO, which present faster electron mobility and lower photoinactivation properties [9,15]. The TiO_2-NT photoanodes produced over flexible titanium wires have also been explored in flexible DSSC for application as a power source in soft electronics [12]. The current studies have been driven to modify TiO_2-NTs with other materials to reduce electron-hole recombination, increase the photo electron lifetime, enhance electron transport, and sensitize the photoanode under visible light to improve DSSC overall efficiency.

7.4 PEC WATER-SPLITTING

Besides solar cells, PEC water-splitting systems have also explored the photocatalytic properties of TiO_2-NT films for hydrogen production. This fuel is a potential alternative to fossil fuels. The PEC devices operate similarly to a water electrolysis process but utilize a photochemical process as an additional drive force. The photogenerated charge carriers in the anode are driven to the electrode/electrolyte interface to produce gaseous O_2 and H_2. The oxygen evolution reaction (OER) occurs in the anode interface, which has protons as a side product, together with O_2. The protons migrate towards the cathode, where they are reduced to H_2 via hydrogen evolution reaction (HER) [1]. The efficiency of this process depends on the capability of light absorption, mobility of charges, number of active sites on the surface, and electron-proton recombination in the photoanode that should be minimized. Since the OER in the anode is the limiting step of the reaction mechanism, most strategies for enhancing PEC efficiency lie majority in improving the photoanode' properties instead of the cathode'. Therefore, recent studies focus on improving the photocurrent density and PEC overall efficiency by developing electrodes with specific bandgap properties, crystallinity, and morphology. The porous vertically aligned structure of TiO_2-NT is an attractive property for a more efficient electrolyte diffusion in DSSCs [8]. 3D hierarchical structures obtained by anodization of Ti meshes showed that the inner diameter and the length of TiO_2 nanotubes influenced the overall performance of the PEC cell. And thinner oxide layers presented better efficiency [16]. TiO_2-NT photoanodes have demonstrated improved performance compared to conventional catalysts in powder form deposit in carbon substrate [17]. Similar to the DSSC, the strategies to change band structure, sensitization under visible light, and reducing charge recombination of TiO_2-NT involve doping the oxide or combining it with other compounds like ZnO, NiO, W, and Cu [9,18,19]. Those approaches aim to enhance the photoanode efficiency by tuning TiO_2-NT surface properties.

7.5 FUEL CELLS

A fuel cell is also considered a promising sustainable technology for electricity generation. It is composed of anode and cathode electrodes immersed in proper electrolyte and connected by an external resistance. The fuel (H_2, CH_3OH, HCOOH, organic pollutant, etc.) is oxidized on the anode surface, while an electron acceptor (O_2, $[Fe(CN)_6]^{3-}$, $S_2O_8^{2-}$, NO_3^-) is reduced on the cathode generating electricity and products. The fuel cells are classified according to the fuel or electrolyte: proton-exchange membrane fuel cell (PEMFC), direct formic acid fuel cell (DFAFC), direct methanol fuel cell (DMFC), microbial fuel cell (MFC), etc. [1,14]. Currently, this technology is in the development stage, and many efforts have been made to produce efficient and economical fuel cells. However, the high cost of the electrodes is still a barrier to the wide utilization of this technology. One of the strategies for reducing costs is improving the electrodes by using porous nanomaterials with high corrosion resistance, stability, and durability as a support for the catalysts.

In this sense, the TiO_2-NT films are the most explored among the anodic oxides in fuel cell electrodes as catalyst support to provide a high surface area and increase the electrode stability. Some studies report its decoration with Pt, Pd, Ni, and NiO in PEMFC, DMFC, and DFAFC in both anode and cathode electrodes [20–22]. When combined with TiO_2-NT, an improvement in corrosion resistance and long-term stability of the material were observed. Besides, the quantity of expensive materials like Pt and Pd is small, reducing electrode costs. Bare or combined with other materials to enhance its properties, TiO_2-NT electrodes have also been explored as bioanodes, cathodes, and photoanodes in MFCs and photo-assisted MFC systems [14,23,24]. Another advantage of TiO_2-NT, observed in MFCs, is the presence of terminal hydroxyl groups on the surface that could enhance the bacteria attachment to the oxide surface of bioanodes [25].

7.6   SUPERCAPACITORS AND LI-ION BATTERIES

In energy storage systems, the TiO_2-NT properties have also been investigated in supercapacitors and Li-ion batteries (LIB) [1]. A supercapacitor usually refers to an electrochemical capacitor, which operates via a double-layer mechanism or charge transfer reactions occurring on the electrodes' surface. They can be symmetric or asymmetric, depending on the electrode type. The electrodes consist of an active material bounded to a current collector. The larger the surface area of the active layer, the higher the storage charge capacity on its surface. Therefore, a large surface area of the vertically aligned TiO_2-NT film combined with its fast ion diffusion can favor fast redox reactions in pseudocapacitive electrodes. Additionally, the anodization allows the fabrication of binder-free electrodes since the oxide has rigidly adhered to the metal substrate, which can act directly as the current collector [1].

The first studies of pseudocapacitance behavior and stability of TiO_2-NT/Ti electrodes showed promising results. The 3D-nanostructured oxide demonstrated great potential in supercapacitor electrodes as an active layer or support for other materials like polymers and MnO_2. When combined with a conductive polymer like polyaniline, the charge transfer rates, electronic ability, and long-term stability significantly increased [26,27]. The large surface area with open channels for Li^+ intercalation and the low volume expansion is an advantage of TiO_2-NT anodes for lithium-ion batteries (LIB), increasing stability and safety. The TiO_2-NT anodes also exhibited fast kinetics and enhanced capacity retention. The low voltage for Li insertion (1.7 V vs. Li/Li^+) also contributes to battery safety since it prevents the formation of dendrites on the electrode surface responsible for the short circuit during overload. Current studies have demonstrated the TiO_2-NT modification by doping or combining with other materials like carbon-nanotubes improved LIB performance by increasing capacity retention, cycling behavior, electronic conductivity, and long-term stability [28,29]. The large size channels of TiO_2-NT anodes have also been an attraction for developing sodium-ion batteries (SIB) [1].

7.7   ANODIC OXIDES FROM TRANSITION METALS OF FOURTH PERIOD: FE_2O_3, NIO, CU_xO, AND ZNO

In the chemical periodic table, the transition metal oxides with relatively close atomic weight (Fe_2O_3, NiO_x, CuO_x and ZnO) in the fourth period have shown particular interest in energy device assembly using anodic oxides. In this series, hematite (Fe_2O_3) obtained by iron anodization was applied in the field of energy devices for PEC water-splitting applications for H_2 production [30,31]. Its preparation has been reported by anodization of high-purity iron substrates in an EG + water + fluoride electrolyte, at voltages ~50 V and short anodizing times (15 min), followed by an annealing procedure under Ar atmosphere (500 °C, 1 h) to form the hematite phase. Despite its low conductivity, this anodic oxide has a suitable bandgap to absorb visible-light irradiation. To improve PEC water-splitting performance and oxide conductivity, doping ions can be added to the electrolyte during the hematite anodization. An increase in IPCE (incident photon-to-current efficiency) was observed when Sn^{2+} dopant ions were added to the Fe_2O_3 photoanode. The enhanced performance was ascribed to an improvement in the energy transfer rate [30].

Similar anodization methods are carried out for anodic oxide preparation based on Ni. NiO mesoporous structure, for instance, is prepared by anodization of Ni foils in $H_3PO_4 + NH_4F$ at low voltages and short times (~ 3 V, 5 min), whereas NiO nanopetals film is produced using oxalic acid electrolyte (50 V, 10 min) [5,32]. Post-treatment as annealing under air atmosphere (at 400–550 °C for 20–45 min) is used for oxide crystallization. Due to its high electrocatalytic performance, reduced cost, and good stability, this synthetic route was used to prepare the mesoporous NiO anodes for a DMFC, using Ni as an alternative to the Pt-based catalyst [32]. As a result, the anode demonstrated good performance as electro-active material with a high current peak for methanol oxidation towards CO_2, which was attributed to its high grain boundary in the mesoporous structure that favored fast reaction kinetics. In supercapacitors, NiO nanopetals electrode was applied in asymmetric supercapacitor as the positive pole and a carbon electrode as the negative pole, separated by cellulose paper and KOH solution as the electrolyte [5]. Despite the low specific energy density and the power density compared to other supercapacitor electrodes based on Ru, RuO, MnO_2, and Fe_3O_4, the NiO electrode demonstrated outstanding cycling stability and the advantages of using a low-cost, environmentally friendly, and simple fabrication procedure for NiO/Ni electrodes.

The following chemical element in the fourth period after the nickel is copper, a highly abundant, non-toxic, and ideal for scale-up fabrication processes. Cupric (Cu_2O) and cuprous (CuO) oxides have been exhibiting great potential in PEC devices and can be produced by anodization [11]. With a bandgap suitable for absorption in the visible-light region (Cu_2O band gap over 2.1 eV and CuO band gap equal 1.44 eV) [33–37], the photoactivity of the p-type semiconductors based on copper oxides has also been explored as photocathodes in PEC water-splitting applications.

Regarding the nanostructured oxide synthesis, Cu foil and Cu-sputtered FTO are used as a substrate for anodization in NaOH or KOH solutions for times varying from 3 to 20 min. The heat treatment is carried out in an Ar atmosphere at 550–600 °C for 4 h. Since copper oxides are susceptible to photo corrosion, additional modification steps are included in their synthesis to protect the oxide. Some reports described materials like Al-doped ZnO, TiO_2, Ru_xO, and C-coating employed to protect the Cu layers [33,34]. An increase of 25% in photoactivity under visible light irradiation was observed using the 3D electrode based on Cu anodizing compared to the planar electrodes of similar architecture. Cu_2O nanowires embedded with C-coated demonstrated an improvement in the charge transfer processes at the electrode/electrolyte interface, increasing the photocurrent response and the photo-corrosion protection [33].

The use of copper oxide/hydroxide in supercapacitors is also described. $Cu(OH)_2$ nanorods obtained from the anodization of Cu foil in alkaline electrolyte were explored as a flexible electrode (positive electrode) in an asymmetric supercapacitor. The overall performance of this device delivered a high capacitance and better stability compared to other pseudocapacitive materials reported in the literature showing the great potential for cost-effective commercial devices with non-toxic materials such as copper [6]. Another possibility is the application of CuO as part of Sn-based anode materials in LIBs motivated by the idea that the volume change ratio during Li$^+$ intercalation in Sn/Cu$_x$O anode would be lower than the typical Sn anodes. Remarkably, Sn/Cu$_x$O anode demonstrated a reversible discharge capacity of 772.5 mA g^{-1} and stable cycle retention after 100 cycles, proving the advantages of Cu$_x$O nanowires as active components in LIBs anodes [38].

Zinc oxide nanowires in DSSC and PEC applications were reported in some studies [15,39,40]. ZnO presents a wide bandgap (3.37 eV) and absorption edge in the UV region, making this oxide an alternative for TiO$_2$ photoanode application since it exhibits higher electron mobility [40]. The standard synthetic route to prepare Zn/ZnO follows a procedure of potentiostatic anodization (5–10 V) for short periods (10–30 min) [39,40]. A post-thermal treatment (300 °C for 1 h) is required to crystallize the amorphous oxide in a wurtzite hexagonal structure. Furthermore, this n-type semiconductor possesses other properties that make it attractive for PEC devices, such as high photostability, thermal stability, low cost, non-toxicity, and biodegradability [40]. An anode based on Zn/ZnO-hexagonal pyramid array was also explored in a Zn-ion battery [41]. The structure was synthesized by pulsed anodization method in an NH$_4$Cl + H$_2$O$_2$ solution, followed by thermal treatment (300 °C for 1 h). This battery type is considered a green energy system using less-flammable electrolytes and low-toxic materials. This anodized electrode inhibited the growth of Zn dendrites, which is reported to decrease the electrode stability and demonstrated excellent performance for coulombic efficiency, long-term cyclability (~99% efficiency after 1000 cycles at 9 A g^{-1}), and improved corrosion resistance [41].

7.8 NANOSTRUCTURED WO$_3$, NB$_2$O$_5$, TA$_2$O$_5$, AND SNO$_2$

More recently, other anodic oxides widely explored in other fields are also being applied in energy applications, such as the W, Nb, Ta, and Sn oxides. Among these, the anodic oxides from W, Nb, and Ta metals presented similar properties to TiO$_2$. These metals belong to the group known as "valve metals", which also comprises Al and Zr [7]. The anodic oxides obtained from the anodization of these two metals are extensively applied in photonic crystals, sensors, biosensors, and biomaterials [4]. Still, they are not so commonly explored as an active layer in energy devices, exerting secondary roles in the energy field. For instance, the nanoporous anodic alumina (Al$_2$O$_3$) is used as a template or scaffold for synthesizing other nanostructures. The templates are removed from the material after synthesis, the scaffold remains as structural material [42–44]. ZrO$_2$ nanotubes films were explored as a solid electrolyte in solid oxide fuel cell (SOFC) [45].

The high stability, photo corrosion resistance, and suitable band gap value to absorption under visible-light irradiation make the WO$_3$ an attractive catalyst for PEC water-splitting. The synthetic route for fabricating nanostructured WO$_3$ photoanodes involves the anodization of W foils at 40–50 V for times varying from 30 min to 20 h, using an aqueous- or organic-based electrolyte containing fluoride ions [46]. The annealing is usually performed at 400–500 °C for 2 or 3h. On the other hand, a method for WO$_3$ nanosheet arrays is described using a two-anodization step procedure in H$_2$SO$_4$ solution at 15 V for 2 h (first step) and 24 h (second step) [47]. This methodology is usually applied to produce highly ordered nanostructures. In this case, the oxide layer formed in the first step is removed by ultrasonication in water and the marked substrate acts as a guide for the oxide pattern formation in the second anodization step [7].

The WO_3-NT and nanoporous WO_3 films prepared by anodizing W foil demonstrated a remarkable photoactivity for water splitting under UV and visible-light irradiation. The excellent performance of this semiconductor is attributed to the fast electron-hole separation rate in the 3D-nanostructure, efficient charge transfer/transport processes at the interface, strong interaction between metallic W substrate and oxide layer, and the high crystallinity of WO_3 [46,48,49]. Due to this excellent performance in PEC systems, current studies have been driven to improve the photo-catalytic properties of this semiconductor and investigate the influence of anodizing conditions on the properties of this metal oxide since it was not widely explored in PEC systems like the TiO_2-NT. Due to its high theoretical capacity, WO_3 was also explored as the anode for LIBs. However, it is susceptible to nanoparticle aggregation reducing its capacity. To solve this, WO_3 nanosheets decorated with Ag-NPs were produced and tested by controlling anodizing parameters, demonstrating enhanced electronic conductivity and good rate performance after 150 cycles [47].

The other oxides, Nb_2O_5, Ta_2O_5, and SnO_2, have been explored in rechargeable batteries. The method reported to prepared nanoporous Nb_2O_5 and Ta_2O_5 electrodes is the anodization of Nb or Ta foils in EG + fluoride electrolyte at low voltages (~15 V, 2 h), followed by heat treatment at 400–600 °C under Ar or H_2 atmosphere for 1–2 h [50,51]. On the other hand, the mesoporous SnO was produced by anodization of tin foil in oxalic acid at low voltage and a short time (10 V, 20 min) [52]. Before the heat treatment, the as-prepared SnO samples were soaked in water at room temperature for different intervals (2–168 h) and then heated at different temperatures (40–100 °C) for 2 h.

The high storage capacity of Nb_2O_5 in LIBs motivated the study of this semicon-ductor as an intercalation anode in Na-ion batteries (SIB). An amorphous nanoporous Nb_2O_5 array enriched by hydrogen via post-thermal treatment, with a uniform pore diameter of 15–20 nm, exhibited excellent performance as a binder-free anode in SIBs [50]. The high reversible capacity and long stability of the amorphous Nb_2O_5 (400 °C) were comparable to the crystalline one (annealed at 600 °C) and other high-performance materials. The good performance was credited to the enhanced higher activity of this semiconductor promoted by the presence of non-metal elements and oxygen vacancies. Aiming for the development of a flexible and durable 3D cathode, the nanostructured Ta_2O_5 was explored in LIBs motivated by the nanoporous struc-ture of the anodic oxide, which tends to facilitate the Li^+ intercalation, increasing the capacity and reversibility of the battery [51]. In preliminary tests performed in a coin cell, the cathode reached a theoretical predicted capacity similar to the LIB system, demonstrating good stability (8000 cycles at a 5 C rate). The outstanding result was ascribed to the oxygen deficiency and the nanoporous architecture of the Ta_2O_5, increasing the electrode transport processes. In searching for a better storage cap-ability electrode for SIBs, a water-immersion method was developed to crystallize the as-anodized mesoporous SnO_2 in the rutile phase under low temperatures. The system exhibited an enhanced performance with a capacity of up to 514 mA h g^{-1} after 100 cycles at 0.1 C, which was attributed to the high surface area with a larger number of reactive sites; the oxide crystallinity that improved the conductivity; and the fast diffusion of Na^+ ions into the short diffusion length available for Na^+ ions transport.

7.9 FINAL REMARKS

The strategies for developing both energy devices revealed remarkable results concerning nanostructured anodic oxides. TiO_2-NT and nanostructured WO_3 are the most explored anodic oxides in energy applications under different architectures. In DSSC and PEC devices, TiO_2-NT anodes stand out, although WO_3 electrodes have also demonstrated outstanding performance in PEC devices. Fe_2O_3, ZnO, and Cu_xO nanostructures also exhibited promising results as photocatalysts in PEC water-splitting. The unidirectional orientation and nanoporous structures enhanced the charge transfer and mass transport processes, while the large specific surface area favored the diffusion and adsorption/desorption of species. Regarding the other nanostructured oxides, NiO had been explored as alternative electrodes to the expensive Pt and Pd electrodes in fuel cells, as well as the positive electrode in asymmetric supercapacitors. The opened 3D nanostructure of Nb_2O_5 and Ta_2O_5 anodized films is an attractive property for Li^+ and Na^+ intercalation processes in rechargeable batteries. Combined with other materials, the anodic oxides can contribute to developing green energy storage systems with high efficiency, long-term stability, and safety. And as a consequence, they can provide significant improvement for the industrial and commercial sectors.

ABBREVIATION LIST

DFAFC: direct formic acid fuel cell; DMFC: direct methanol fuel cell; DSSC: dye-sensitized solar cells; EG: ethylene glycol; FTO: fluorine-doped tin oxide; HER: hydrogen evolution reaction; IPCE: incident photon-to-current efficiency; LIB: Lithium-ion batteries; MFC: microbial fuel cell; NP: nanoparticles; NT: nanotubes; OER: oxygen evolution reaction; PEC: photoelectrochemical; PEMFC: proton membrane exchange fuel cell; PV: photovoltaic; SIB: sodium-ion batteries; SOFC: solid oxide fuel cell.

REFERENCES

1. Santos JS, Araujo PDS, Pissolitto YB, et al. The use of anodic oxides in practical and sustainable devices for energy conversion and storage. *Materials (Basel)*. 2021;14(2).
2. Giziński D, Brudzisz A, Santos JS, et al. Nanostructured anodic copper oxides as catalysts in electrochemical and photoelectrochemical reactions. *Catalysts*. 2020;10(11):1338.
3. Fu Y, Mo A. A Review on the electrochemically self-organized titania nanotube arrays: synthesis, modifications, and biomedical applications. *Nanoscale Research Letters*. 2018;13(1):187.
4. Ruiz-Clavijo A, Caballero-Calero O, Martín-González M. Revisiting anodic alumina templates: From fabrication to applications. *Nanoscale*. 2021;13(4):2227–2265.
5. Cheng G, Bai Q, Si C, et al. Nickel oxide nanopetal-decorated 3D nickel network with enhanced pseudocapacitive properties. *RSC Advances*. 2015;5:15042–15051.

6. Chen J, Xu J, Zhou S, et al. Facile and scalable fabrication of three-dimensional Cu(OH)2 nanoporous nanorods for solid-state supercapacitors. *Journal of Materials Chemistry A.* 2015;3(33):17385–17391.

7. Trivinho-Strixino F, Santos JS, Souza Sikora M. *3 – Electrochemical Synthesis of Nanostructured Materials.* In: Da Róz AL, Ferreira M, de Lima Leite F, et al., editors. Nanostructures: William Andrew Publishing; 2017. p. 53–103. https://doi.org/10.1016/B978-0-323-49782-4.00003-6

8. Lee AR, Kim J-Y. Highly ordered TiO2 nanotube electrodes for efficient Quasi-Solid-State Dye-Sensitized solar cells. *Energies.* 2020;13(22).

9. Navarro-Gázquez PJ, Muñoz-Portero MJ, Blasco-Tamarit E, et al. Original approach to synthesize TiO2/ZnO hybrid nanosponges used as Photoanodes for Photoelectrochemical applications. *Materials.* 2021;14(21).

10. Siddiqui S-ET, Rahman MA, Kim J-H, et al. A review on recent advancements of Ni-NiO nanocomposite as an Anode for high-performance Lithium-Ion Battery. *Nanomaterials.* 2022;12(17).

11. Giziński D, Brudzisz A, Santos JS, et al. Nanostructured Anodic copper oxides as catalysts in electrochemical and photoelectrochemical reactions. *Catalysts.* 2020;10(11).

12. Xiao B-C, Lin L-Y. Substrate diameter-dependent photovoltaic performance of flexible fiber-type dye-sensitized solar cells with TiO2 Nanoparticle/TiO2 nanotube array photoanodes. *Nanomaterials.* 2020;10(1).

13. Stepniowski WJ, Misiolek WZ. Review of fabrication methods, Physical properties, and applications of Nanostructured copper oxides formed via electrochemical oxidation. *Nanomaterials.* 2018;8(6).

14. Santos JS, Tarek M, Sikora MS, et al. Anodized TiO2 nanotubes arrays as microbial fuel cell (MFC) electrodes for wastewater treatment: An overview. *Journal of Power Sources.* 2023;564:232872. https://doi.org/10.1016/j.jpowsour.2023.232872

15. Miles DO, Lee CS, Cameron PJ, et al. Hierarchical growth of TiO2 nanosheets on anodic ZnO nanowires for high efficiency dye-sensitized solar cells. *Journal of Power Sources.* 2016;325:365–374.

16. Saboo T, Tavella F, Ampelli C, et al. Water splitting on 3D-type meso/macro porous structured photoanodes based on Ti mesh. *Solar Energy Materials and Solar Cells.* 2018;178:98–105.

17. Stoll T, Zafeiropoulos G, Tsampas MN. Solar fuel production in a novel polymeric electrolyte membrane photoelectrochemical (PEM-PEC) cell with a web of titania nanotube arrays as photoanode and gaseous reactants. *International Journal of Hydrogen Energy.* 2016;41(40):17807–17817.

18. Rasheed MA, Rahimullah R, Uddin SK, et al. Role of temperature and NiO addition in improving photocatalytic properties of TiO2 nanotubes. *Applied Nanoscience.* 2019;9(8):1731–1742.

19. Momeni MM, Ghayeb Y, Ezati F. Fabrication, characterization and photoelectrochemical activity of tungsten-copper co-sensitized TiO2 nanotube composite photoanodes. *Journal of Colloid and Interface Science.* 2018;514:70–82.

20. Haskul M, Ülgen AT, Döner A. Fabrication and characterization of Ni modified TiO2 electrode as anode material for direct methanol fuel cell. *International Journal of Hydrogen Energy.* 2020;45(7):4860–4874.

21. Manikandan M, Vedarajan R, Kodiyath R, et al. Pt decorated free-standing TiO2 nanotube arrays: Highly active and durable electrocatalyst for oxygen reduction and methanol oxidation reactions. *Journal of Nanoscience and Nanotechnology.* 2016;16(8):8269–8278.

22. Pisarek M, Kędzierzawski P, Andrzejczuk M, et al. TiO2 nanotubes with Pt and Pd nanoparticles as catalysts for electro-oxidation of formic acid. *Materials*. 2020;13(5):1195.

23. Deng L, Dong G, Zhang Y, et al. Lysine-modified TiO2 nanotube array for optimizing bioelectricity generation in microbial fuel cells. *Electrochimica Acta*. 2019;300:163–170.

24. Feng H, Liang Y, Guo K, et al. TiO2 nanotube arrays modified titanium: a stable, scalable, and cost-effective bioanode for microbial fuel cells. *Environmental Science & Technology Letters*. 2016;3(12):420–424.

25. Guo T, Wang C, Xu P, et al. Effects of the structure of TiO2 nanotube arrays on its catalytic activity for microbial fuel cell. *Global Challenges*. 2019;3(5):1800084.

26. Chen J, Xia Z, Li H, et al. Preparation of highly capacitive polyaniline/black TiO2 nanotubes as supercapacitor electrode by hydrogenation and electrochemical deposition. *Electrochimica Acta*. 2015;166:174–182.

27. Zhang J, Li Y, Zhang Y, et al. The enhanced adhesion between overlong TiNxOy/MnO2 nanoarrays and Ti substrate: Towards flexible supercapacitors with high energy density and long service life. *Nano Energy*. 2018;43:91–102.

28. Madian M, Ummethala R, Naga AOAE, et al. Ternary CNTs@TiO2/CoO Nanotube Composites: Improved Anode Materials for High Performance Lithium Ion Batteries. *Materials*. 2017;10(6).

29. Rahman MA, Wong YC, Song G, et al. Improvement on electrochemical performances of nanoporous titania as anode of lithium-ion batteries through annealing of pure titanium foils. *Journal of Energy Chemistry*. 2018;27(1):250–263.

30. Lv X, Rodriguez I, Hu C, et al. Modulated anodization synthesis of Sn-doped iron oxide with enhanced solar water splitting performance. *Materials Today Chemistry*. 2019;12:7–15.

31. Lucas-Granados B, Sánchez-Tovar R, Fernández-Domene RM, et al. Influence of electrolyte temperature on the synthesis of iron oxide nanostructures by electrochemical anodization for water splitting. *International Journal of Hydrogen Energy*. 2018;43(16):7923–7937.

32. Wang L, Zhang G, Liu Y, et al. Facile synthesis of a mechanically robust and highly porous NiO film with excellent electrocatalytic activity towards methanol oxidation. *Nanoscale*. 2016;8(21):11256–11263.

33. Shi W, Zhang X, Li S, et al. Carbon coated Cu2O nanowires for photo-electrochemical water splitting with enhanced activity. *Applied Surface Science*. 2015;358:404–411.

34. Luo J, Steier L, Son M-K, et al. Cu2O nanowire photocathodes for efficient and durable solar water splitting. *Nano Letters*. 2016;16(3):1848–1857.

35. Zoolfakar AS, Rani RA, Morfa AJ, et al. Nanostructured copper oxide semiconductors: A perspective on materials, synthesis methods and applications. *Journal of Materials Chemistry C*. 2014;2(27):5247–5270.

36. Stepniowski WJ, Misiolek WZ. 13-Nanostructured anodic films grown on copper: A review of fabrication techniques and applications. In: Sulka GD, editor. *Nanostructured Anodic Metal Oxides*. Amsterdam: Elsevier; 2020. p. 415–452.

37. Giziński D, Mojsilović K, Brudzisz A, et al. Controlling the morphology of barrel-shaped nanostructures grown via CuZn electro-oxidation. *Materials*. 2022;15(11):3961.

38. Kim M, Choi I, Kim JJ. Facile electrochemical synthesis of heterostructured amorphous-Sn@CuxO nanowire anode for Li-ion batteries with high stability and rate-performance. *Applied Surface Science*. 2019;479:225–233.

39. Faid AY, Allam NK. Stable solar-driven water splitting by anodic ZnO nanotubular semiconducting photoanodes. *RSC Advances*. 2016;6(83):80221–80225.

40. Batista-Grau P, Sánchez-Tovar R, Fernández-Domene RM, et al. Formation of ZnO nanowires by anodization under hydrodynamic conditions for photoelectrochemical water splitting. *Surface and Coatings Technology.* 2020;381:125197.

41. Kim JY, Liu G, Shim GY, et al. Functionalized Zn@ZnO hexagonal pyramid array for dendrite-free and ultrastable zinc metal anodes. *Advanced Functional Materials.* 2020;30(36):2004210.

42. Kwon C-W, Lee J-I, Kim K-B, et al. The thermomechanical stability of micro-solid oxide fuel cells fabricated on anodized aluminum oxide membranes. *Journal of Power Sources.* 2012;210:178–183.

43. Kwon H-C, Kim A, Lee H, et al. Parallelized nanopillar perovskites for semitransparent solar cells using an anodized aluminum oxide scaffold. *Advanced Energy Materials.* 2016;6(20):1601055.

44. Wei Q, Fu Y, Zhang G, et al. Rational design of novel nanostructured arrays based on porous AAO templates for electrochemical energy storage and conversion. *Nano Energy.* 2019;55:234–259.

45. Buyukaksoy A, Fürstenhaupt T, Birss VI. First-time electrical characterization of nanotubular ZrO2 films for micro-solid oxide fuel cell applications. *Nanoscale.* 2015;7(18):8428–8437.

46. Lu H, Yan Y, Zhang M, et al. The effects of adjusting pulse anodization parameters on the surface morphology and properties of a WO3 photoanode for photoelectrochemical water splitting. *Journal of Solid State Electrochemistry.* 2018;22(7):2169–2181.

47. Yang X, Qi Y, Liu Y, et al. Self-Assembly growth of 3D WO3 framework with interpenetrated nanosheets as binder-free anode for lithium ion batteries. *Journal of The Electrochemical Society.* 2017;164(12):A2783–A2789.

48. Li L, Zhao X, Pan D, et al. Nanotube array-like WO3/W photoanode fabricated by electrochemical anodization for photoelectrocatalytic overall water splitting. *Chinese Journal of Catalysis.* 2017;38(12):2132–2140.

49. Das PK, Arunachalam M, Seo YJ, et al. Functional blocking layer of twisted tungsten oxide nanorod grown by electrochemical anodization for photoelectrochemical water splitting. *Journal of The Electrochemical Society.* 2020;167(6):066501.

50. Ni J, Wang W, Wu C, et al. Highly reversible and durable Na storage in niobium pentoxide through optimizing structure, composition, and nanoarchitecture. *Advanced Materials.* 2017;29(9):1605607.

51. Xia S, Ni J, Savilov SV, et al. Oxygen-deficient Ta2O5 nanoporous films as self-supported electrodes for lithium microbatteries. *Nano Energy.* 2018;45:407–412.

52. Bian H, Dong R, Shao Q, et al. Water-enabled crystallization of mesoporous SnO2 as a binder-free electrode for enhanced sodium storage. *Journal of Materials Chemistry A.* 2017;5(45):23967–23975.

8 Green and Bio-waste-based Materials for Energy Production, Conversion, Storage, and Hybrid Technologies

Valentin Romanovski, Alexandr Dubina, Ali Akbari Sehat, Xintai Su, and Dmitry Moskovskikh

8.1 INTRODUCTION

People widely consume large amounts of energy in various applications of their livelihoods. Energy can be generated from renewable sources (e.g., solar, wind, sea tides, biomass, geothermal energy) and non-renewable sources (nuclear and fossil fuels such as oil, natural gas and coal). Fossil fuels are the most used source of energy production in the world. However, energy derived from fossil fuels cannot meet the growing global demand for energy due to limited supplies and negative environmental impacts. According to some prospects (Hatfield-Dodds et al., 2017), by 2050, fossil fuel production will increase by about 53%. Global energy demand is also estimated to grow by about 65% by 2040 (Kambo and Dutta, 2015). It is expected that the extensive exploitation of fossil fuels can lead to the depletion of non-renewable resources of the Earth's crust. In addition, energy generation processes from fossil fuels also release greenhouse gases (e.g. carbon dioxide, methane, nitrous oxide, sulfur oxides, nitrogen oxides, fly ash, dioxins and furans), which can contribute to environmental pollution and exacerbate global changing of the climate. As a result of some global meetings, such as the G7 Leaders Declaration Summit and the United Nations Summit, the importance of sustainable development and commitment to best practices in natural resource management were emphasized (Hatfield-Dodds et al., 2017). Due to these global challenges, efforts must be made to develop renewable energy technologies to reduce dependence on fossil energy. Renewable energy is sustainable, clean and does not pollute the environment. There are different types of renewable energy sources and related energy sectors (Rahman et al., 2022): *i*) wind – wind energy; *ii*) sunlight – solar energy; *iii*) energy of moving water – hydropower; *iv*) plants (biomass), waste – bioenergy; *v*) the heat of the Earth – geothermal energy.

DOI: 10.1201/9781003318859-8

The share of renewable energy sources is 13% (Popp et al., 2021), while the share of bioenergy is about 77% in the balance of renewable energy sources, hydropower is 15%. A variety of organic waste forms the raw material base of bioenergy from agriculture and forestry, food, woodworking and other industries, municipal (Hurynovich et al., 2021) and household waste, various plants and crops are specially grown for energy processing, as well as algae and microorganisms (Cui et al., 2021).

Modern ways of extracting renewable energy from biomass involve using highly efficient conversion systems of various capacities (from small household appliances to industrial conversion plants) to obtain energy from the so-called "modern biomass". Depending on the type of primary biomass, two main conversion approaches are used: biochemical/chemical and thermochemical to obtain a range of final bioenergy products through processes (Okolie et al., 2022): *i*) thermochemical impact on biomass with the production of synthesis gas and the subsequent production of heat and/or electricity; *ii*) anaerobic digestion of biomass with the production of biogas and the subsequent production of heat and electricity/biomethane as a gas motor fuel; *iii*) biochemical/chemical transformation of starch- and sugar-containing raw materials into bioethanol and fat- and oil-containing raw materials into biodiesel for the production of liquid transport and aviation fuels; *iv*) producing biodiesel from algae biomass by catalytic hydroprocessing and "green" gasoline based on catalytic cracking of oils. Another fraction – carbohydrates – is less energy efficient than lipids but can be successfully used to produce various fuel types, for example, ethanol, butanol, and aviation fuel. Residues and wastes of biomass after extraction can be processed by anaerobic fermentation into biogas or by the thermochemical method – into synthesis gas, as well as into associated valuable products, pharmaceutical substances, dietary supplements, animal feed, etc. (on the principles of biorefining).

It is noteworthy that the direct transportation to consumers, supply, and storage of renewable energy (mainly solar, wind, and geothermal) have restrictions, for example, restrictions in the emission of greenhouse gases. These are limiting factors for the broader use of renewable energy sources. After producing renewable energy, it must be converted into electricity and stored in energy storage devices (Denholm 2006). Performance improvements in energy storage devices are important to maximize renewable energy storage. To date, energy storage devices have been fabricated for renewable energy storage, such as supercapacitors with higher energy/power density and longer lifetimes (Lee and Lee 2018; Purushothaman et al., 2017). Currently, there is extensive research into supercapacitor devices to improve renewable energy storage to solve the problem of increasing the residual energy. One of the most critical aspects of sustainable development and green technologies is reducing waste generation. This can be achieved through the introduction of low-waste and waste-free technologies, mainly by increasing the degree of processing of raw materials (Osipova et al., 2022). A simpler approach involves recycling waste. There are many examples of this approach, for example, waste can be used to produce building materials (gypsum (Kamarou 2020; 2021a; Romanovski et al., 2022a), anhydrite (Kamarou 2021b, 2022; Romanovski et al., 2021), binders (Kamarou 2021c), building blocks (Ademati et al., 2022) and slabs (Akinwande 2022a), composites (Akinwande 2022b,

2022c), pigments (Zalyhina et al., 2021a, 2021b), metal extractions (Romanovskaia et al., 2021).

As a rule, the ideal constituent material for supercapacitor devices should be obtained by a simple, inexpensive, environmentally friendly synthesis method. The use of organic wastes is promising as a carbon matrix, and industrial wastes can be used as modifiers. Below are examples of waste that can be used to produce materials for energy production, conversion, storage and hybrid technologies. Today all carbon derivatives can be obtained from biomass because biomass is widely available as a renewable feedstock and requires simple synthesis (Romanovski 2021). So far, studies have reviewed the potential of biomass-derived materials (Zhang et al., 2015) and biomass-derived nanomaterials. Currently, researchers are focusing on low-cost carbon electrode materials to develop energy storage devices, including high-energy-density supercapacitors and lithium-ion batteries. The attention is on various types of natural carbon sources used to synthesize graphene and carbon products/derivatives to manufacture supercapacitors with high electrochemical characteristics (Matsukevich et al., 2021b, 2022). However, along with natural carbon sources, many organic wastes of anthropogenic origin can also be considered.

There are various types of supercapacitors, such as pseudocapacitors, symmetrical, asymmetric and hybrid capacitors (Wang and Dai, 2013). Hybrid capacitors are electrochemical capacitors that achieve high power density by combining the characteristics of a supercapacitor and a battery (Wang and Dai, 2013). Supercapacitors can be divided into two categories: *i*) symmetrical supercapacitors are those that have a similar type of material on the positive and negative electrodes; *ii*) Asymmetric supercapacitors are capacitors that use different types of materials on the positive and negative electrodes, such as a supercapacitor having one EDLC (Electrochemical Double Layer Capacitor) electrode and another pseudocapacitor electrode (Gonzalez et al., 2016; Sundriyal et al., 2018). A hybrid electrochemical capacitor (HEC) is a type of asymmetric capacitor, a supercapacitor, in which one electrode is either an EDLC or pseudo capacitance type, and the other is a battery. These hybrid capacitors can store energy comparable to lithium-ion batteries. HECs outperform EDLCs and pseudocapacitors in terms of high cyclic stability, high energy and power. It is related to the potential window of the symmetric supercapacitor, which is limited to only 1 V due to water splitting. However, in an asymmetric supercapacitor, the potential window goes beyond 2 V with classical aqueous electrolytes (Li et al., 2018).

Classification of various supercapacitor technologies, based on (Raza et al., 2018): *i*) electric double-layer capacitor: activated carbon, graphene, carbon nanotubes; *ii*) pseudocapacitor: conducting polymers, metal oxides; *iii*) hybrid capacitors: composite hybrids, asymmetric hybrids, and battery type of hybrids. The objectives of this study were to *i*) sources of secondary raw materials (wastes) for green and biowaste-based materials for energy production, conversion, storage, and hybrid technologies, *ii*) estimate different methods for the synthesis of green and biowaste-based materials for energy production, conversion, storage and hybrid technologies, *iii*) provide examples of the use of waste sources and green methods and approaches for their synthesis.

8.2 SOURCES OF SECONDARY RAW MATERIALS (WASTES) FOR GREEN AND BIO-WASTE-BASED MATERIALS FOR ENERGY PRODUCTION, CONVERSION, STORAGE, AND HYBRID TECHNOLOGIES

8.2.1 WASTE SOURCES OF CARBON

According to estimates, about 2.24 billion tons of solid waste enter the biosphere annually with an increase to 3.88 billion expected by 2050, of which 50–60% are organic compounds (Max-Ikechebelu et al., 2022). Waste occupies vast land areas, and significant funds are spent on transportation and storage. In addition, they lose valuable components that could be involved in economic and production activities. In industrialized countries, up to 5 tons (by dry weight) of waste per person are accumulated yearly. The share of waste processed into valuable products (utilized, recycled) in the total mass is 10–20%. According to environmental legislation, in many countries, wastes containing more than 5–10% organic matter should not be stored but should be recycled. Below we consider organic waste of natural origin suitable for energy processing, as well as the primary source of carbon for energy production, conversion, storage, and hybrid technologies.

Coniferous and hardwood wood and its components: cellulose and hemicellulose, lignin, and bark. Wood makes up the bulk of the total primary production of organic matter. Of the approximately 200 billion tons (dry matter) of annual biomass growth globally, about 120 billion tons is wood (Idowu et al., 2021). About 2.73 billion tons of wood is harvested annually in the world (Sustainability, 2022), of which 45% is used for fuel, 40% for the production of lumber, 15% is processed in the chemical and biotechnological industries (mainly in the form of waste from the woodworking and pulp and paper industries).

Wood contains polysaccharides (cellulose, hemicellulose) and lignin. Coniferous wood contains approximately 60% hexoses, and hardwood wood contains approximately 45% hexoses (Irle et al., 2012). Along with cellulose, the most essential component of wood is lignin. It is one of the most complex heterogeneous polymers with a three-dimensional structure, consisting mainly of phenylpropane units and methoxyl groups. Lignin contains 58–70% carbon, 5–7% hydrogen, and 24–30% oxygen (Glasser, 1985). Phenylpropane units are connected by bonds extremely resistant to enzymes: carbon-carbon C–C and ether C–O–C. The structure of lignin belongs to polyphenols and is similar to humic acids. Under natural conditions, lignin is structurally associated with polysaccharides, encrusting the cellulose of plant fibres. Lignin is the main component that determines the properties of lignocellulosic materials. It is one of the most stable organic polymers in nature, is slowly biodegradable and impairs the digestibility of cellulose and hemicellulose by ruminants. Lignin is decomposed by microorganisms belonging to the ligninolytic group (Phukongchai et al., 2022).

Technical lignin. It is formed during the chemical processing of wood at hydrolysis-yeast enterprises. When processing crushed vegetable raw materials by chemical percolation hydrolysis with dilute sulfuric acid at elevated pressure and temperature, wood hydrolysates are selected at the outlet of the hydrolysis apparatus. Ethanol is obtained from sugars obtained from wood hydrolysates during anaerobic fermentation of hexoses, and fodder yeast is obtained from hexoses and pentoses or pentoses after fermentation of hexoses and ethanol distillation. After hydrolysis, a waste remains – technical (hydrolysis) lignin. The presence of sulfuric acid in hydrolysis lignin limits the possibility of its utilization. Part of the technical lignin is used as a fuel, additive to cement and fillers for plastics, soil stabilizers on roads, to obtain sorbents, etc.

Agricultural waste (straw, corn cob, rice husk, sunflower husk, etc.). The lignin content of these wastes is low; the proportion of cellulose or other polysaccharides reaches 90% (Sah et al., 2022). Due to the low bulk density of these wastes, it is unprofitable to transport them; therefore, small plants are built for their processing near the sources of raw materials.

Sugar industry waste – molasses, beet pulp, filtration sediment (defecate), refined molasses, beet "battle," and beet tails. Molasses contains 70–75% solids and up to 50% sucrose. It contains about 2% nitrogenous substances, 6% ash, amino acids (primarily aspartic and glutamine), betaine (nitrogenous base, formed by decarboxylation of amino acids), amines, amides, and vitamins. The pulp contains 6.0–7.5% dry matter, including 0.2–0.4% sugar.

Waste of alcohol, winemaking and brewing industries – primary stillage (grain-potato, molasses) – is formed after the distillation of alcohol, secondary – after growing fodder yeast on grain-potato or molasses stillage, lute water, head (ether-aldehyde) fraction, fusel oil, grape pomace and combs, cream of tartar, grain waste, grain alloy, brewer's grains, hop grains, malt sprouts, protein sludge, sedimentary yeast, fermentation carbon dioxide, etc. Barda contains organic acids – up to 20% of dry matter (DM), reducing substances – up to 7% of DM, amino acids – up to 3% of DM, glycerin and other alcohols – up to 6% of DM, solid-phase (yeast biomass, part of crushed malt).

Fruit and vegetable production waste – fruits, fruit stones, substandard juices, pomace, etc.

Waste from the meat, poultry and fish industries, as well as the leather and fur industries – protein and fat-containing waste, collagen-containing (greaves, fascia, skin, vein) and keratin-containing (horns, hooves, wool, bristles, low-value feathers, fluff, hair) raw materials, meat and bone meal, fish meal, chitin, etc. Chitin is the main component of the cell wall of fungi, the supporting skeleton of many invertebrates, including crustaceans; is a polymer chain consisting of glucose residues in which acetylated amino groups replace the hydroxyl groups at the second carbon atom.

Oil and fat industry waste – oilcake and meal, sunflower pulp, phosphatide concentrates, soap stocks obtained as a result of the neutralization of vegetable oils and fats, waste generated during refining, hydrogenation and interesterification of vegetable oils and fats, tar.

Tobacco industry waste is generated during tobacco cultivation and fermentation in tobacco factories.

Wastes of essential oil production – from the processing of grain essential oil, flower and herbaceous raw materials, the production of absolute oils, and the processing of essential oils.

Food acid production waste – mycelium, calcium citrate filtrate, gypsum sludge, and lime sediment from lactic acid production. Calcium citrate filtrate is formed during filtration of suspensions neutralized with lime milk of fermented molasses solutions in a volume of 7 m^3 per 1 ton of crystalline citric acid with surface and about 15 m^3 with deep production methods; Contains 10% solids, of which organic matter is 80%, amino acids – up to 6%.

Dairy industry waste – whey (cottage cheese, cheese, casein), skimmed milk (reverse), buttermilk. The milk whey remaining in the production of cheese, cottage cheese or casein, depending on its production method, is divided into cheese, curd and technical (casein). One of the energy-efficient ways to utilize whey is its methanogenic fermentation to produce biogas.

Animal husbandry waste (manure and bedding of cattle and pigs, gull – liquid manure, chicken manure). From one head of cattle, 45 kg/day of manure is formed, pigs – 4.5 kg/day, poultry – 0.1 kg/day. For animal waste processing, low-temperature or high-temperature drying and biotechnological methods, or a combination thereof, are most often used. With low-temperature drying, only a dehydrated, but not disinfected, product is obtained. It retains some pathogens and viable weed seeds. Such a product is undesirable to use in the fields as a fertilizer. In addition, freeze dryers are difficult to manufacture, expensive to purchase, and uneconomical to operate. It is economically and environmentally more efficient to use biotechnological methods for processing these wastes.

Waste from treatment facilities (sedimentary sludge and activated sludge). In terms of volume, the amount of activated sludge and precipitation generated annually is comparable to the amount of industrial and domestic waste. In European countries, the total amount of sludge in 2015 reached almost 9 million tons of dry matter. Households without a centralized sewerage network accumulate 70–80 litres of sludge per inhabitant per year. Precipitation and excess activated sludge are formed in the process of mechanical, biological and physico-chemical (reagent) wastewater treatment. The processing and disposal costs of sludge can be up to 50% of all wastewater treatment costs (Hurynovich et al., 2021). The cost of sludge processing is high due to its high humidity (90–99.7%), high content of harmful substances and pathogenic microorganisms. The elemental composition of the dry matter of sediments varies widely (% wt. of the dry matter of the sediment): C – 35.4–87.8; H – 4.5–8.7; S – 0.2–2.7; N – 1.6–8; O – 7.6–35.4. Sediments contain compounds of silicon, aluminium, iron, oxides of calcium, magnesium, potassium, sodium, zinc, nickel, chromium, etc., about two times less proteins, and 2.5–to 3 times more carbohydrates than in activated sludge. Sludge from biological wastewater treatment typically has a pH of 6.5–8.0. The content of trace elements in sediments is mg/kg of dry matter: boron up to 15, cobalt 2–120, manganese 60–750, copper 50–3200, molybdenum 0.5–10, zinc 40–5000. The sludge formed in municipal wastewater treatment plants is a good source of carbon-based materials for different kinds of environmental applications (Bakhsh et al., 2022).

Bottom silt and sediments. In terms of composition and properties, silts and bottom sediments of canals, ponds, and special reservoirs are closest to the sediments formed during biological wastewater treatment. Bioremediation, bioleaching, phytoremediation and other methods have been proposed for biological processing.

Municipal solid waste. The world's urban population has proliferated since 1950, from 746 million to 3.9 billion in 2014 (Bongaarts 2006). According to the UN, this figure is expected to increase to 9.7 billion by 2050, with almost 90 per cent of this increase in urban areas in Africa and Asia. Today, the world volumes of municipal solid waste are estimated at 2 billion tons per year. Unlike projections for world population and urbanization trends, there are currently no UN projections for future per capita waste generation. However, there is a general understanding that the amount of waste will increase substantially. The drivers of the increase in waste are the increased consumption of goods by the growing urban population, changes in lifestyle and the growing wealth of the growing middle class. According to the International Energy Agency, the waste generation rate in low-income countries will more than double over the next twenty years. Regardless of the accuracy of these forecasts, such colossal volumes of waste will pose a huge challenge for many local city governments, which are already struggling to manage the current amount of waste. An urban dweller in a developing country or country with an economy in transition produces an average of 100 to 400 kg of MSW per year. The most modern, economically and environmentally sound option for processing MSW is their preliminary separation and sorting. In this case, an organic component with the prospect of using it to produce carbon and an inorganic component, predominantly containing metals, can be isolated. The organic waste listed above, in addition to direct use for energy production, can be used to store the energy already received in supercapacitors in the form of activated carbon for the manufacture of electrodes.

8.2.2 WASTE SOURCES OF METALS

Industrial wastes can also be used as sources of inorganic substances (Figure 8.1). For example, wastewater treatment and industrial wastewater treatment. In this case, wastewater treatment is the cleanest and low-component. They do not contain additional fillers. Spent ion-exchange resins can be considered as organic-granulated wastes of the water treatment process (Romanovskii and Martsul, 2009, 2012). Among the inorganic wastes of the water treatment process, one can single out: *i*) sediments formed during the coagulation of surface waters (Romanovski et al., 2021a). These wastes contain mainly calcium carbonate, but also FeOOH and $Al(OH)_3$. *ii*) sediments formed during the purification of washing water from iron removal filters (during the purification of groundwater from aquifers) (Romanovskii and Khort 2017; Ramanouski and Andreeva, 2012; Romanovski 2020). In these wastes, the main share is FeOOH. Good sources of metals are different mining slags (Smorokov et al., 2022a, 2022b, 2022c) and other waste streams of different industries (Romanovski et al., 2022b).

Wastes from electroplating industries may be of particular interest (Martsul et al., 2012a, 2012b, 2013). In their composition, they can be considered as a source of iron, nickel, chromium, zinc, lead, cadmium, and other metals. The content of metals in them strongly depends on the type of electroplated coating. These wastes are generated by tens of tons per year at large machine-building enterprises. In this case, it is possible to separate their removal and purification with the production of precipitates containing a narrow range of necessary metals. Also, in the enterprises in many processes, metals and alloys are used, which form waste, such as metal shavings and gas cleaning dust at metallurgical enterprises. Calcium, phosphorus, potassium and other elements contained in the waste of electroplating industries can play a synergistic role in the properties of the resulting substances.

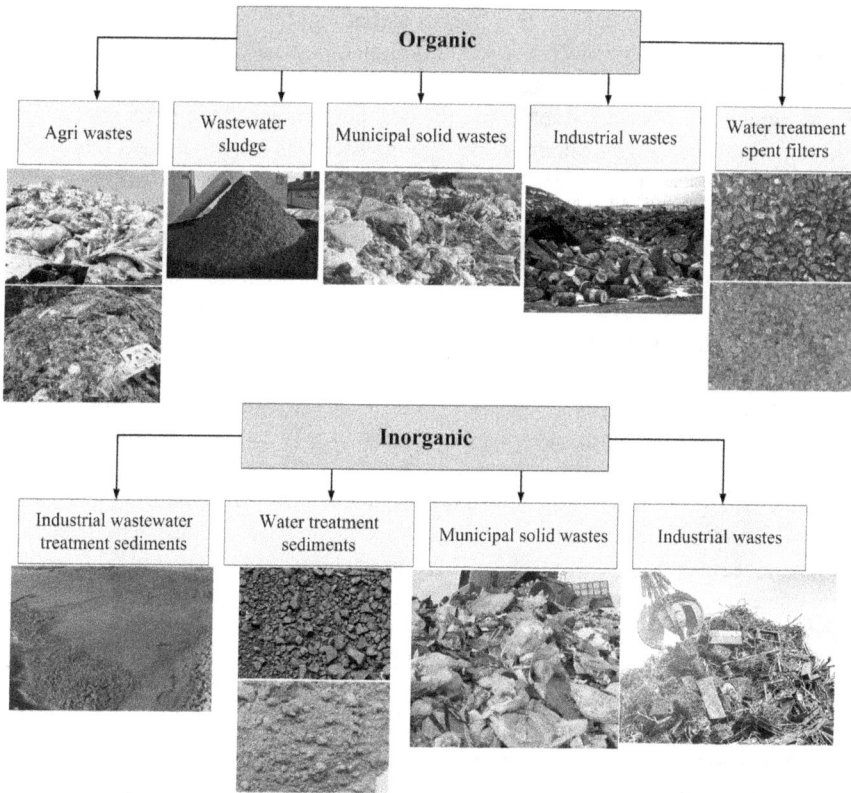

FIGURE 8.1 Organic and inorganic wastes as sources for the production of materials for energy production, conversion, storage, and hybrid technologies.

8.3 METHODS FOR THE SYNTHESIS OF GREEN AND BIO WASTE-BASED MATERIALS FOR ENERGY PRODUCTION, CONVERSION, STORAGE, AND HYBRID TECHNOLOGIES

As shown in section 8.2.1, almost any biomass and biowaste can be used as a carbon source, such as wood, plant leaves, straw, fruit shells, animal tissue, agricultural waste, etc. Typical biomass consists of hemicellulose, lignin, cellulose, oil or fat, protein, chitosan, etc. Various precursors differ in composition, microstructure, and metallic and nonmetallic elements, all of which affect the properties of carbon products.

The ultimate performance of produced carbon in energy applications is highly dependent on structural and textural characteristics, e.g. particle size, porosity, morphology and physico-chemical properties, which are determined by a selection of both organic precursors and processing conditions (e.g. process temperature, heating rate, exposure and chemical or physical activators used). Pyrolysis and hydrothermal carbonization (HTC) are the two main thermal treatment methods for carbonizing any sustainable carbonaceous resource. Inorganic components could modify further carbon materials to increase the properties of obtained materials for energy production, conversion, storage and hybrid technologies. During the HTC process, water molecules help to convert the waste biomass into carbonaceous products. This process produces porous carbons with a high density of surface oxygen functional groups, which are critical for achieving the best performance in various applications of obtained materials. Hydrothermal processing methods can be divided into hydrothermal carbonization (180–250 °C) for the production of biochar, hydrothermal liquefaction (200–370 °C with a pressure of 4 and 20 MPa) for bio-fuel production and hydrothermal gasification (near-critical temperatures up to 500 °C) to produce hydrogen-rich gas. This method has made significant progress in obtaining carbon dots, including low toxicity, selective detection of target species, water-solubility, fluorescence, biocompatibility, etc. (Meng et al., 2019).

This process is used to convert carbonaceous feedstocks into functionalized porous carbon materials. The purity of the phase and the density of functional groups on the carbon surface can be changed by selecting the source of raw materials and adjusting the pyrolysis temperatures. In general, pyrolysis is a common strategy for carbon raw materials subjected to a heating process (200–1000 °C) in an inert gas atmosphere. Depending on the conditions of pyrolysis, it can be: *i*) according to the heating rate: slow, convection, fast, instant; *ii*) by temperature: high temperature and low temperature; *iii*) by reactor type; *iv*) according to the type of medium in the reactor. Gasification can be distinguished as a type of pyrolysis. In this case, ammonia, CO, and CO_2 can be used to increase the yield of the gaseous phase. After the pyrolysis process, activation of the resulting carbon materials is usually required. Physical and chemical activation processes based on various activating agents can create high-porosity carbon materials and increase the specific surface area (Keppetipola et al., 2021). Physical activation is a two-step process in which the biomass is first pretreated at a suitable temperature (400–450 °C) in an inert atmosphere and carbonized again at high temperatures using air (600 °C), steam or CO_2 (up to 1200 °C) as activators. Chemical activation mainly involves the use of agents such as potassium hydroxide (KOH), potassium carbonate (K_2CO_3), sodium hydroxide (NaOH), phosphoric acid (H_3PO_4), calcium chloride, zinc chloride and iron salts.

Physical activation consists of two stages: the first stage includes the carbonization of the feedstock, followed by activation in the second stage with CO_2 and/or steam (Mui et al., 2010; Song et al., 2012). Heat treatment usually occurs in the temperature range of 600 to 900 °C in either an argon- or nitrogen-inert atmosphere. CO_2 is considered the preferred choice for physical activation due to ease of handling, control of various parameters, and low reaction rate. Depending on the experimental conditions during high-temperature activation, the micropores initially formed by CO_2 activation can rupture to form larger pores falling into the mesopore size range with increasing exposure time. The CO_2 activation process helps selectively remove the most reactive carbon atoms from biochar (Liu et al., 2015). On the other hand, steam activation offers advantages such as low cost, high quantity, environmentally friendly process, and avoiding post-activation procedures (Jin et al., 2014). Through physical activation, biochar is additionally activated using steam, carbon dioxide, air or a mixture of gases as an activating agent at temperatures around 350–1000 °C, which leads to an increase in carbon porosity.

Chemical activation involves impregnating carbon sources with chemical agents such as acids, bases, or salts before or after carbonization, resulting in nanoporous carbon materials. The temperature range used in chemical activation depends on the choice of precursor and the nature of the catalyst (450–900 °C). The activating agent helps to remove residual water from the raw material by acting as a dehydrating agent and also helps as an oxidizing agent. Both processes affect the decomposition of the precursors and the rearrangement of the resulting carbon atoms into an aromatic structure. Chemical activation has the added benefit of introducing functional groups such as -COOH, -NH or -OH onto the porous carbon surface. However, the crystallinity of the sample after chemical activation decreases due to continuous dehydration and oxidation by the activating agent, which creates many defective areas along the carbon walls of the final product. Several chemical agents such as potassium hydroxide (KOH), zinc chloride ($ZnCl_2$), calcium carbonate ($CaCO_3$), hydrogen peroxide (H_2O_2) and potassium permanganate ($KMnO_4$), were used as activating agents during chemical activation. As a rule, the activating agent is adsorbed and dissociated in the active sites of carbon surfaces, forming gases, which, in turn, can be adsorbed on the active sites, thereby inhibiting them. The above reaction is reversible and active sites are continuously formed while creating porosity.

The authors of Cheng et al., (2018) proposed a microwave carbonization method. The authors note a significant reduction in run time and increased energy efficiency as an advantage of microwave carbonization. Compared to the hydrothermal method, the heating time can even be reduced to a few minutes. It is evident from many works that the hydrothermal method using microwaves is highly efficient in the production of such biochars compared to classical approaches such as pyrolysis and hydrothermal methods. Another approach, Molten Salt Synthesis (MSS), as its name suggests, introduces kinds of inorganic salts, such as metal halides or salts involving oxides, for biomass-derived 3D carbon materials. In particular, salts with a low melting point (no more than 1000 °C) such as zinc (potassium, sodium) chloride, carbonates and nitrates are common salt agents used in MSS, creating a liquid flow phase for sufficient mass transfer and contact area on the solid-liquid interface. It was suggested in Wang et l., (2018) that, during the MSS process, the biomass is annealed in the air rather than in an inert gas atmosphere; this, to some extent, improved the electrochemical properties of pre-prepared porous carbon sheets for supercapacitors since oxygen was involved in the introduction of porosity and the formation of pseudocapacitance.

Obtaining *graphene materials* for energy production, conversion, storage, and hybrid technologies refers to all three generations of these materials. In addition to graphene, there are other graphene-like carbon materials. These are multilayer graphene, graphene quantum dots (GQD), graphene nanoribbons (GNR), nanogrids, nanosheets, graphene oxides (GO), and reduced graphene oxide (rGO). Graphene oxide (GO) has many properties of graphene, but unlike hydrophobic graphene, GO is soluble in water. The process of obtaining graphene from inexpensive graphite is as follows: first, graphene oxide (GO), highly soluble in water, is obtained, then thermal, chemical or electrochemical methods can be used to obtain reduced graphene oxide (rGO). Other forms of carbon existence appear when graphene is saturated (hydrated) with hydrogen: these are Graphane – fully hydrated graphene, in which CH bonds appear on two sides of the graphene, and Graphone – half-hydrated graphene with CH bonds on one side. At the same time, Graphane has high stability, is a dielectric and can be used in nanoelectronics and in graphene-hydride fuel cells. The process of obtaining Graphane is reversible and consists of treating graphene with cold plasma (argon with hydrogen) at low pressure and temperature. This makes it possible to create regions with dielectric properties in conductive graphene and use them in nanoelectronics and biotechnologies for drug delivery.

Graphone has anti-magnetic properties. Fluorographene (FG) is regarded as the thinnest Teflon-type insulator, and at the same time, it is not chemically inert. FG is able to form various compounds of graphene, for example, with acids. FG will find its use in almost all of the above applications. FG is obtained by fluorination of graphene, and mechanical or chemical splitting of graphite fluoride. Graphene quantum dots (GQDs) are pieces of graphene smaller than 10 nm. They have many of the properties of graphene due to the attachment of various chemical groups along the edges of the particles and get new properties. Graphyne and Graphdiyne are two allotropic forms of carbon in which graphene forms structures with different bonds. Both are semiconductors created by methods of synthetic organic chemistry. A separate area of application of graphene is the creation of new materials based on it with unique properties due to doping of graphene with various elements. This makes it possible to develop materials for various applications, for example, for energy, sensors, photovoltaics, nanoelectronics, catalysts, supercapacitors, magnetic materials, biomedicine, etc. Non-metals (N, B, S, P, Se, O, Si, I) and metals (Mn, Co, Ni, Al, Ti, Pd, Ru, Rh, Pt, Au, Ag).

Graphene can be obtained in the following ways: *i*) micromechanical separation at room temperature to obtain particles larger than 1000 microns; *ii*) obtaining from a colloidal suspension at room temperature with obtaining particles with sizes less than 0.1 microns; *iii*) chemical cleavage at a temperature of about 100 °C to obtain particles with a size of about 100 microns; *iv*) chemical vapour deposition (CVD) at a temperature of about 1000 °C to obtain particles with a size of about 1000 microns; *v*) epitaxial growth at temperatures above 1100 °C to obtain particles with a size of about 50 µm. Synthesis of graphene quantum dots also includes both bottom-up (molecular carbonization and electron beam irradiation (EBI) methods), and top-down strategies (e.g. chemical oxidation, hydrothermal, electrochemical oxidation, chemical vapour deposition (CVD) methods, and pulsed laser ablation (PLA) technique, or a combination).

Doping is a process that can improve the electrical properties of a material by introducing a new element (dopant) with desired properties into the material's structure metals (e.g. sodium, lithium, zinc), and non-metals (e.g. nitrogen, phosphorus, and sulfur). Doping of biomass-derived carbon nanofibers can produce defects and pores on the surface of the carbon nanofiber. It is advantageous to produce a material with a high porous surface for use as an electrode in supercapacitor devices, as this can improve energy storage performance by providing large entry paths for electrolyte ions and helping to increase active volume. Doping of a carbon nanofiber can also increase electrons for strong π-electron delocalization, improving the electrical conductivity and catalytic activity of the carbon nanofiber. The range of obtained materials may include: *i*) metal-metal oxide systems (Me-MeO, MeO-MeO, MeO) consisting of individual oxides (ZnO, WO_3, etc.), complex oxides (e.g. $ZnFe_2O_4$); *ii*) metal-organic (Me-Organic, MeO-organic, Me-Carbon, MeO-Carbon, etc.); *iii*) and other composite materials.

Among them are mainly distinguished: *i*) physical methods: surface disaggregation (milling, HEBM etc. (Rogachev 2019; Cao et al., 2022)); laser ablation; sputtering; physical vapour deposition; etc. *ii*) chemical methods: sol-gel (Abdol Aziz et al., 2019); chemical vapour deposition; colloidal; spray pyrolysis; combustion methods; solvothermal synthesis (Liang et al., 2019), microwave heating of chemical reagents (Shad et al., 2020), etc. *iii*) biological methods: in bacteria; in fungi; in algae; yeast; plants extracts; using enzymes and biomolecules; etc. In recent years, among the chemical methods, self-sustaining exothermic methods for the synthesis of nano-oxide materials have been increasingly used. The combustion of solutions is a phenomenon of propagation of a self-sustaining reaction through an aqueous solution or gel (Aruna and Mukasyan, 2008; Mukasyan et al., 2019; Khaliullin et al., 2020). This process allows the synthesis of various nanomaterials, including simple and complex oxides, metals and their alloys. These materials are used in many important areas such as nanoelectronics, energy, catalysis, optics and bioceramics. These methods make it possible to obtain nanoscale metals (Sdobnyakov et al., 2020; Romanovskiy et al., 2018; Romanovski et al., 2020) and oxides (Glinskaya et al., 2021; Matsukevich et al., 2021a, 2021b, 2022), as well as metal-graphene structures (Khort et al., 2020a, 2020b) in one stage. Also, this method can be effectively used for carbon (Propolsky et al., 2020) and inorganic materials modification. These methods are characterized by simplicity, high reaction rate, and short preparatory operations and synthesis times, avoiding the formation of waste, wastewater, and exhaust of poisonous gases compared to others (Romanovski et al., 2021b), and are excellent for scaling.

8.4 CONCLUSIONS AND FUTURE PERSPECTIVES

Among the prospects for the development of materials for energy production, conversion, storage and hybrid technologies, the following areas can be distinguished, the purpose of which is to create promising materials that meet the goals of sustainable development:

- improvement and the possibility of regulating the composition, structure and properties of carbon-containing materials;
- simplification of technical approaches to regulating the structure, composition and properties of carbonaceous materials;
- new approaches and improvement of existing approaches to the ligation of carbon-containing materials – development of the third generation of carbon materials for energy production, conversion, storage and hybrid technologies – hybrid and nanostructured materials;
- further development in the study of carbon quantum dots and graphene solar cells with quantum dots;
- development of new, less aggressive electrolytes;

- increase in current density;
- microfluidic batteries;
- two-phase batteries using immiscible redox electrolytes, which are separated according to thermodynamic principles;
- getting rid of membranes in expensive ion-selective membranes in the energy conversion reactor;
- the stability and safety of the materials used and the final products;
- development of hybrid systems;
- use of secondary material resources to obtain materials for energy production, conversion, storage and hybrid technologies;
- new "green" approaches to alloying carbon materials.

REFERENCES

Abdol Aziz, R.A., S.F. Abd Karim, and N.A. Rosli. 2019. The effect of ph on zinc oxide nanoparticles characteristics synthesized from banana peel extract. In *Key Engineering Materials*. Trans Tech Publications Ltd. 797:271–279.

Ademati, A.O., A.A. Akinwande, O.A. Balogun, and V. Romanovski. 2022. Optimization of bamboo-fiber-reinforced composite-clay bricks for development of low-cost farm-settlements towards boosting rural agri-business in Africa. *Journal of Materials in Civil Engineering*. https://doi.org/10.1061/(ASCE)MT.1943-5533.0004489

Akinwande, A.A., O.A. Balogun, V. Romanovski, H. Danso, M. Kamarou, and A.O. Ademati. 2022a. Mechanical performance and Taguchi optimization of kenaf fiber/cement-paperboard composite for interior application. *Environmental Science and Pollution Research*. https://doi.org/10.1007/s11356-022-19449-8

Akinwande, A.A., O.A. Balogun, V. Romanovski, A.O. Ademati, and Y.V.Adetula 2022b. Recycling of synthetic waste wig fiber in the production of cement-adobe for building envelope: Physio-hydric properties. *Environmental Science and Pollution Research*. https://doi.org/10.1007/s11356-022-18649-6

Akinwande, A.A., O.A. Balogun, and V. Romanovski. 2022c. Modeling, multi-response optimization, and performance reliability of green metal composites produced from municipal wastes. *Environmental Science and Pollution Research*. https://doi.org/10.1007/s11356-022-20023-5

Aruna, S.T., and A.S. Mukasyan. 2008. Combustion synthesis and nanomaterials". *Current Opinion in Solid State Materials Science* 12:44.

Bakhsh, E.M., Sh.B. Khan, K. Akhtar, E.Y. Danish, T.M. Fagieh, Ch. Qiu, Y. Sun, V. Romanovski, and X. Su. 2022. Simultaneous preparation of humic acid and Mesoporous Silica from municipal sludge and their adsorption properties for U(VI). *Colloids and Surfaces A: Physicochemical and Engineering Aspects*. https://doi.org/10.1016/j.colsurfa.2022.129060

Bongaarts, J. 2006. United nations department of economic and social affairs, population division world mortality report 2005. *Population and Development Review*, 32(3):594–596.

Cao, A., M. Zhang, X. Su, V. Romanovski, and S. Chu. 2022. In situ fabrication of NiS2-decorated graphitic carbon Nitride/Metal-Organic framework nanostructures for photocatalytic H2 evolution. *ACS Applied Nano Materials*. https://doi.org/10.1007/s11356-022-20023-5

Cheng, Y., B. Li, Y. Huang, Y. Wang, J. Chen, Y. Wei, Y. Feng, D. Jia, and Y. Zhou. 2018. Molten salt synthesis of nitrogen and oxygen enriched hierarchically porous carbons derived from biomass via rapid microwave carbonization for high voltage supercapacitors. *Applied Surface Science*, 439:712–723. https://doi.org/10.1016/j.apsusc.2018.01.006

Cui, X., J. Yang, Z. Wang, and X. Shi. 2021. Better use of bioenergy: A critical review of co-pelletizing for biofuel manufacturing. *Carbon Capture Science & Technology*, 1:100005.

Denholm, P. 2006. Improving the technical, environmental and social performance of wind energy systems using biomass- based energy storage. *Renew Energy*, 31:1355e70.

Glasser, W.G. 1985. Lignin. In *Fundamentals of Thermochemical Biomass Conversion* (pp. 61–76). Springer, Dordrecht.

Glinskaya, A., G. Petrov, and V. Romanovski. 2021. Crystal structure, physicochemical and sensory properties of solid solutions $Bi_{1-x}La_xFe_{1-x}Co_xO_3$ (x = 0, 0.05, 0.1). *Journal of Materials Science: Materials in Electronics*, 32:22579–22587.

Gonzalez, A., E. Goikolea, J.A. Barrena, and R. Mysyk. 2016. Review on supercapacitors: Technologies and materials. *Renewable* and *Sustainable. Energy Reviews*, 58:1189e1206.

Hammadi, O.A. 2020. Effects of extraction parameters on particle size of titanium dioxide nanopowders prepared by physical vapor deposition technique. *Plasmonics*, 15(6):1747–1754. https://doi.org/10.1007/s11468-020-01205-8

Hatfield-Dodds, S., H. Schandl, D. Newth, M. Obersteiner, Y. Cai, T. Baynes, J. West, and P. Havlik. 2017. Assessing global resource use and greenhouse emissions to 2050, with ambitious resource efficiency and climate mitigation policies. *Journal of Cleaner Production*, 144:403e14. https://doi.org/10.1016/j.jclepro.2016.12.170

Hurynovich, A., M. Kwietniewski, and V. Romanovski. 2021. Evaluation of the possibility of utilization of sewage sludge from wastewater treatment plant – case study. *Desalination and Water Treatment*, 227:16–25.

Idowu, I.A., K. Hashim, A. Shaw, and L.J. Nunes. 2021. Enhancing the fuel properties of beverage wastes as non-edible feedstock for biofuel production. *Biofuels*, 1–8. DOI: 10.1080/17597269.2021.1923934

Irle, M.A., M.C. Barbu, R. Réh,, L. Bergland, and R.M. Rowell. 2012. *Handbook of Wood Chemistry and Wood Composites*. In Handbook of Wood Chemistry and Wood Composites, ed. Roger M. Rowell. Boca Raton: CRC Press, 06 Sep 2012, Routledge Handbooks Online.

Jin, Z., X. Yan, Y. Yu, and G. Zhao. 2014. Sustainable activated carbon fibers from liquefied wood with controllable porosity for high-performance supercapacitors, *Journal of Materials Chemistry. A*, 2:11706–11715. https://doi.org/10.1039/C4TA01413H

Kamarou, M., N. Korob, V. Romanovski. 2021a. Structurally controlled synthesis of synthetic gypsum derived from industrial wastes: Sustainable approach. *Journal of Chemical Technology and Biotechnology*, 96(11):3134–3141.

Kamarou, M., N. Korob, A. Hil, D. Moskovskikh, and V. Romanovski. 2021b. Low-energy technology for producing anhydrite in the $CaCO_3 - H_2SO_4 - H_2O$ system derived from industrial wastes. *Journal of Chemical Technology and Biotechnology*, 96(7):2065–2071.

Kamarou, M., N. Korob, W. Kwapinski, V. Romanovski. 2021c. High-quality gypsum binders based on synthetic calcium sulfate dihydrate produced from industrial wastes. *Journal of Industrial and Engineering Chemistry*, 100:324–332. https://doi.org/10. 1016/ j. jiec 2021. 05. 006

Kamarou, M., M. Kuzmenkov, N. Korob, W. Kwapinski, and V. Romanovski. 2020. Structurally controlled synthesis of calcium sulphate dihydrate from industrial wastes of spent sulphuric acid and limestone, *Environmental Technology & Innovation*, 17:100582.

Kamarou, M., D. Moskovskikh, H.L. Chan, H. Wang, T. Li, A.A. Akinwande, and V. Romanovski. 2022. Low energy synthesis of anhydrite cement from waste lime mud.

Journal of Chemical Technology and Biotechnology, 98(3):789–796. https://doi.org/10.1002/jctb.7284

Kambo, H.S., and A. Dutta. 2015. A comparative review of biochar and hydrochar in terms of production, physico-chemical properties and applications. *Renewable Sustainable Energy Reviews,* 45:359e78.

Keppetipola, N.M., C. Olivier, T. Toupance, L. Cojocaru. 2021. Biomass-derived carbon electrodes for supercapacitors and hybrid solar cells: towards sustainable photo-supercapacitors, Sustain. *Energy Fuels,* 5:4784–4806. https://doi.org/10.1039/D1SE00954K

Khaliullin, S.M., V.D. Zhuravlev, V.G. Bamburov, A.A. Khort, S.I. Roslyakov, G.V. Trusov, and D.O. Moskovskikh. 2020. Effect of the residual water content in gels on solution combustion synthesis temperature. *Journal of Sol-Gel Science and Technology,* 93(2):251–261. 10.1007/s10971-019-05189-8

Khort, A., V. Romanovski, D. Leybo, and D. Moskovskikh. 2020a. CO oxidation and organic dyes degradation over graphene-Cu and graphene-CuNi catalysts obtained by solution combustion synthesis. *Scientific Reports,* 10:16104.

Khort, A., V. Romanovski, V. Lapitskaya, T. Kuznetsova, K. Yusupov, D. Moskovskikh, Y. Haiduk, K. Podbolotov. 2020b. Graphene@metal nanocomposites by solution combustion synthesis. *Inorganic Chemistry,* 59:6550–6565.

Lee, S.-H., and T.-H. Lee. 2018. High performance hybrid supercapacitors with LiNi1/3Mn1/3Co1/3O2/activated carbon cathode and activated carbon anode. *International Journal of Hydrogen Energy,* 43:15365e9.

Li, B., J. Zheng, H. Zhang, L. Jin, D. Yang, H. Lv, C. Shen, A. Shellikeri, Y. Zheng, and R. Gong. 2018. Electrode materials, electrolytes, and challenges in nonaqueous lithium-ion capacitors. *Advanced Materials,* 30(17):1705670.

Liang, S., Shi, S., Zhang, H., Qiu, J., Yu, W., Li, M., Gan, Q., Yu, W., Xiao, K., Liu, B. and Hu, J. 2019. One-pot solvothermal synthesis of magnetic biochar from waste biomass: Formation mechanism and efficient adsorption of Cr (VI) in an aqueous solution. Science of the Total Environment, 695:133886.

Liu, W.J., Jiang, H. and Yu, H.Q. 2015. Development of biochar-based functional materials: toward a sustainable platform carbon material. Chemical Reviews, 115(22):12251–12285.

Martsul, V.N., A.V. Likhacheva, L.A. Shibeka, O.S. Zalygina, V.I. Ramanouski, and V.V. Khodin. 2012b. Inventory of galvanic sludge and deposits of the treatment facilities formed at the enterprises of the Republic of Belarus. *Proceedings of BSTU. Chemistry and Technology of Inorganic Substances.* 3:71–77.

Martsul, V.N., O.S. Zalygina, A.V. Likhacheva, and V.I. Romanovski. 2013. Treatment of electroplating shop wastewater at enterprises of the Republic of Belarus. *Proceedings of BSTU. Chemistry and Technology of Inorganic Substances.* 3:57–62.

Martsul, V.N., O.S. Zalygina, L.A. Shibeka, A.V. Likhacheva, and V.I. Ramanouski. 2012a. Some applications of galvanic manufacture waste proceedings of BSTU. *Chemistry and Technology of Inorganic Substances,* 3:66–70.

Matsukevich, I., Y. Lipai, V. Romanovski. 2021a. Cu/MgO and Ni/MgO composite nanoparticles for fast, high-efficiency adsorption of aqueous lead(II) and chromium(III) ions. *Journal of Materials Science,* 56(8):5031–5040.

Matsukevich, I., A. Kulak, V. Palkhouskaya, V. Romanovski, J.H. Jo, Y. Aniskevich, and S.G. Mohamed. 2021b. Comparison of different synthesis methods for $Li_2MTi_3O_8$ (M – Co, Cu, Zn) as electrode materials. *Journal of Chemical Technology & Biotechnology,* 97(4) (2022) 1021–1026. https://doi.org/10.1002/jctb.6992

Matsukevich, I.V., A.I. Kulak, V.I. Popkov, V.I. Romanovski, M.G. Fayedc, and S.G. Mohamed. 2022. Lithium cobalt titanate with the spinel structure as an anode material for lithium Ion batteries. *Inorganic Materials*, 58(2). 160–164. https://doi.org/10.1134/S002016852 2010083

Max-Ikechebelu, J.O., O.A. Falusi, and O.M. Bolaji, 2022. Physiochemical parameters of composited cow and goat waste as mitigation to municipal waste. *Science*, 10(4), 165–181.

Meng, W., X. Bai, B. Wang, Z. Liu, S. Lu, B. Yang. 2019. Biomass-derived carbon dots and their applications. *Energy & Environmental Materials*, 2:172–192. https://doi.org/ 10.1002/eem2.12038

Mui, E.L.K., W.H. Cheung, M. Valix, and G. McKay. 2010. Mesoporous activated carbon from waste tyre rubber for dye removal from effluents. *Microporous and Mesoporous Materials*, 130:287–294. https://doi.org/10.1016/j.micromeso.2009.11.022

Mukasyan, A.S., S. Roslyakov, J.M. Pauls, L.C. Gallington, T. Orlova, X. Liu, M. Dobrowolska, J.K. Furdyna, and K.V. Manukyan. 2019. Nanoscale metastable ε-Fe3N ferromagnetic materials by self-sustained reactions. *Inorganic Chemistry*, 58(9):5583–5592. https:// doi.org/10.1021/acs.inorgchem.8b03553

Okolie, J.A., E.I. Epelle, M.E. Tabat, U. Orivri, A.N. Amenaghawon, P.U. Okoye, and B. Gunes, 2022. Waste biomass valorization for the production of biofuels and value-added products: A comprehensive review of thermochemical, biological and integrated processes. *Process Safety and Environmental Protection*, 159:323–344.

Osipov, V.V., Y.A. Kotov, M.G. Ivanov, O.M. Samatov, V.V. Lisenkov, V.V. Platonov, A.M. Murzakaev, A.I. Medvedev, and E.I. Azarkevich. 2006. Laser synthesis of nanopowders. *Laser Physics*, 16(1):116–25. https://doi.org/10.1134/S1054660X0 6010105

Osipova, E., V. Shevchuk, A. Stromski, and V. Romanovski. 2022. Intensification of potash ore flotation by the introduction of industrial oils. *Journal of Chemical Technology and Biotechnology*, 97(1):312–318. https://doi.org/10.1002/jctb.6945

Phukongchai, W., W. Kaewpradit, and F. Rasche, 2022. Inoculation of cellulolytic and ligninolytic microorganisms accelerates decomposition of high C/N and cellulose rich sugarcane straw in tropical sandy soils. *Applied Soil Ecology*, 172:104355.

Popp, J., S. Kovács, J. Oláh, Z. Divéki, and E. Balázs, 2021. Bioeconomy: Biomass and biomass-based energy supply and demand. *New Biotechnology*, 60:76–84.

Propolsky, D., E. Romanovskaia, W. Kwapinski, V. Romanovski. 2020. Modified activated carbon for deironing of underground water. *Environmental Research*, 182:108996.

Purushothaman, K.K., I.M. Babu, and B. Saravanakumar. 2017. Hierarchical mesoporous Cox Ni1AxO as advanced electrode material for hybrid supercapacitors. *International Journal of Hydrogen Energy*, 42:28445e52.

Rahman, A., O. Farrok, and Haque, M.M. 2022. Environmental impact of renewable energy source based electrical power plants: Solar, wind, hydroelectric, biomass, geothermal, tidal, ocean, and osmotic. *Renewable and Sustainable Energy Reviews*, 161:112279.

Ramanouski, V.I., and N.A. Andreeva. 2012. Purification of washing waters of iron removal stations. *Proceedings of bstu. Chemistry and Technology of Inorganic Substances*, 3:62–65.

Raza, W., F. Ali, N. Raza, Y. Luo, K.-H. Kim, J. Yang, S. Kumar, A. Mehmood, and E.E. Kwon. 2018. Recent advancements in supercapacitor technology. *Nano Energy*, 52:441e473.

Rogachev, A.S. 2019. Mechanical activation of heterogeneous exothermic reactions in powder mixtures. *Russian Chemical Reviews*, 88(9):875. https://doi.org/10.1070/RCR4884

Romanovskaia, E., V. Romanovski, W. Kwapinski, and I. Kurilo. 2021. Selective recovery of vanadium pentoxide from spent catalysts of sulfuric acid production: Sustainable approach. *Hydrometallurgy*, 200:105568.

Romanovski, V. 2020. New approach for inert filtering media modification by using precipitates of deironing filters for underground water treatment. *Environmental Science and Pollution Research*, 27:31706–31714.

Romanovski, V. 2021. Agricultural waste based-nanomaterials: Green technology for water purifications. In *Aquananotechnology*, 1st Edition. Applications of Nanomaterials for Water Purification. Editors: Kamel Abd-Elsalam Muhammad Zahid. Elsevier. 577–595.

Romanovskii, V.I., A.A. Khort. 2017. Modified anthracites for deironing of underground water. *Journal of Water Chemistry and Technology*, 39:299–304.

Romanovskii, V.I., and V.N. Martsul. 2009. Distribution of heteroatoms of synthetic ion exchangers in pyrolysis products. *Russian Journal Applied Chemistry*, 82:836–839.

Romanovskii, V.I., and V.N. Martsul. 2012. Functional group distribution over the surface and in the bulk of particles of spent Ion exchangers in the course of mechanochemical destruction. *Russian Journal of Applied Chemistry*, 85:371–376.

Romanovski, V., A. Yu. Kolosov, A.A. Khort, V.S. Myasnichenko, K.B. Podbolotov, K.G. Savina, D.N. Sokolov, E.V. Romanovskaia, and N. Yu. Sdobnyakov. 2020. Features of Cu-Ni nanoparticle synthesis: Experiment and computer simulation. *Physical and Chemical Aspects of the Study of Clusters Nanostructures and Nanomaterials*, 12:293–309.

Romanovski, V., A. Klyndyuk, and M. Kamarou. 2021a. Green approach for low-energy direct synthesis of anhydrite from industrial wastes of lime mud and spent sulfuric acid. *Journal of Environmental Chemical Engineering*, 9(6), 106711.

Romanovski, V., E. Romanovskaia, D. Moskovskikh, K. Kuskov, V. Likhavitski, F.A. Mehmet, S. Beloshapkin, I. Matsukevich, A. Khort. 2021b. Recycling of iron-rich sediment for surface modification of filters for underground water deironing. *Journal of Environmental Chemical Engineering*, 9(4):105712.

Romanovski, V., L. Zhang, X. Su, A. Smorokov, and M. Kamarou. 2022a. Gypsum and high quality binders derived from water treatment sediments and spent sulfuric acid: Chemical engineering and environmental aspects. *Chemical Engineering Research and Design*, 184:224–232.

Romanovski, V., X. Su, L. Zhang, A. Paspelau, A. Smorokov, A.A. Sehat, A.A. Akinwande, N. Korob, and Kamarou, M. 2022b. Approaches for filtrate utilization from synthetic gypsum production. *Environmental Science and Pollution Research*, 30(12):33243–33252, 1–10.

Romanovskiy, V.I., A.A. Khort, K.B. Podbolotov, N.Yu. Sdobnyakov, V.S. Myasnichenko, and D.N. Sokolov. 2018. One-step synthesis of polymetallic nanoparticles in air environment. *Izvestiya Vysshikh Uchebnykh Zavedeniy Khimiya Khimicheskaya Tekhnologiya*, 61:43–48.

Sah, M.K., S. Mukherjee, B. Flora, N. Malek, and S.N. Rath. 2022. Advancement in "Garbage In Biomaterials Out (GIBO)" concept to develop biomaterials from agricultural waste for tissue engineering and biomedical applications. *Journal of Environmental Health Science and Engineering*, 20(2):1015–1033, 1–19.

Sdobnyakov, N., A. Khort, V. Myasnichenko, K. Podbolotov, E. Romanovskaia, A. Kolosov, D. Sokolov, and V. Romanovski. 2020. Solution combustion synthesis and Monte Carlo simulation of the formation of CuNi integrated nanoparticles. *Computational Materials Science*, 184:109936.

Shad, S., Belinga-Desaunay-Nault, M.F.A., Bashir, N. and Lynch, I. 2020. Removal of contaminants from canal water using microwave synthesized zero valent iron nanoparticles. *Environmental Science: Water Research & Technology*, 6(11):3057–3065.

Smorokov, A., A. Kantaev, D. Bryankin, A. Miklashevich, M. Kamarou, and V. Romanovski. 2022a. Low-temperature desiliconization of activated zircon concentrate by NH4HF2 solution. *Minerals Engineering*, 189:107909.

Smorokov, A., A. Kantaev, D. Bryankin, A. Miklashevich, M. Kamarou, and V. Romanovski. 2022b. Low-temperature method for desiliconization of polymetallic slags by ammonium bifluoride solution. *Environmental Science and Pollution Research*, 30(11):30271–30280, 1–10.

Smorokov, A., A. Kantaev, D. Bryankin, A. Miklashevich, M. Kamarou, and V. Romanovski. 2022c. A novel low-energy approach of leucoxene concentrate desiliconization by ammonium bifluoride solutions. *Journal of Chemical Technology & Biotechnology*, 98(3):726–733.

Song, X.H., R. Xu, A. Lai, H.L. Lo, F.L. Neo, and K. Wang. 2012. Preparation and characterization of mesoporous activated carbons from waste tyre. *Asia-Pacific Journal of Chemical Engineering*, 7(3):474–478. https://doi.org/10.1002/apj.544

Sundriyal, S., V. Shrivastav, H. Kaur, S. Mishra, and A. Deep. 2018. High-performance symmetrical supercapacitor with a combination of a ZIF-67/rGO composite electrode and a redox additive electrolyte. *ACS Omega*, 3(12):17348e17358.

Sustainability. *Is There Enough Biomass to Fuel the World? Part I*. www.mr-sustainability.com/stories/2020/7/26/is-there-enough-biomass-to-fuel-the-world-1 Date of access 20.12.2022.

Wang, C., D. Wu, H. Wang, Z. Gao, F. Xu, and K. Jiang. 2018. A green and scalable route to yield porous carbon sheets from biomass for supercapacitors with high capacity. *Journal of Materials Chemistry A*, 6:1244–1254. https://doi.org/10.1039/C7TA07579K

Wang, H., H. Dai. 2013. Strongly coupled inorganicnano-carbon hybrid materials for energy storage. *Chemistry Society Reviews*, 42(7):3088e3113.

Wang, H., Q. Gao, and J. Hu. 2009. High hydrogen storage capacity of porous carbons prepared by using activated carbon. *Journal of the American Chemistry Society*, 131:7016–7022. https://doi.org/10.1021/ja8083225

Xu, Y., G. Jian, Y. Zhu, M.R. Zachariah, and Ch. Wang. 2014. Superior electrochemical performance and structure evolution of mesoporous Fe2O3 anodes for lithium-ion batteries. *Nano Energy*, 3:26–35. https://doi.org/10.1016/j.nanoen.2013.10.003

Zalyhina, V., V. Cheprasova, V. Belyaeva, V. Romanovski. 2021a. Pigments from spent Zn, Ni, Cu, and Cd electrolytes from electroplating industry. *Environmental Science and Pollution Research*, 28:32660–32668.

Zalyhina, V., V. Cheprasova, and V. Romanovski. 2021b. Pigments from spent chloride-ammonium zinc plating electrolytes. *Journal of Chemical Technology & Biotechnology*, 96(10):2767–2774.

Zhang, L., Z. Liu, G.Cui, and L. Chen. 2015. Biomass-derived materials for electrochemical energy storages. *Progress in Polymer Science*, 43:136e64.

9 Metal–organic Framework-based Materials for Energy Production, Conversion, and Storage

Athira A R and Xavier T S

9.1 INTRODUCTION

Metal–organic frameworks (MOFs), or coordination polymers, are highly ordered porous crystalline structures with metal ions and organic linkers (ligands). The Revolution of MOFs established the relationship between coordination chemistry and material science. Diversity in the choice of metal ions and ligands outburst developing novel MOFs with vivid topology and tuneable size-shape pores [1–3]. The choice of metal ions includes transition metal oxides, alkaline earth metals, lanthanides, and ligands like multidentate molecules with N or O donor atoms; for example, pyridyl, polyamines, carboxylate, etc., are widely used for constructing MOFs [4–6]. The hybrid crystalline network structures originated from organic and inorganic constituents via molecular self-assembly pioneered in the late 1990s. So far, more than 90000 different MOFs have been developed and are still growing daily [7].

MOFs connect the inorganic vertices and organic linkers through coordinate bonds and form a network structure (nano micro MOF). Because of this hollow network structure, MOFs possess a great internal surface area of more than 7800 m^2/g [8]. In addition, MOFs exhibit unique structure diversity corresponding to the rigid periodic arrangement of their organic-inorganic components, which were not accessible in conventional porous materials like zeolites. Also, compared to other porous materials such as active carbon, and mesoporous silica, MOFs possess superior chemical tailorability due to their diversified functional groups associated with their framework [9]. The tuneable porosity with complex sorption performances is the signature of MOFs, making them suitable for vivid applications. The porous surface permits the easy diffusion of size-shape selective guest molecules into their bulk. Once it is diffused, it undergoes chemical transformations which step to designing customized MOF structures by their final application [10]. Hence, MOFs have recently become a research hotspot due to their high surface area, porous nature, controllable morphology,

 DOI: 10.1201/9781003318859-9

structural stability, and multi-functionalities [11, 12]. MOFs have the potential as heterogeneous catalysts, although applications have not been commercialized. Their high surface area, tunable porosity, and diversity in metal and functional groups make them attractive in catalysis applications. Besides, MOFs are known for their structural flexibility in response to external stimuli like temperature changes, mechanical forces etc., which paved the development of flexible MOFs; they possess wide application in gas storage and separation [13]. The availability of recent literature indicates the progress in the development of MOFs [14, 15]. Most concentrate on the different synthesis strategies for tailoring various morphologies accountable for diverse applications [16, 17, 18]. It is also reported that MOFs can be used as versatile precursors for developing morphology-tuned porous materials like porous carbon, metal oxides, metal carbides, metal sulfides, metal phosphides etc. [19]. Tang et al. reported the growth of 1D nanocarbon rods and 2D graphene nanoribbons by the controlled carbonization of MOF precursors [20]. Lu et al. reported porous Co_3O_4 concave nanocubes using the calcination of CO-based MOFs as the precursor at an optimized temperature of 300°C. However, properly controlling calcination temperature is vital in designing porous metal oxide structures with high porosity and large surface area [21].

Selecting appropriate metal ions and ligands resulted in the development of MOFs with specific physiochemical properties. Apart from their porosity, various pore shapes and sizes can be tailored from the opposite choice of organic ligands. Hence an easy tuning of desired network structure is possible with the development of MOFs. In the early stages of development, MOFs were used in applications such as the separation and storage of gas, catalysis, and liquid purification owing to their microporosity [22]. In addition to these properties, high surface area, unique morphology, good chemical stability, and cage-like structure, later on, its application extended to sensor detection, energy generation, conversion, and storage [23]. Recently, reports revealed the potentiality of MOFs in biological, environmental, and food antibacterial applications [24]. This chapter emphasizes the developments of MOFs and their derivatives in energy applications. The surge to develop innovative energy technologies explored designing porous materials that can store energy carriers or facilitate fast mass and electron transport. Extensive studies have been done to identify the best materials for photo and electrochemical energy applications. Compared with traditional porous materials, MOFs possess a hybrid inorganic nature with a uniform dispersion of components, tunable topologies and pore size, highly porous nature, and multifunctional ties, enabling them to excel amazingly in energy applications. One of the most promising applications of MOFs in energy is gas storage. MOFs can store large amounts of gas in their pores, making them an attractive alternative to traditional gas storage methods. For example, MOFs have been used to store hydrogen gas for fuel cells, which have the potential to power cars and other vehicles without producing harmful emissions. Additionally, MOFs have been used to store natural gas, which could provide a safer and more efficient method of transporting natural gas.

MOFs have also shown promise in catalysis, which is the process of accelerating chemical reactions. Their porous structure can provide numerous advantages, such as abundant charge carrier pathways, ample reaction active sites and rich adsorption and desorption channels during the photocatalytic process. MOFs can be designed to have specific catalytic properties, making them useful in various applications, such as converting carbon dioxide into useful chemicals or producing hydrogen from water. Additionally, MOFs can be used as support for catalysts, improving their performance and stability. Furthermore, MOFs have been explored for energy conversion and storage applications [25, 26]. MOFs can be used as electrodes in batteries and supercapacitors, providing a high surface area for energy storage. Additionally, MOFs have been used in photocatalysis, which uses light to initiate chemical reactions. MOFs can be designed to have specific photocatalytic properties, making them useful in a range of applications, including water splitting for hydrogen production and carbon dioxide reduction.

In conclusion, MOFs have great potential in energy applications due to their unique properties, such as high surface area, tunable pore size, and chemical stability. MOFs have been explored for gas storage, catalysis, and energy conversion and storage applications and have shown promising results in each area. As research on MOFs continues, they have the potential to revolutionize the way we store and use energy.

9.2   SYNTHESIS OF MOFS AND THEIR COMPOSITES

Inspired by the structural benefits, many researchers attempted to design MOFs with diverse topology and porosity; hence synthesis of MOFs has attracted wide attention during the last two decades [27, 28]. Integrating various functional groups is an effective strategy for improving MOF performances [29]. Traditional synthesis procedures adopted for the development of MOFs and their composites are shown in figure 9.1 and are described as follows.

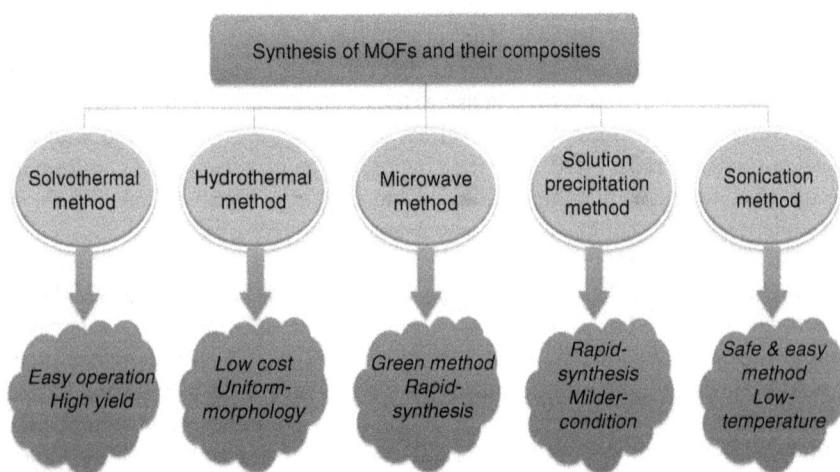

FIGURE 9.1   Conventional routes to synthesis of MOFs and their composites.

9.2.1 Solvothermal Method

The solvothermal method is one of the easy routes to prepare MOFs and their composites owing to its simple procedure, pressure-temperature control, high yield, etc. [30]. However, selecting a suitable solvent is critical since it behaves as the structure-directing agent and the synthesis media. In general, organic solvents such as N, N-dimethylformamide (DMF), N, N-diethyl formamide (DEF), dimethyl sulfoxide (DMSO), toluene, etc., are employed for the synthesis of MOFs and their composites. DMF is a widely accepted solvent because of its high solubility and boiling point. DEF is the second prior solvent but is expensive for large-scale production. For example, Mckinstry et al. reported the solvothermal synthesis of MOF5 using the synthetic solvent DMF and zinc nitrate ($ZnNO_3$) and terephthalic acid (H_2BDC) as precursors [31]. The literature demonstrates that the topology-tuned MOFs can be synthesized through a solvothermal route. Maka et al. reported linkers' effect on tuning in MOF's porosity. Subtle variations in the structure of likers generated in MOFs with different topologies and significant absorption of gas with varying capacities [32].

Solvothermal procedures are generally time-consuming; sometimes, synthesizing MOFs takes several hours to a day. To reduce the duration, conventional microwave-assisted synthesis strategies were adopted with the solvothermal process, reducing the reaction duration to several minutes [33, 34, 35]. Ni et al. reported the microwave-assisted solvothermal synthesis of MOF; they completed the reaction in 30 s to 2 min. The method uses microwave to nucleate the crystal growth, further enhancing the reaction [36]. Even though the solvothermal route is facile, organic solvent toxicity and un-eco-friendliness obstruct the large-scale production of MOFs. Also, the complete removal of solvents is unattainable. A solvent such as DEF is expensive, inhibiting the large-scale production of MOFs through solvothermal.

9.2.2 Hydrothermal Method

The hydrothermal (HT) method is one of the best facile methods to fabricate MOFs and their composites. In the HT method, the reaction conditions, such as reaction temperature, the reaction duration, precursors, solvents, and pH of the components, can be easily adaptable. Lee et al. reviewed the HT synthesis of zeolitic imidazolate framework (ZIF) structures, ZIF-8 and different MOFs using organic ligands 1,4-benzene dicarboxylic acid 1,3,5-benzene tricarboxylic acid I connection with different metal nodes such as Zn^{2+}, Cr^{3+}, Al^{3+}, Fe^{3+} and Zr^{3+} [37]. Through properly controlling the synthesis parameters, the HT method is very much appreciable for developing size-controlled MOF structures for specific applications [38, 39, 40].

Even though the HT method is efficient, it still faces some difficulties in developing application-oriented MOF nanocomposite structures through the HT method. Some noticeable drawbacks are the long reaction time and low yield. Also, the total mechanism behind the formation of the composite structures cannot be understood [41].

9.2.3 Microwave Method

The microwave method is a powerful strategy for developing functional nanomaterials. This procedure rapidly transfers electromagnetic radiations ranging from 0.2 to 300

GHz to the reaction medium. Compared to conventional procedures, the microwave method is a faster and safer route [42]. By altering the synthesis conditions such as microwave power level, irradiation time, temperature, solvent concentration etc., specific morphology-oriented structures can be developed using this method within a few minutes. Choi reported the development of MOF-5 through microwave procedure; the group achieved specific uniform cubic MOF- crystals through the optimized reaction conditions. Also, they noted a sharp deterioration in physiochemical properties through prolonged microwave radiation [43]. Han et al. reported the pillared Ni-MOF electrode materials through 30 min microwave irradiation.

Further annealing of the obtained MOF material offered higher specific capacitance and rate performance. The microwave method can rapidly achieve highly porous material with an enhanced surface area [44]. Layered MOF structures can be developed using the microwave route. Double hydroxide layers of Zn(II)are bridged by terephthalic acid, and the interlayer spacing can be tuned by varying the pH of the precursor solution. The storage performance of the layered Zn-TPA MOF as an efficient electrode for energy storage was reported by Hirai et al. [45].

9.2.4 SOLUTION PRECIPITATION METHOD

The solution precipitation or coprecipitation method is well known for developing inorganic materials. It is a fast and easier procedure in which simultaneous precipitation of cations occurs during the reaction. Compared to the conventional hydrothermal method, it is easy to develop MOF structures and composites because of their easy manipulation and mind experimental conditions. Generally, the HT method requires hours to days to synthesize MOFs; Zhuang et al. reported the synthesis of the HKUST-1 framework using the facile coprecipitation method within minutes. Ethanol and methanol were considered better solvents for developing well-defined crystalline materials. The formation of dense surface–mounted MOF films, called SURMOFs, is also possible using this method [46]. Room temperature synthesis of MOFs is advisable since the harmful organic solvents can be partially substituted with water. For developing MOFs at room temperature conditions, some amines are to be added along with the joint solution of metal precursor and the ligand to make an abrupt change in the pH of the solution. This method is preferable for the synthesis of MOF-5. Even though the exact mechanism is unclear, the amine causes deprotonation to the ligand, predisposing it to react with the metal ion to the solution [47]. Park et al. reported the uniform-sized dodecahedral structures of ZIF-67 samples obtained via coprecipitation. The product material exhibited good dispersibility and better photocatalytic activity [48]. Hence, high-quality MOFs and their composites with desired morphology can be easily tailored by carefully modifying the synthesis conditions like stirring speed, reaction temperature, and reaction time.

9.2.5 SONICATION METHOD

The sonication method (ultrasonication) is one of the effective procedures for synthesizing MOFs and their composite since it is cost-effective, has environmental friendliness, operational safety, and low reaction temperature. During sonication, high-frequency energy is transferred to the reaction solution, and the generated high energy and high-pressure lead to the formation of nanocrystals. The first report on the synthesis of MOFs by sonication method was reported in 2008 by Qui et al. The work demonstrated the rapid synthesis of microporous $Zn_3(BTC)_2 \cdot 12\ H_2O$, for the selective sensing of organoamenes [49]. Ultrasonication effectively synthesizes MOF/MXene hybrid nanosheets as electrodes for supercapacitors [50]. High energy during ultrasonication breaks the van der Waals force between the interlayers and supports to enhance the effective surface area of the product material [51].

9.3 MOFS IN ENERGY APPLICATIONS

9.3.1 MOFs in Energy Generation: Photocatalysis Application of MOF-based Composites

The depletion of fossil fuels and the urge for clean energy research motivated the development of facile and environmentally friendly energy generation strategies. Conversion of solar and wind energy is the most promising solution for generating a clean energy environment. Sun is the ultimate source of nonconventional energy; in this aspect, researchers are encouraged to develop clean energy from the sun. One of the facile and green techniques for converting solar energy to chemical energy is photocatalysis; the material support for photocatalysis is known as photocatalysts. Photocatalysis is a process that involves the use of a photocatalyst to initiate a chemical reaction when exposed to light. During photocatalysis, hydrogen is evolved from water in the presence of a photocatalyst. So the role of photocatalysts is pretty important and can be considered a gem in producing clean energy. Several materials have been used as photocatalysts, but lack of electronic conductivity, low apparent quantum efficiency, rapid recombination of charge carriers, poor cycle stability and chemical stability hinders the use of conventional photocatalysts like metal oxides, metal phosphide, and metal sulphide-based materials. The construction of superior functional materials with outstanding overall performance is highly desirable. Due to exceptional stability, large surface area, and superior photocatalytic activity, the MOF has attracted much research interest in solar energy generation.

MOFs are used as photocatalysts to enhance the conversion efficiency of solar cells. Owing to their synthetic tunability and viability, MOFs can be used to design new atomically precise cluster-based catalysts. However, the poor electronic conductivity, deprived light absorption capacity, short-term cycle durability and rapid recombination of photo-generated electron-hole pairs impede the extensive application of MOFs in energy generation. Numerous strategies have been adopted to optimize the photocatalytic properties of various MOF structures [52]. Recently, there has been growing interest in using metal–organic framework (MOF)-based composites as photocatalysts for various applications (Figure 9.2) [53–56].

(a) *Metal ions-doped MOFs composites*: Doping with photocatalytic metal ions (noble or non-noble metal ions) is an effective strategy to promote the photocatalytic activity of MOFs since the injected ions significantly alter the surface electronic

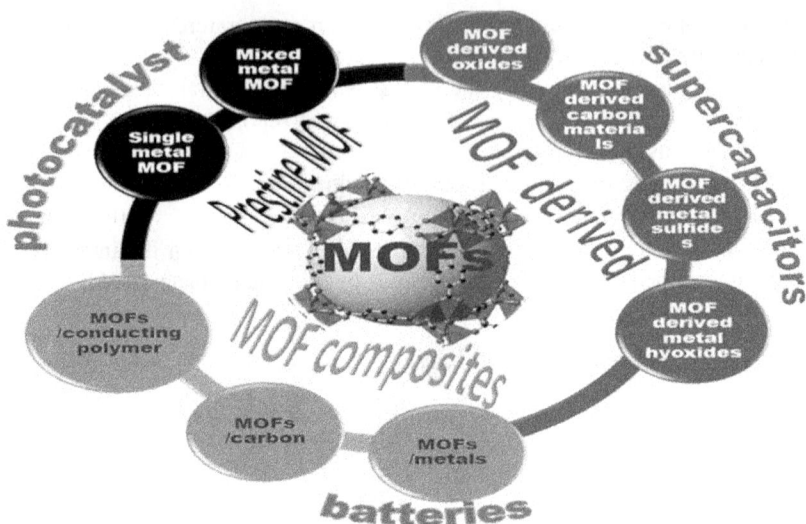

FIGURE 9.2   MOFs and their composites for various energy applications.

properties of the MOFs. Also, these impurity ions can behave like electron traps, further enhance the electron-hole separation, and provide enormous active sites for photocatalytic activity [57]. Because of noticeable electronic conductivity and high catalytic performance, noble metal ions like Ir, Pt, Au, Ag, and Ru are superior co-catalysts for preparing MOF-based composites [58–61]. Non-noble metal ions such as Co^{3+}, Fe^{3+}, Cu^{2+}, Ti^{4+}, and Zr^{4+} are commonly used as co-catalysts; due to extensive thermal and structure stability, Fe^{3+} attained considerable attention [62, 63]. But the scarcity and high cost of noble metal ions and poor cycle stability of non-noble metal ion co-catalysts limit their large-scale production; hence, research focuses on simultaneously incorporating both noble and non-noble co-catalysts.

(b) *Metal oxide-MOF composite photocatalysts*: Because of their low cost and excellent physiochemical properties, metal oxides such as TiO_2, Fe_2O_3, $BiVO_4$, CuO, Cu_2O etc. have been utilized as co-catalysts [64–66]. Owing to the suitable bandgap and ease of mass production, TiO_2 is considered a better candidate for compositing with MOFs [67, 68].

(c) *Metal sulphide-MOF composite photocatalyst*: Metal sulfides possess narrow bandgap, small band structure, high electronic conductivity, and excellent photocatalytic activity. Taking care of these advantages, they are incorporated with MOFs to enhance their overall photocatalytic performance. CdS, MoS_2, Bi_2S_3, Ni_4S_3, and CuS have been widely studied metal sulfides. Of many of them, CdS is a pioneer owing to its narrow bandgap (~2.4 eV) and excellent light absorption capability [69, 70].

(d) *Metal phosphide-MOF composite photocatalyst*: Utilizing metal phosphides incorporated MOFs recently gained much research interest due to the co-catalyst's high catalytic activity. Metal phosphides also possess high electronic conductivity, and their abundance and eco-friendliness are the advantages of using them as co-catalysts for developing MOF-based photocatalysts [71].

(e) *Carbon derivative-MOF composite photocatalysts*: Graphene, carbon nanotubes (CNT), carbon dots (CDs), graphitic carbon, graphene oxide (GO), etc., are some of the greatly studied carbon derivatives, with large surface area, high conductivity, unique morphology, etc. These salient features make them superior for catalytic application. Combining these materials with MOFs opened a new era for efficient, clean energy generation [72, 73]. CNT/MOFs possess unique hollow morphology beneficial for easy trapping of charge carriers and support to improve the photocatalytic activity [74]. CDs have had research interest recently due to their bio-compatibility and low cost. The large surface functional groups, high aqueous solubility, and exceptional electron transfer capacity make CDs viable co-catalysts for preparing efficient CD-MOF composite photocatalysts [75]. Graphene and GO are well known for their enhanced surface area, high thermal stability, and high electron mobility. Compared to graphene, it possesses better light harvesting ability and excellent optical properties, allowing it a superior candidate for catalysis. Furthermore, GO is enriched with various surface functional groups comprising epoxy, hydroxyl, carboxy etc., remarkably enhancing its photocatalytic activity [76]. Hence GO proved to be a superior co-catalyst for boosting the photocatalytic performance of the GO-MOF composite.

In addition to these conventional materials, highly porous aerogels are utilized as co-catalysts for enhancing the catalytic activities of MOFs [77, 78]. MXene, perovskites, and quantum dots are some of the rarely studied co-catalysts for enhancing the photocatalytic performance of MOFs [79]. Majevsky et al. reviewed the rational modification of organic linkers and the possibility of incorporating photosensitizers or functional groups into MOF structures to modify them for solar fuel generation applications [80]. The evolution of clean energy fuel hydrogen from water in the presence of photocatalysts has been a hot topic in recent research. Several reports are available on MOF-based photocatalysts to produce hydrogen from water under visible light irradiation. The photocatalytic activities of MOF-based photocatalysts for solar fuel production can be categorized into three: (i) photocatalytic water splitting to hydrogen fuel, (ii) photocatalytic carbon dioxide reduction to hydrocarbon fuel, and (iii) photocatalytic nitrogen fixation to high energy fuel carriers such as ammonia [81].

Furthermore, environmental remediation is one of the main applications of MOF-based photocatalysts. These materials can degrade organic pollutants in water and air, such as dyes and pesticides. Kampouri et al. reported the first report on the MOF system used for duel photocatalytic activity- hydrogen generation and dye degradation using MIL-125-NH_2 under visible light irradiation. The work investigated how the metal-based co-catalysts affect the hydrogen generation of MIL-125- NH_2 and the result verified that the Ni_2P/ MIL-125- NH_2 system exhibited a high hydrogen evolution rate of 1230 µmol $h^{-1} g^{-1}$. Also, when the electro donor is replaced with rhodamine B(RhB), the organic pollutant undergoes degradation, which proves the compatibility of MIL-125-NH_2 for sustainable wastewater remediation [82]. Yang et al. reported a pillared-paddlewheel type MOF, CCNU-1, capable of absorbing visible light from 200–800 nm, exposed enhanced hydrogen generation rate (4680 µmol $h^{-1} g^{-1}$) while decorated with Pt nanoparticles and L-ascorbic acid as a sacrificial reagent, highlighting the promising future of Pt/CCNU-1 MOFs in solar-to-chemical energy conversion [83]. Li et al. reported the development of polymer (PSS)/MOF hybrid membranes in nanofluidic systems that will control the nanofluidic transport. Adjusting the PSS content, tuning the cation selectivity and rectifying the ion current is possible. Incorporating these membranes with a salinity-gradient-driven device, the high power density of 2.87 W/m2 can be achieved and render their application in the field of energy conversion [84].

9.3.2 Carbon Capture

Carbon capture is another important application of MOF composites in energy generation. The excessive carbon emissions caused by burning fossil fuels contribute significantly to global warming. MOF composites can be used to capture carbon dioxide and reduce carbon emissions. These composites have high selectivity and adsorption capacity, making them a promising material for carbon capture. Because of their high surface area, microporosity and high absorption capacity, functionalized MOFs were used as a potential candidate for carbon capture [85].

9.3.3 MOF Composites in Energy Storage: Supercapacitors and Batteries

Pursuing sustainable storage systems has become one of the greatest issues of concern due to the intermittent nature of renewable energy sources. The need for effective and reliable energy storage technologies has become a worldwide research target to meet the energy crisis in the modern era [85]. Evolution of electric transportation- the development of hybrid electric vehicles (HEVs), plugin hybrid vehicles (PHEVs), and full electric vehicles (FEV) boosted the demand for electrical energy storage (EES) systems with high power as well as energy densities. Recent research showed that batteries and supercapacitors (SCs) are at the forefront of EES devices, which play a vital role in future energy applications [86]. However, they are not enough to solve all energy crises owing to their limitations. Batteries possess high energy density; however, their poor power density inhibits them from wide energy storage areas. SCs exhibit excellent power density compared to batteries, but their poor cycle stability and rate performance impede them from long-range storage applications. One effective solution to overcome these limitations is to develop efficient materials such as electrodes or electrolytes for both batteries and SCs. Great research efforts have been devoted to developing competent electrode materials for next-generation batteries and SCs and improving the utilization of clean energy. MOFs are one of the emerging candidates for developing EES devices. Several reports summarized the importance of MOFs in energy storage, especially hydrogen storage, electrode materials for fuel cells, batteries, and SCs.

9.3.3.1 MOFs as Electrodes for Supercapacitors

In recent years, supercapacitors or ultracapacitors attracted much attention in electrochemical energy storage because of their ability to deliver a bundle of energy over a short time than conventional capacitors or batteries. They possess better rate performance and appreciable cycle stability over thousands of charge-discharge cycles. The basic charge storage mechanism, SCs are broadly classifieds into electric double-layer capacitors (EDLC), pseudocapacitors (reversible- faradaic), and hybrid capacitors. The storage mechanisms were based on the materials used for fabricating electrode materials. In general storage performance of an SC can be defined by the adsorption/ desorption of electrolyte ions into the electrode materials. In EDLC, the energy is stored by the charge separation between the electrode and electrolyte (non-faradaic process). This double layer, or the Helmholtz layer, is formed by applying an electric potential across the electrodes of an SC. Because of the instant charge accumulation on the electrode surface, EDLC performs a fast charge-discharge rate and high stability due to the absence of faradaic redox reactions. More precisely, only the physical adsorption of electrolyte ions over the electrode surface contributes to the charge/ energy storage in EDLC [87].

In comparison with EDLCs, the energy storage mechanism in pseudocapacitors is different. It occurs by reversible faradaic redox reactions; reduction and oxidation occur on the electroactive material, the charge carriers cross the double layer, and a faradaic current is generated [88]. Because of this faradaic charge storage process, pseudocapacitance possesses higher storage performance than EDLC but less cycle stability [89]. Hybrid supercapacitors are developed by combing the advantages of EDLC and pseudocapacitance mechanism, utilizing the advantages of high cycle stability and power densities of EDLC and remarkable specific capacitance and energy densities of pseudocapacitors [11]. Because of their porous morphology, tailored pore size, and enhanced surface area, MOFs are apposite electrode materials for supercapacitors [12]. A surplus of literature exhibits the feasibility of MOFs in supercapacitor applications [86,90–92]. A variety of nanomaterial crystals with varying organic functionalities have been developed and tested for their adaptability as an electrode for supercapacitor applications [93,94]. It proves that metal–organic nanocrystals have apparent electrochemical performance compared to their metal ion counterparts. MOFs can be applied in three ways to develop electrode materials: (i) Pristine MOFs can be used as such electrodes; the charge is stored through the physisorption of the electrolyte ions into the internal surface of the electrode or by the reversible redox reaction of the metal centres; (ii) Converting MOFs to the corresponding metal oxide by preserving the electrons through charge transfer between the electrolyte and the electrode; (iii) pyrolyzing the MOFs to corresponding porous carbon and enhancing the capacitance through improving the conductivity [86]. Presine MOFs offer fast diffusion of electrolyte ions during the charging-discharging process, which is the basic requirement for developing efficient electrodes for energy storage. However, the pristine MOFs lack conductivity. Incorporating active cations into the MOF structure can enhance the conductivity and capacitance via the pseudocapacitive mechanism.

MOF-derived metal oxides are another class of effective electrode materials. Transition metal oxides such as RuO_2, MnO_2, Co_3O_4, NiO, Fe_3O_4 etc., are well known for pseudocapacitor electrodes. They possess better capacitance properties; faradaic process-governed pseudocapacitors offer much higher capacitance than EDLC. MOF-derived metal oxides offer a large accessible internal surface area, providing many diffusion sites for active ions, critical for effective electrochemical activity [95,96].

Co_3O_4 is an attractive electrode for SC because of its favourable pseudocapacitive properties [97]. In supercapacitors, MOFs have been used as electrodes directly or as precursors for developing porous electrodes. Xu et al. reviewed the recent progress in MOFs in SC applications as precursors for electrodes, also the recycling of organic ligands to drastically reduce the cost of production by the MOF hydrolyzing strategy [98]. Carbon materials developed from MOF precursors offer the privileged structures of MOF, viz enlarged surface area, porous structure, etc. MOF-derived carbon materials possess enhanced surface area, and pore size allows fast ion transport and enhances the interaction between electrolyte ions and the electrodes. Hence MOF-derived carbon electrodes are promising for SC applications. Properly selecting the MOF precursors can tailor the morphology and compositions of the product carbon composite. The low conductivity of MOFs restricts the wide application in SCs. Several combinations of metal oxides or carbon composites were incorporated with MOFs to address the issue. Liu et al. report the SC electrodes using MOF-derived materials, broadly on MOF-derived porous carbon and MOF-derived metal oxides/sulfides. The durability and effectiveness of these materials as electrodes were briefly reviewed [99]. Choudhari et al. reviewed the conducting polymer incorporated MOF composites in SC applications. The ion dynamics in the electrode materials were briefly studied using nuclear magnetic resonance spectroscopy (NMR) [100]. Polyaniline (PANI) incorporated more than 87.6% PANI/MOFs composite electrode capacitance after the 6000 charges discharging test, showing the system's excellent cycle stability [101]. The effective ion diffusion in polymer/MOF composites provides better storage performance in SC applications [102, 103].

Various MOF nanocrystals with different functionalities, metal ions, have been synthesized and analyzed for their viability as an electrode in SC applications. These procedures or composites possess enhanced electrochemical performance. To modify the conductivity of MOFs, effective electrodeposition strategies were adopted. Zhu et al. reported the fabrication of ZnO@MOF@PANI through combined electrodeposition and hydrothermal procedures. The composite material showed better rate capacity and capacitance retention [104]. Liu et al. reported the electrodeposition of MOF/polypyrrole composite on carbon cloth as a positive electrode for SC showed remarkable storage performance [105]. Due to the high electrical conductivity, surface area, and excellent electrochemical behaviour, reduced graphene oxide (rGO) is better for EDLC applications. The incorporation of rGO into MOF structures is proven to be a better material as electrodes for SCs. The excellent synergy between MOF and rGO offers better electrochemical characteristics favourable for energy storage [106]. The doping of metal ions further enhances electrochemical performances. Ni-doped ZIF67 prepared in the presence of rGO exhibits high specific capacitance. Ni doping and concentration of rGO have been optimized to attain the desired redox activity and electrical conductivity. Also, substituting Ni ions with Co ions shows better specific capacitance by inducing additional pseudocapacitance [107]. Nanoporous metal oxides developed from hybrid MOF structures (ZIF67 and ZIF8) offer better electrochemical performance. rGO incorporated hybrid MOF electrodes rGO electrodes were used to fabricate asymmetric SC, which offers an enhanced working potential window for the SC with a higher energy density of 12.4 Wh/kg [108].

Recently, *inorganic pseudocapacitance materials incorporated* MOF electrodes gained much attention in SC applications. MOF-derived metal oxides/sulphides gained much attention in recent research due to the high demand for efficient pseudocapacitive materials. Since MOFs are metal ions and ligands combinations, higher concentrations of metal-containing groups can be used as metal sources. Nanoporous metal oxides provide tunable porosity and crystallinity, which can be easily synthesized through thermal decomposition. A simple thermal decomposition method can develop another metal oxide or carbon-based composite for supercapacitors [109].

9.3.3.2 MOFs as Electrodes in Batteries

Li-ion batteries (LIBs) have found great commercial importance for the energy supply of portable electronic devices. But for large-scale energy storage applications such as electric vehicles, the LIBs fail to meet the energy requirements due to their low power density even though they have a long life span and specific energy. The working mechanism behind the LIBs is the generation of Li ions through the oxidation of the Li-based anode, and then migrate to the cathode during discharge and reverse during the charging. Hence, new materials were to be developed as electrodes to reach beyond the levels of conventional LIB. MOFs, composite and MOF derivatives are better options for developing anode, cathode, and electrolyte materials for next-generation LIBs.

Figure 9.3 explores the next-generation batteries other than LIBs, where the cathodes are made up of porous MOFs and their derivatives such as MOF composites, MOF-derived oxides and MOF-derived sulfides. Graphite is the conventional anode material used for the fabrication of batteries. Yet their low capacity ($372\ mAhg^{-1}$)

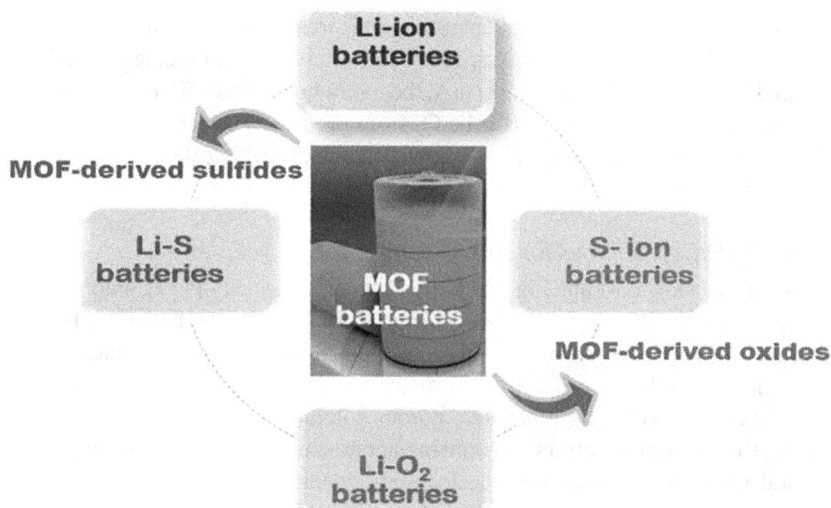

FIGURE 9.3 MOF-derived batteries for next-generation energy storage.

and rate performance paved the development of alternate anode materials [86]. MOF structures have an inherent porous structure which effectively encapsulates Li ions in its interior and is reversibly inserted/extracted, leading to the insertion-type electrodes. More impressive electrode materials can be developed by properly selecting the metal centres and organic ligands. Primary investigations highlighted pristine MOFs' inefficiency since they never favour easy reversible Li-ion transport due to their lower cycling capacity. However, the better discharge capacity suggested the possibility of modified MOFs as anode materials. Further studies showed that intercalated MOFs are preferable to prestine MOFs [110].

(a) MOFs for LIBs

In recent years, researchers have explored various strategies to optimize the electrochemical performance of MOFs, including structural modifications, doping, and hybridization with other materials. MOF-based electrodes have demonstrated high specific capacity, excellent rate performance, and long cycle life, making them a promising alternative to conventional LIB electrode materials. However, further research is needed to optimize the synthesis and performance of MOFs for use in commercial LIBs.

(b) MOF-derived Oxides for LIBs

Oxides are desirable electrode materials due to their theoretical capacity and safety [111,112]. MOF-derived oxides exhibit excellent LIB-specific capacity and long cycle life [113]. Prussian blue analogue (PBA, $Co_3[Co(CN)_6]_2$) sintering produced porous Co_3O_4 nanocages with 1465 mAh g^{-1} stable capacity after 50 cycles at 300 mA g^{-1} [114]. Nanocages performed well electrochemically because of their porous shell, tiny size, and large surface area. Fang et al. [115] used liquid-phase deposition with MOF precursors to create porous metal oxide nanosheets on 3D substrates for LIBs with high-rate capability and long-term cyclic stability. Fe_2O_3 could be an anode material with its high theoretical capacity, low treatment cost, and non-toxicity. Zhang et al. [116] annealed Prussian blue (PB) microcubes to create hierarchically structured Fe_2O_3 micro boxes. At 200 mA, the micro boxes had 950 mAh g^{-1}. Xu et al. [60] produced spindle-like porous α-Fe_2O_3 from $Fe_3O(H_2O)2Cl(BDC)_3 \cdot nH_2O$ (MIL-88-Fe) with a specific capacity of 911 mAh g^{-1} over 50 cycles at 0.2 C. α-Fe_2O_3 has 424 mAh g^{-1} at 10 C.

(c) MOF-derived Sulfides for LIBs

Because of their low cost and high theoretical specific capacity [117], sulphides have recently gained favour as an electrode material for lithium-ion batteries. However, their functional-specific capacity and cycle performance are constrained by their low electronic conductivity and poor structural stability throughout the charge/discharge process. Developing innovative porous sulphide composites, such as porous sulfide/carbon materials, offers a promising approach. Carbon materials improve the structural integrity and increase the electronic conductivity of electrode materials, while the porous structure allows for volume variations during lithiation/de-lithiation. Reduced graphene oxide (rGO)@CoSx and CoSx-rGO-CoSx composites were created by thermally sulfurizing MOFs/graphene oxide (GO) precursors [118]. These composites demonstrated high specific capacities, outstanding rate capabilities, and long cycle life for LIBs, demonstrating MOFs as a promising precursor for high-performance sulphides.

(d) SIBs

Sodium-ion batteries are an emerging technology that has gained significant attention as an alternative to traditional lithium-ion batteries due to their low cost and abundant resources. Sodium-ion batteries use sodium ions instead of lithium ions to transfer charges between the electrodes. Sodium is much more abundant and less expensive than lithium, making it a more sustainable and cost-effective option for energy storage. One of the key components of a sodium-ion battery is the cathode material. Various materials have been tested for use as cathodes in sodium-ion batteries, including transition metal oxides, phosphates, and polyanions. One of the most promising materials for use as a cathode is $Na_3V_2(PO_4)_3$, or NVP, which has a high specific capacity and excellent cycling stability. In addition to cathode materials, anode materials are an important consideration in sodium-ion batteries. Carbon-based materials have been used as anodes in sodium-ion batteries but tend to have low capacity and poor cycling stability. Recently, alternative anode materials, such as metal sulfides and oxides, have shown promise in sodium-ion batteries due to their high specific capacity and good cycling stability. Recent developments in sodium-ion batteries have focused on improving their energy density, power density, and cycling stability. Researchers have developed new cathode materials, such as P2-type $Na_{2/3}[Ni1/3Mn_{2/3}]O_2$, with a high specific capacity and good cycling stability. Other advancements have focused on improving the electrolyte, which transfers ions between the cathode and anode. Researchers have developed new electrolyte materials, such as $NaPF_6$ in a carbonate solvent, showing improved cycling stability and rate capability.

(e) MOFs for SIBs

As a result of the extensive availability and inexpensive cost of sodium resources, sodium-ion batteries, also known as sodium-ion batteries (SIBs), have arisen as competitive alternatives to lithium-ion batteries, also known as LIBs. They are among the next generation's most promising energy storage technologies [119,120] and share the same working principles and components as LIBs. However, compared to lithium-ion insertion materials, the exploration of sodium-ion insertion materials is restricted due to the poor kinetics caused by the larger radius (102 pm) and heavier atomic mass of Na+ than those of Li+ [121,122,123]. This is because sodium ions have a larger radius than lithium ions. Because the diffusion of Na+ demands a greater tunnel size, organic compounds, which have structural diversity and flexibility, are considered viable candidates for use as SIBs [124].

(f) MOF-derived Metal Oxides for SIBs

Low electronic conductivity and instability in conversion reactions are disadvantages of metal oxide anodes used in LIBs. The charge/discharge process for SIBs is slower for sodium ions (0.102 nm) than lithium ions (0.059 nm), and the volume change is higher. Similarly to how LIBs are handled, transition metal oxides, including Fe_2O_3

[125], Co_3O_4 [126], and CuO [127], have been employed as anodes for SIBs. The speed of the conversion reaction, the electronic conductivity, and the ability to adapt to the volume change can all be enhanced by treatments such as porous nanostructure engineering and carbon modifications. NiO/Ni/graphene hollow ball-in-ball nanostructure was produced using Ni-based metal–organic frameworks (Ni-MOFs) [128], and MOF precursors have been used to prepare metal oxides for SIBs. The NiO/Ni/graphene anode had a consistent cycle life (0.2% specific capacity fading per cycle) and good rate performance (207 mAh g^{-1} at 2 A g^{-1}) in SIBs. NiO/Ni/graphene has a novel hierarchical hollow ball-in-ball structure that can support the volume change of active materials across multiple cycles. The highly conductive graphene made fast electron transport possible, which also contributed to creating a stable solid electrolyte interface (SEI) film.

(g) MOF-derived Carbon Materials for SIBs

Nanocasting silica as a template improves mesoporous carbon Na+ storage kinetics and performance [129]. Direct pyrolysis of MOFs can produce porous carbon with a narrow pore size distribution and large surface area [130]. Qu et al. [131] pyrolyzed 2-methylimidazole zinc salt to create microporous carbon (ZIF-C) with a well-proportioned pore size of 0.5 nm (ZIF-8). CMK-3 was less reversible than ZIF-C. The microporous ZIF-C also reduced irreversible capacity loss during the initial cycle and improved Na+ storage following subsequent cycles. ZIF-promising C's electrochemical performance and template-free synthesis show MOF-derived carbon materials' promise as SIB anodes. Despite MOFs' widespread use in new-generation SIBs, it's crucial to understand the electrochemical reaction process, energy storage mechanism, and innovative structure-Na^+ storage performance interaction. To create SIBs and design superior electrode materials, the effects of porous structure, specific area, pore volume, and heteroatom doping on Na^+ storage capability for MOF-derived carbon materials should be explored. $LiCoO_2$-graphite LIBs have 491 Wh L^{-1} specific energy and are frequently employed in electric automobiles [132]. Silicon hybridization in the graphite anode can boost this specific energy. Some of the nanomaterials above perform well electrochemically in laboratory studies for LIBs and SIBs, but the synthesis process must be simplified for large-scale production.

(h) Li-O_2 Batteries

Lithium-oxygen (Li-O2) batteries, also known as lithium-air batteries, are a promising technology for energy storage due to their high theoretical energy density. Li-O_2 batteries work by using lithium metal as the anode and oxygen as the cathode, which reacts to produce lithium oxide and release energy. This reaction is reversible, allowing the battery to be recharged. One of the main challenges of Li-O_2 batteries is the limited cycle life caused by the formation of insoluble lithium peroxide during discharge. This can lead to the formation of a passivation layer that blocks the cathode, reducing the battery's performance. Researchers have been exploring different materials and designs for Li-O_2 batteries to overcome this issue. One recent development in Li-O_2 battery technology is using carbon-based cathodes, which can promote the formation of soluble lithium superoxide instead of lithium peroxide.

This can improve the battery's efficiency and cycle life. Another approach involves using catalysts to enhance the reaction kinetics and prevent the formation of passivation layers. For example, researchers have used metal-based catalysts such as cobalt and manganese to improve the performance of Li-O_2 batteries. Other materials studied for Li-O_2 batteries include solid electrolytes and polymer-based electrolytes. Solid electrolytes can improve the safety and stability of the battery by preventing the growth of dendrites, which can cause short circuits. Polymer-based electrolytes can also enhance the battery's performance by improving the transport of ions and preventing the formation of passivation layers.

(i)   MOFs for Li-O_2 Batteries

MOFs have attracted much attention as potential catalysts for increasing the cycling performance of Li-O_2 batteries [133,134,135] and [136]. MOFs have a high surface area, accessible metal sites, and a flexible structure. MOFs were shown to raise the concentration of O_2 in micropores, which resulted in a concentration that was 18 times higher than the concentration of pure oxygen at 273 K under ambient pressure. This discovery was made by Wu and colleagues [134]. The Mn-MOF-74/ Super P cathode displayed a capacity over four times greater than that of the Super P cathode when measured under 1 atmosphere of oxygen at ambient temperature. Due to the high surface area and open metal sites of Mn-MOF-74, the concentration of O_2 molecules within the micropores was improved, which led to the outstanding electrochemical performance of Li-O_2 batteries. This was determined to be the cause of this improvement.

(j)   MOF-derived Materials for Li-O_2 Batteries

Several other kinds of MOF derivatives, including metal oxides, carbon materials, metal oxide/carbon composites, and sulphides, have been investigated to possibly improve the intrinsic catalysis of Li-O_2 batteries [136,137,138]. Metal oxides and sulphides derived from MOFs are well known for having a large surface area, a hierarchical porous structure, and efficient channels for mass transport. These characteristics and the fact that they can help improve the catalytic activity of Li-O_2 batteries have led to their widespread use. For instance, Co-Mn-O nanocubes were synthesized from MOFs and used as cathode catalysts for Li-O_2 batteries. This resulted in high catalytic activity, a low overpotential, and improved cycle life with a limited capacity of 500 mAh g^{-1} up to 100 cycles [136]. However, the capacity of the batteries was still limited to 500 mAh g^{-1}.

Additionally, mesoporous cobalt sulphide nanosheets were utilized as a bifunctional electrocatalyst for rechargeable Li-O_2 batteries [138]. These nanosheets demonstrated an initial discharge capacity of roughly 5917 mAh g^{-1} and a high reversibility of 95.72%. MOF-derived sulphides are considered interesting candidates for high-performance Li-O_2 batteries due to their enormous surface area and hierarchically porous structure.

(k) MOFs for Li-S Batteries

Sulfur is well-suited to porous metal–organic framework (MOF) materials[139]. Tarascon et al. [140] proposed sulphur host materials for Li-S batteries using MOFs in 2011. MIL-100(Cr) with a large specific surface area (1485 m^2g^{-1}) improved Li-S battery capacity retention, and XPS demonstrated weak interaction between the oxygenated framework and lithium polysulfides. Since then, sulphur encapsulation with MOFs has gained attention. Qian et al. employed MOFs HKUST-1 [141] and ZIF-8 [142] as sulphur hosts for Li-S batteries. HKUST-1's adequate pore space and open Cu_2^+ site ensured robust sulphur confinement, resulting in the steady cycle life of the HKUST-1 composite with 40% sulphur content and 500 mAh g^{-1} capacity after 170 cycles. Xiao et al. [143] investigated the sulphur storage mechanism of MOFs and found that a new Ni-MOF, Ni6(BTB)$_4$(BP)$_3$ (BTB = benzene-1,3,5-tri benzoate and BP= 4,4'-bipyridyl), could host sulphur. The sulphur composite retained about 89% capacity after 100 cycles at 0.1 C. The mesopores (2.8 nm) and micropores (1.4 nm) in Ni-MOF and strong contacts between the polysulfide base and Lewis acidic Ni(II) core prevented soluble polysulfides from migrating out of the pores. Theoretical computations showed a high coordination force on soluble $Li_2S_8/Li_2S_6/Li_2S_4$. This study suggested new ways to construct Li-S battery systems via Lewis acid–base interactions. Li et al. [144] tested sulphur composites with different MOFs for electrochemical performance. Small MOF hosts increased sulphur usage, and small apertures with functions in the open framework had an affinity for polysulfide anions through Lewis acid–base interactions, allowing stable cycling. They also found that decreasing ZIF-8 particle size (<20 nm: >950 mAh g^{-1} at 0.5 C) increased capacity, while about 200 nm (75% over 250 cycles at 0.5 C) improved cycling performance [145].

(l) MOF-derived Carbon Materials for Li-S Batteries

Due to low electron conductivity, MOFs/S composites have low sulphur content (<50%) and require a lot of conductive chemicals to make sulphur cathodes. Pyrolysis of MOFs in an inert atmosphere produces hierarchical carbons with micro-, meso-, and macropores, high specific surface areas, and huge pore volumes. Porous MOF-derived carbon sulphur hosts increase composite conductivity and reduce lithium polysulfide dissolution. We pioneered sulphur hosting using MOF-derived carbon compounds [147]. One-step pyrolysis of MOF-5 produced hierarchically porous carbon nanoplates (HPCN) with an average thickness of 50 nm, a high specific surface area of 1645 m^2 g^{-1}, and a huge pore volume of 1.18 cm^3 g^{-1}. After 50 cycles at 0.5 C, the HPCN/S composite discharged 730 mAh g^{-1} after high specific capacity and good cycling performance. HPCN's porous structure and conductive carbon framework improved electron transport and inhibited lithium polysulfide diffusion. The HPCN/S composite's residual mesopore channels enabled quick Li^+ diffusion and electrolyte infiltration. Kumar et al. [148] pyrolyzed four types of zinc-containing MOFs to create carbon materials with hierarchical pore architectures. They discovered that MOF-derived carbon materials' pore size distribution and volume were critical for Li-S battery electrochemical performance. Lou et al. [149] synthesized microporous carbon polyhedrons from ZIF-8 as conductive hosts to load sulphur for Li-S batteries and discovered that sulphur content and preparation technique significantly affected battery performance. The composite with sulphur incorporated in micropores had stable cycle life in both DOL/DME and EC/DEC electrolytes, but sulphur outside the micropores had poor performance with various electrolytes.

9.4   CONCLUSION

Metal–organic frameworks (MOFs) are a form of porous substance held together by organic ligands, metal ions, or clusters. They are excellent candidates for various energy-related applications, including energy production, conversion, and storage, because of their high surface area, variable porosity, and simplicity of functionalization. In catalysis, where they can be employed as heterogeneous catalysts for various processes, including CO_2 reduction, water splitting, and methane activation, MOFs have some of the most promising uses. MOFs' high surface area and pore size distribution can be adjusted to increase their effectiveness in these applications as photocatalysts for solar energy conversion. MOFs are being researched for energy storage uses, including hydrogen and methane storage in addition to catalysis and photocatalysis. Researchers strive to create materials that can retain large volumes of these gases with little loss or deterioration over time by manipulating MOFs' pore size and surface chemistry. As MOFs are also being investigated as excellent materials for energy storage systems like batteries and supercapacitors, they can perform better thanks to their large surface area and distinctive electrical characteristics, resulting in faster charging times and higher energy densities.

Overall, MOFs are appealing candidates for various applications due to their distinct characteristics. The continued research and development in this area are encouraging for the future of sustainable energy production, conversion, and storage, even if there are still obstacles to be solved, such as MOF stability and scalability.

REFERENCES

1. Kitagawa, S. (2014). Metal–organic frameworks (MOFs). *Chemical Society Reviews*, *43*(16), 5415–5418.
2. Wang, S., McGuirk, C. M., d'Aquino, A., Mason, J. A., & Mirkin, C. A. (2018). Metal–organic framework nanoparticles. *Advanced Materials*, *30*(37), 1800202.
3. Xie, L. S., Skorupskii, G., & Dincă, M. (2020). Electrically conductive metal–organic frameworks. *Chemical Reviews*, *120*(16), 8536–8580.
4. Chai, L., Pan, J., Hu, Y., Qian, J., & Hong, M. (2021). Rational design and growth of MOF-on-MOF heterostructures. *Small*, *17*(36), 2100607.
5. Xian, J. Y., Xie, X. X., Huang, Z. Y., Liu, Y. L., Song, H. Y., Chen, Z. Q., ... & Zheng, S. R. (2023). Structure and properties of a Mixed-Ligand Co-MOF that was synthesized in Situ from a single Imidazole–Pyridyl–Tetrazole Trifunctional Ligand. *Crystal Growth & Design*, *23*(3), 1448–1454.
6. Subramaniyam, V., Ravi, P. V., & Pichumani, M. (2022). Structure coordination of solitary amino acids as ligands in metal-organic frameworks (MOFs): A comprehensive review. *Journal of Molecular Structure*, *1251*, 131931.

7. MOF Metal Organic Framework – Definition, fabrication and use (nanowerk.com), www.nanowerk.com/mof-metal-organic-framework.php

8. De Villenoisy, T., Zheng, X., Wong, V., Mofarah, S. S., Arandiyan, H., Yamauchi, Y., ... & Sorrell, C. C. (2023). Principles of design and synthesis of metal derivatives from MOFs. *Advanced Materials*, *35*, 2210166.

9. Li, B., Wang, H., & Chen, B. (2014). Microporous metal–organic frameworks for gas separation. *Chemistry–An Asian Journal*, *9*(6), 1474–1498.

10. James, S. L. (2003). Metal-organic frameworks. *Chemical Society Reviews*, *32*(5), 276–288.

11. Mageto, T., de Souza, F. M., Kaur, J., Kumar, A., & Gupta, R. K. (2023). Chemistry and potential candidature of metal-organic frameworks for electrochemical energy storage devices. *Fuel Processing Technology*, *242*, 107659.

12. Xu, G., Nie, P., Dou, H., Ding, B., Li, L., & Zhang, X. (2017). Exploring metal organic frameworks for energy storage in batteries and supercapacitors. *Materials Today*, *20*(4), 191–209.

13. Schneemann, A., Bon, V., Schwedler, I., Senkovska, I., Kaskel, S., Fischer, R.A. (2014). Flexible metal–organic frameworks. *Chemical Society Reviews*, *43*(16), 6062–96. Flexible MOF's (berkeley.edu)

14. Kaneti, Y. V., Tang, J., Salunkhe, R. R., Jiang, X., Yu, A., Wu, K. C. W., & Yamauchi, Y. (2017). Nanoarchitectured design of porous materials and nanocomposites from metal-organic frameworks. *Advanced Materials*, *29*(12), 1604898.

15. He, Y., Wang, Z., Wang, H., Wang, Z., Zeng, G., Xu, P., ... & Zhao, Y. (2021). Metal-organic framework-derived nanomaterials in environment related fields: Fundamentals, properties and applications. *Coordination Chemistry Reviews*, *429*, 213618.

16. Zhang, A., Zong, H., Fu, H., Wang, L., Cao, X., Zhong, Y., ... & Liu, J. (2022). Controllable synthesis of nickel doped hierarchical zinc MOF with tunable morphologies for enhanced supercapability. *Journal of Colloid and Interface Science*, *618*, 375–385.

17. Ray, A., Sultana, S., Paramanik, L., & Parida, K. M. (2020). Recent advances in phase, size, and morphology-oriented nanostructured nickel phosphide for overall water splitting. *Journal of Materials Chemistry A*, *8*(37), 19196–19245.

18. Feng, L., Wang, K. Y., Powell, J., & Zhou, H. C. (2019). Controllable synthesis of metal-organic frameworks and their hierarchical assemblies. *Matter*, *1*(4), 801–824.

19. Nano/Micro Metal-Organic Frameworks, Springer Nature Singapore Pte Ltd, ISBN 978-981-16-4071-1 (eBook)

20. Tang, J., & Yamauchi, Y. (2016). MOF morphologies in control. *Nature Chemistry*, *8*(7), 638–639.

21. Lü, Y., Zhan, W., He, Y., Wang, Y., Kong, X., Kuang, Q., ... Zheng, L. (2014). MOF-templated synthesis of porous co3o4 concave nanocubes with high specific surface area and their gas sensing properties. *ACS Applied Materials & Interfaces*, *6*(6), 4186–4195. doi:10.1021/am405858v

22. Li, B., Wang, H., & Chen, B. (2014). Microporous metal–organic frameworks for gas separation. *Chemistry–An Asian Journal*, *9*(6), 1474–1498.

23. Dolgopolova, E. A., Rice, A. M., Martin, C. R., & Shustova, N. B. (2018). Photochemistry and photophysics of MOFs: Steps towards MOF-based sensing enhancements. *Chemical Society Reviews*, *47*(13), 4710–4728.

24. Shen, M., Forghani, F., Kong, X., Liu, D., Ye, X., Chen, S., & Ding, T. (2020). Antibacterial applications of metal–organic frameworks and their composites. *Comprehensive Reviews in Food Science and Food Safety*, *19*(4), 1397–1419.

25. Ye, Z., Jiang, Y., Li, L., Wu, F., & Chen, R. (2021). Rational design of MOF-based materials for next-generation rechargeable batteries. *Nano-Micro Letters*, *13*, 1–37.

26. Liu, J., Zhu, D., Guo, C., Vasileff, A., & Qiao, S. Z. (2017). Design strategies toward advanced MOF-derived electrocatalysts for energy-conversion reactions. *Advanced Energy Materials*, *7*(23), 1700518.

27. Stock, N., & Biswas, S. (2012). Synthesis of metal-organic frameworks (MOFs): routes to various MOF topologies, morphologies, and composites. *Chemical Reviews*, *112*(2), 933–969.

28. Khan, M. S., & Shahid, M. (2022). Synthesis of metal-organic frameworks (MOFs): routes to various MOF topologies, morphologies, and composites. In *Electrochemical Applications of Metal-Organic Frameworks* (pp. 17–35). Elsevier.

29. Wang, H., Zhu, Q. L., Zou, R., & Xu, Q. (2017). Metal-organic frameworks for energy applications. *Chem*, *2*(1), 52–80.

30. Ahmed, I., & Jhung, S. H. (2014). Composites of metal–organic frameworks: Preparation and application in adsorption. *Materials Today*, *17*(3), 136–146.

31. McKinstry, C., Cathcart, R.J., Cussen, E.J., Fletcher, A.J., Patwardhan, S.V. Sefcik, J. (2016). Scalable continuous solvothermal synthesis of metal organic framework (MOF-5) crystals. *Chemical Engineering Journal*, *285*, 718–725.

32. Maka, V. K., Tamuly, P., Jindal, S., & Moorthy, J. N. (2020). Control of In-MOF topologies and tuning of porosity through ligand structure, functionality and interpenetration: Selective cationic dye exchange. *Applied Materials Today*, *19*, 100613.

33. Xiang, Z., Cao, D., Shao, X., Wang, W., Zhang, J., & Wu, W. (2010). Facile preparation of high-capacity hydrogen storage metal-organic frameworks: A combination of microwave-assisted solvothermal synthesis and supercritical activation. *Chemical Engineering Science*, *65*(10), 3140–3146.

34. Nguyen, V. H., Nguyen, T. D., & Van Nguyen, T. (2020). Microwave-assisted solvothermal synthesis and photocatalytic activity of bismuth (III) based metal–organic framework. *Topics in Catalysis*, *63*, 1109–1120.

35. Zhang, Z., Tao, C. A., Zhao, J., Wang, F., Huang, J., & Wang, J. (2020). Microwave-assisted solvothermal synthesis of UiO-66-NH2 and its catalytic performance toward the hydrolysis of a nerve agent simulant. *Catalysts*, *10*(9), 1086.

36. Ni, Z., & Masel, R. I. (2006). Rapid production of metal– organic frameworks via microwave-assisted solvothermal synthesis. *Journal of the American Chemical Society*, *128*(38), 12394–12395.

37. Lee, Y. R., Kim, J., & Ahn, W. S. (2013). Synthesis of metal-organic frameworks: A mini review. *Korean Journal of Chemical Engineering*, *30*, 1667–1680.

38. Qian, Y., Zhang, F., & Pang, H. (2021). A review of MOFs and their composites-based photocatalysts: synthesis and applications. *Advanced Functional Materials*, *31*(37), 2104231.

39. Chen, X., Peng, X., Jiang, L., Yuan, X., Yu, H., Wang, H., ... & Xia, Q. (2020). Recent advances in titanium metal–organic frameworks and their derived materials: Features, fabrication, and photocatalytic applications. *Chemical Engineering Journal*, *395*, 125080.

40. Cao, X., Tan, C., Sindoro, M., & Zhang, H. (2017). Hybrid micro-/nano-structures derived from metal–organic frameworks: Preparation and applications in energy storage and conversion. *Chemical Society Reviews*, *46*(10), 2660–2677.

41. Heidari, M., Dutta, A., Acharya, B., & Mahmud, S. (2019). A review of the current knowledge and challenges of hydrothermal carbonization for biomass conversion. *Journal of the Energy Institute*, *92*(6), 1779–1799.

42. Collins Jr, M. J. (2010). Future trends in microwave synthesis. *Future Medicinal Chemistry*, 2(2), 151–155.

43. Choi, J. S., Son, W. J., Kim, J., & Ahn, W. S. (2008). Metal–organic framework MOF-5 prepared by microwave heating: Factors to be considered. *Microporous and Mesoporous Materials*, 116(1–3), 727–731.

44. Han, X., Tao, K., Ma, Q., & Han, L. (2018). Microwave-assisted synthesis of pillared Ni-based metal–organic framework and its derived hierarchical NiO nanoparticles for supercapacitors. *Journal of Materials Science: Materials in Electronics*, 29, 14697–14704.

45. Hirai, Y., Nishiumi, N., Sun, H., Matsushima, Y., Masuhara, A., Shiroishi, H., & Yoshida, T. (2018, July). Microwave synthesis of Zn-Terephthalate MOF as anode material for redox batteries. In *Electrochemical Society Meeting Abstracts 4dms18* (No. 4, pp. 233–233). The Electrochemical Society, Inc.

46. Zhuang, J. L., Ceglarek, D., Pethuraj, S., & Terfort, A. (2011). Rapid room-temperature synthesis of metal–organic framework HKUST-1 crystals in bulk and as oriented and patterned thin films. *Advanced Functional Materials*, 21(8), 1442–1447.

47. Getachew, N., Chebude, Y., Diaz, I., & Sanchez-Sanchez, M. (2014). Room temperature synthesis of metal organic framework MOF-2. *Journal of Porous Materials*, 21, 769–773.

48. Park, H., Reddy, D. A., Kim, Y., Ma, R., Choi, J., Kim, T. K., & Lee, K. S. (2016). Zeolitic imidazolate framework-67 (ZIF-67) rhombic dodecahedrons as full-spectrum light harvesting photocatalyst for environmental remediation. *Solid State Sciences*, 62, 82–89.

49. Qiu, L. G., Li, Z. Q., Wu, Y., Wang, W., Xu, T., & Jiang, X. (2008). Facile synthesis of nanocrystals of a microporous metal–organic framework by an ultrasonic method and selective sensing of organoamines. *Chemical Communications*, 31, 3642–3644.

50. Qu, Y., Shi, C., Cao, H., & Wang, Y. (2020). Synthesis of Ni-MOF/Ti3C2Tx hybrid nanosheets via ultrasonific method for supercapacitor electrodes. *Materials Letters*, 280, 128526.

51. Zhou, X., Jin, H., Xia, B. Y., Davey, K., Zheng, Y., & Qiao, S. Z. (2021). Molecular cleavage of metal-organic frameworks and application to energy storage and conversion. *Advanced Materials*, 33(51), 2104341.

52. Alfonso-Herrera, L. A., Torres-Martinez, L. M., & Mora-Hernandez, J. M. (2022). Novel strategies to tailor the photocatalytic activity of metal—organic frameworks for hydrogen generation: a mini-review. *Frontiers in Energy*, 16, 1–13.

53. Subudhi, S., Tripathy, S. P., & Parida, K. (2021). Metal oxide integrated metal organic frameworks (MO@ MOF): Rational design, fabrication strategy, characterization and emerging photocatalytic applications. *Inorganic Chemistry Frontiers*, 8(6), 1619–1636.

54. Zhu, J., Li, P. Z., Guo, W., Zhao, Y., & Zou, R. (2018). Titanium-based metal–organic frameworks for photocatalytic applications. *Coordination Chemistry Reviews*, 359, 80–101.

55. Li, R., Zhang, W., & Zhou, K. (2018). Metal–organic-framework-based catalysts for photoreduction of CO2. *Advanced Materials*, 30(35), 1705512.

56. Wang, W., Xu, X., Zhou, W., & Shao, Z. (2017). Recent progress in metal-organic frameworks for applications in electrocatalytic and photocatalytic water splitting. *Advanced Science*, 4(4), 1600371.

57. Cadiau, A., Kolobov, N., Srinivasan, S., Goesten, M. G., Haspel, H., Bavykina, A. V., ... & Gascon, J. (2020). A titanium metal–organic framework with visible-light-responsive photocatalytic activity. *Angewandte Chemie*, 132(32), 13570–13574.

58. Subramanian, V., Wolf, E., & Kamat, P. V. (2001). Semiconductor– metal composite nanostructures. To what extent do metal nanoparticles improve the photocatalytic activity of TiO2 films?. *Journal of Physical Chemistry B, 105*(46), 11439–11446.

59. Shen, L., Luo, M., Huang, L., Feng, P., & Wu, L. (2015). A clean and general strategy to decorate a titanium metal–organic framework with noble-metal nanoparticles for versatile photocatalytic applications. *Inorganic Chemistry, 54*(4), 1191–1193.

60. Ma, Y., Zhou, K., Dong, H., & Zhou, X. (2021). Effects of adsorbing noble metal single atoms on the electronic structure and photocatalytic activity of Ta3N5. *Journal of Physical Chemistry C, 125*(32), 17600–17611.

61. Han, Y., Xu, H., Su, Y., Xu, Z. L., Wang, K., & Wang, W. (2019). Noble metal (Pt, Au@Pd) nanoparticles supported on metal organic framework (MOF-74) nanoshuttles as high-selectivity CO2 conversion catalysts. *Journal of Catalysis, 370,* 70–78.

62. Ren, X., Wei, S., Wang, Q., Shi, L., Wang, X. S., Wei, Y., ... & Ye, J. (2021). Rational construction of dual cobalt active species encapsulated by ultrathin carbon matrix from MOF for boosting photocatalytic H2 generation. *Applied Catalysis B: Environmental, 286,* 119924.

63. Yu, H., Tian, J., Chen, F., Wang, P., & Wang, X. (2015). Synergistic effect of dual electron-cocatalysts for enhanced photocatalytic activity: RGO as electron-transfer mediator and Fe (III) as oxygen-reduction active site. *Scientific Reports, 5*(1), 13083.

64. Su, Y. (2018). *Copper Oxide Nano Photocatalyst for Wastewater Purification using Visible Light* (Doctoral dissertation, University of Cambridge).

65. Rizzi, G. A., & Girardi, L. (2022). Metal oxides for photoelectrochemical fuel production. In *Tailored Functional Oxide Nanomaterials: From Design to Multi-Purpose Applications*, Edited by C. Maccato and D. Barreca (pp 339–377), Wiley-VCH.

66. Irtem, I. E. (2017). *Production of Solar Fuels By Photoelectrochemical Conversion of Carbon Dioxide* (Doctoral dissertation, Universitat de Barcelona).

67. Kafizas, A., Kellici, S., Darr, J. A., & Parkin, I. P. (2009). Titanium dioxide and composite metal/metal oxide titania thin films on glass: a comparative study of photocatalytic activity. *Journal of Photochemistry and Photobiology A: Chemistry, 204*(2–3), 183–190.

68. Nolan, M. (2011). Surface modification of TiO 2 with metal oxide nanoclusters: a route to composite photocatalytic materials. *Chemical Communications, 47*(30), 8617–8619.

69. Liu, Y., Huang, D., Cheng, M., Liu, Z., Lai, C., Zhang, C., ... & Liang, Q. (2020). Metal sulfide/MOF-based composites as visible-light-driven photocatalysts for enhanced hydrogen production from water splitting. *Coordination Chemistry Reviews, 409,* 213220.

70. Peng, Y., Guo, Z., Yang, J., Wang, D., & Yuan, W. (2014). Enhanced photocatalytic H 2 evolution over micro-SiC by coupling with CdS under visible light irradiation. *Journal of Materials Chemistry A, 2*(18), 6296–6300.

71. Wang, X., Zhang, G., Yin, W., Zheng, S., Kong, Q., Tian, J., & Pang, H. (2022). Metal–organic framework-derived phosphide nanomaterials for electrochemical applications. *Carbon Energy, 4*(2), 246–281.

72. Kaur, R., Kim, K. H., & Deep, A. (2017). A convenient electrolytic assembly of graphene-MOF composite thin film and its photoanodic application. *Applied Surface Science, 396,* 1303–1309.

73. Chronopoulos, D. D., Saini, H., Tantis, I., Zbořil, R., Jayaramulu, K., & Otyepka, M. (2022). Carbon nanotube based metal–organic framework hybrids from fundamentals toward applications. *Small, 18*(4), 2104628.

74. Abdi, J., Banisharif, F., & Khataee, A. (2021). Amine-functionalized Zr-MOF/CNTs nanocomposite as an efficient and reusable photocatalyst for removing organic contaminants. *Journal of Molecular Liquids*, *334*, 116129.

75. Bazazi, S., Hosseini, S. P., Hashemi, E., Rashidzadeh, B., Liu, Y., Saeb, M. R., ... & Seidi, F. (2023). Polysaccharide-based C-dots and polysaccharide/C-dot nanocomposites: fabrication strategies and applications. *Nanoscale*, *15*(8), 3630–3650.

76. Xiang, Q., Yu, J., & Jaroniec, M. (2011). Preparation and enhanced visible-light photocatalytic H2-production activity of graphene/C3N4 composites. *The Journal of Physical Chemistry C*, *115*(15), 7355–7363.

77. Ramasubbu, V., Kumar, P. R., Chellapandi, T., Madhumitha, G., Mothi, E. M., & Shajan, X. S. (2022). Zn (II) porphyrin sensitized (TiO2@ Cd-MOF) nanocomposite aerogel as novel photocatalyst for the effective degradation of methyl orange (MO) dye. *Optical Materials*, *132*, 112558.

78. Chattopadhyay, P. K., & Singha, N. R. (2021). MOF and derived materials as aerogels: Structure, property, and performance relations. *Coordination Chemistry Reviews*, *446*, 214125.

79. Saini, H., Srinivasan, N., Sedajova, V., Majumder, M., Dubal, D. P., Otyepka, M., ... & Jayaramulu, K. (2021). Emerging MXene@ Metal–organic framework hybrids: Design strategies toward versatile applications. *ACS Nano*, *15*(12), 18742–18776.

80. Majewski, M. B., Peters, A. W., Wasielewski, M. R., Hupp, J. T., & Farha, O. K. (2018). Metal–organic frameworks as platform materials for solar fuels catalysis. *ACS Energy Letters*, *3*(3), 598–611.

81. Xiao, J. D., Li, R., & Jiang, H. L. (2023). Metal–Organic framework-based photo-catalysis for solar fuel production. *Small Methods*, *7*(1), 2201258.

82. Kampouri, S., Nguyen, T. N., Spodaryk, M., Palgrave, R. G., Züttel, A., Smit, B., & Stylianou, K. C. (2018). Concurrent photocatalytic hydrogen generation and dye degradation using MIL-125-NH2 under visible light irradiation. *Advanced Functional Materials*, *28*(52), 1806368.

83. Yang, H., Wang, J., Ma, J., Yang, H., Zhang, J., Lv, K., ... & Peng, T. (2019). A novel BODIPY-based MOF photocatalyst for efficient visible-light-driven hydrogen evolution. *Journal of Materials Chemistry A*, *7*(17), 10439–10445.

84. Li, R., Jiang, J., Liu, Q., Xie, Z., & Zhai, J. (2018). Hybrid nanochannel membrane based on polymer/MOF for high-performance salinity gradient power generation. *Nano Energy*, *53*, 643–649.

85. Peng, Y., Zhou, T., Ma, J., Bai, Y., Cao, S., & Pang, H. (2022). Metal-organic frame-work (MOF) composites as promising materials for energy storage applications. *Advances in Colloid and Interface Science*, *307*, 102732.

86. Wang, L., Han, Y., Feng, X., Zhou, J., Qi, P., & Wang, B. (2016). Metal–organic frameworks for energy storage: Batteries and supercapacitors. *Coordination Chemistry Reviews*, *307*, 361–381.

87. Liu, J., Wang, J., Xu, C., Jiang, H., Li, C., Zhang, L., ... & Shen, Z. X. (2018). Advanced energy storage devices: Basic principles, analytical methods, and rational materials design. *Advanced Science*, *5*(1), 1700322.

88. Wang, Y., Song, Y., & Xia, Y. (2016). Electrochemical capacitors: Mechanism, materials, systems, characterization and applications. *Chemical Society Reviews*, *45*(21), 5925–5950.

89. Shen, M., & Ma, H. (2022). Metal-organic frameworks (MOFs) and their derivative as electrode materials for lithium-ion batteries. *Coordination Chemistry Reviews*, *470*, 214715.

90. Bello, M. M., Asghar, A., & Raman, A. A. A. (2019). Metal-organic frameworks for supercapacitors. In *Inorganic Nanomaterials for Supercapacitor Design* (pp. 277–305). CRC Press.

91. Young, C., Kim, J., Kaneti, Y. V., & Yamauchi, Y. (2018). One-step synthetic strategy of hybrid materials from bimetallic metal–organic frameworks for supercapacitor applications. *ACS Applied Energy Materials, 1*(5), 2007–2015.

92. Li, G. C., Hua, X. N., Liu, P. F., Xie, Y. X., & Han, L. (2015). Porous Co3O4 microflowers prepared by thermolysis of metal-organic framework for supercapacitor. *Materials Chemistry and Physics, 168,* 127–131.

93. Salunkhe, R. R., Kaneti, Y. V., & Yamauchi, Y. (2017). Metal–organic framework-derived nanoporous metal oxides toward supercapacitor applications: progress and prospects. *ACS Nano, 11*(6), 5293–5308.

94. Sajjad, M., & Lu, W. (2021). Covalent organic frameworks based nanomaterials: Design, synthesis, and current status for supercapacitor applications: A review. *Journal of Energy Storage, 39,* 102618.

95. Wu, S., Liu, J., Wang, H., & Yan, H. (2019). A review of performance optimization of MOF-derived metal oxide as electrode materials for supercapacitors. *International Journal of Energy Research, 43*(2), 697–716

96. Dai, S., Han, F., Tang, J., & Tang, W. (2019). MOF-derived Co3O4 nanosheets rich in oxygen vacancies for efficient all-solid-state symmetric supercapacitors. *Electrochimica Acta, 328,* 135103.

97. Ren, J., Huang, Y., Zhu, H., Zhang, B., Zhu, H., Shen, S., ... & Liu, Q. (2020). Recent progress on MOF-derived carbon materials for energy storage. *Carbon Energy, 2*(2), 176–202.

98. Xu, B., Zhang, H., Mei, H., & Sun, D. (2020). Recent progress in metal-organic framework-based supercapacitor electrode materials. *Coordination Chemistry Reviews, 420,* 213438.

99. Liu, Y., Xu, X., & Shao, Z. (2020). Metal-organic frameworks derived porous carbon, metal oxides and metal sulfides-based compounds for supercapacitors application. *Energy Storage Materials, 26,* 1–22.

100. Choudhary, R. B., Ansari, S., & Majumder, M. (2021). Recent advances on redox active composites of metal-organic framework and conducting polymers as pseudocapacitor electrode material. *Renewable and Sustainable Energy Reviews, 145,* 110854.

101. Wang, M., Ma, Y., & Ye, J. (2020). Controllable layer-by-layer assembly of metal-organic frameworks/polyaniline membranes for flexible solid-state microsupercapacitors. *Journal of Power Sources, 474,* 228681.

102. Srinivasan, R., Elaiyappillai, E., Nixon, E. J., Lydia, I. S., & Johnson, P. M. (2020). Enhanced electrochemical behaviour of Co-MOF/PANI composite electrode for supercapacitors. *Inorganica Chimica Acta, 502,* 119393.

103. Athira, A. R., Raj, B. B., & Xavier, T. S. (2021). Microwave-Induced polyindole on cobalt MOF-Electrodes for high-performance supercapacitors. *Journal of the Electrochemical Society, 168*(12), 120524.

104. Zhu, C., He, Y., Liu, Y., Kazantseva, N., Saha, P., & Cheng, Q. (2019). ZnO@ MOF@ PANI core-shell nanoarrays on carbon cloth for high-performance supercapacitor electrodes. *Journal of Energy Chemistry, 35,* 124–131.

105. Liu, Y., Xu, N., Chen, W., Wang, X., Sun, C., & Su, Z. (2018). Supercapacitor with high cycling stability through electrochemical deposition of metal–organic frameworks/polypyrrole positive electrode. *Dalton Transactions, 47*(38), 13472–13478.

106. Zha, X., Wu, Z., Cheng, Z., Yang, W., Li, J., Chen, Y., ... & Yang, Y. (2021). High performance energy storage electrodes based on 3D Z-CoO/RGO nanostructures for supercapacitor applications. *Energy*, *220*, 119696.

107. Sundriyal, S., Shrivastav, V., Mishra, S., & Deep, A. (2020). Enhanced electrochemical performance of nickel intercalated ZIF-67/rGO composite electrode for solid-state supercapacitors. *International Journal of Hydrogen Energy*, *45*(55), 30859–30869.

108. Borhani, S., Moradi, M., Kiani, M. A., Hajati, S., & Toth, J. (2017). CoxZn1−×ZIF-derived binary Co3O4/ZnO wrapped by 3D reduced graphene oxide for asymmetric supercapacitor: Comparison of pure and heat-treated bimetallic MOF. *Ceramics International*, *43*(16), 14413–14425.

109. Maiti, S., Pramanik, A., & Mahanty, S. (2014). Extraordinarily high pseudocapacitance of metal organic framework derived nanostructured cerium oxide. *Chemical Communications*, *50*(79), 11717–11720.

110. Mehek, R., Iqbal, N., Noor, T., Amjad, M. Z. B., Ali, G., Vignarooban, K., & Khan, M. A. (2021). Metal–organic framework based electrode materials for lithium-ion batteries: a review. *RSC Advances*, *11*(47), 29247–29266.

111. Reddy, M. V., Subba Rao, G. V. & Chowdari, B. V. R., 2013. Metal oxides and oxysalts as anode materials for Li ion batteries. *Chemical Reviews*, *113*(7), 5364–5457.

112. Liu, S., Wang, Z., Yu, C., Wu, H. B., Wang, G., Dong, Q., Qiu, J., Eychmüller, A. & Lou, X. W., 2013. A flexible TiO2 (B)-based battery electrode with superior power rate and ultralong cycle life. *Advanced Materials*, *25*(25), 3462–3467.

113. Cao, X., Zheng, B., Rui, X., Shi, W., Yan, Q. & Zhang, H., 2014. Metal oxide-coated three-dimensional graphene prepared by the use of metal–organic frameworks as precursors. *Angewandte Chemie International Edition*, *53*(5), 1404–1409.

114. Hu, L., Yan, N., Chen, Q., Zhang, P., Zhong, H., Zheng, X., Li, Y. & Hu, X., 2012. Fabrication based on the kirkendall effect of Co3O4 porous nanocages with extraordinarily high capacity for lithium storage. *Chemistry–A European Journal*, *18*(29), 8971–8977.

115. Fang, G., Zhou, J., Liang, C., Pan, A., Zhang, C., Tang, Y., Tan, X., Liu, J. & Liang, S., 2016. MOFs nanosheets derived porous metal oxide-coated three-dimensional substrates for lithium-ion battery applications. *Nano Energy*, *26*, 57–65.

116. Zhang, L., Wu, H. B., Madhavi, S., Hng, H. H. & Lou, X. W., 2012. Formation of Fe2O3 microboxes with hierarchical shell structures from metal–organic frameworks and their lithium storage properties. *Journal of the American Chemical Society*, *134*(42), 17388–17391.

117. Jin, R., Zhou, J., Guan, Y., Liu, H. & Chen, G., 2014. Mesocrystal Co 9 S 8 hollow sphere anodes for high performance lithium ion batteries. *Journal of Materials Chemistry A*, *2*(33), 13241–13244.

118. Yin, D., Huang, G., Zhang, F., Qin, Y., Na, Z., Wu, Y. & Wang, L., 2016. Coated/sandwiched rGO/CoSx composites derived from metal–organic frameworks/GO as advanced anode materials for lithium-ion batteries. *Chemistry–A European Journal*, *22*(4), 1467–1474.

119. Ong, S. P., Chevrier, V. L., Hautier, G., Jain, A., Moore, C., Kim, S., Ma, X. & Ceder, G., 2011. Voltage, stability and diffusion barrier differences between sodium-ion and lithium-ion intercalation materials. *Energy & Environmental Science*, *4*(9), 3680–3688.

120. Pan, H., Lu, X., Yu, X., Hu, Y. S., Li, H., Yang, X. Q. & Chen, L., 2013. Sodium storage and transport properties in layered Na2Ti3O7 for room-temperature sodium-ion batteries. *Advanced Energy Materials*, *3*(9), 1186–1194.

121. Lee, K. T., Ramesh, T. N., Nan, F., Botton, G. & Nazar, L. F., 2011. Topochemical synthesis of sodium metal phosphate olivines for sodium-ion batteries. *Chemistry of Materials*, *23*(16), 3593–3600.

122. Ellis, B. L. & Nazar, L. F., 2012. Sodium and sodium-ion energy storage batteries. *Current Opinion in Solid State and Materials Science*, *16*(4), 168–177.

123. Song, M. K., Park, S., Alamgir, F. M., Cho, J. & Liu, M., 2011. Nanostructured electrodes for lithium-ion and lithium-air batteries: The latest developments, challenges, and perspectives. *Materials Science and Engineering: R: Reports*, *72*(11), 203–252.

124. Bennett, T. D., Goodwin, A. L., Dove, M. T., Keen, D. A., Tucker, M. G., Barney, E. R., Soper, A. K., Bithell, E. G., Tan, J. C. & Cheetham, A. K., 2010. Structure and properties of an amorphous metal-organic framework. *Physical Review Letters*, *104*(11), 115503.

125. Zhang, N., Han, X., Liu, Y., Hu, X., Zhao, Q. & Chen, J., 2015. 3D porous γ-Fe2O3@ C nanocomposite as high-performance anode material of Na-ion batteries. *Advanced Energy Materials*, *5*(5), p.1401123.

126. Jian, Z., Liu, P., Li, F., Chen, M. & Zhou, H., 2014. Monodispersed hierarchical Co 3 O 4 spheres intertwined with carbon nanotubes for use as anode materials in sodium-ion batteries. *Journal of Materials Chemistry A*, *2*(34), 13805–13809.

127. Lu, Y., Zhang, N., Zhao, Q., Liang, J. & Chen, J., 2015. Micro-nanostructured CuO/ C spheres as high-performance anode materials for Na-ion batteries. *Nanoscale*, *7*(6), 2770–2776.

128. Zou, F., Chen, Y. M., Liu, K., Yu, Z., Liang, W., Bhaway, S. M., Gao, M. & Zhu, Y., 2016. Metal organic frameworks derived hierarchical hollow NiO/Ni/graphene composites for lithium and sodium storage. *ACS Nano*, *10*(1), 377–386.

129. Wenzel, S., Hara, T., Janek, J. & Adelhelm, P., 2011. Room-temperature sodium-ion batteries: Improving the rate capability of carbon anode materials by templating strategies. *Energy & Environmental Science*, *4*(9), 3342–3345.

130. Hu, M., Reboul, J., Furukawa, S., Torad, N. L., Ji, Q., Srinivasu, P., Ariga, K., Kitagawa, S. & Yamauchi, Y., 2012. Direct carbonization of Al-based porous coordination polymer for synthesis of nanoporous carbon. *Journal of the American Chemical Society*, *134*(6), 2864–2867.

131. Qu, Q., Yun, J., Wan, Z., Zheng, H., Gao, T., Shen, M., Shao, J. & Zheng, H., 2014. MOF-derived microporous carbon as a better choice for Na-ion batteries than mesoporous CMK-3. *RSC Advances*, *4*(110), 64692–64697.

132. Choi, J. W. & Aurbach, D., 2016. Promise and reality of post-lithium-ion batteries with high energy densities. *Nature Reviews Materials*, *1*(4), 1–16.

133. Morozan, A. & Jaouen, F., 2012. Metal organic frameworks for electrochemical applications. *Energy & Environmental Science*, *5*(11), 9269–9290.

134. Wu, D., Guo, Z., Yin, X., Pang, Q., Tu, B., Zhang, L., Wang, Y. G. & Li, Q., 2014. Metal–organic frameworks as cathode materials for Li–O2 batteries. *Advanced Materials*, *26*(20), 3258–3262.

135. Zhang, Z., Chen, Y., He, S., Zhang, J., Xu, X., Yang, Y., Nosheen, F., Saleem, F., He, W. & Wang, X., 2014. Hierarchical Zn/Ni-MOF-2 nanosheet-assembled hollow nanocubes for multicomponent catalytic reactions. *Angewandte Chemie International Edition*, *53*(46), 12517–12521.

136. Zhang, J., Wang, L., Xu, L., Ge, X., Zhao, X., Lai, M., Liu, Z. & Chen, W., 2015. Porous cobalt–manganese oxide nanocubes derived from metal organic frameworks as a cathode catalyst for rechargeable Li–O 2 batteries. *Nanoscale*, *7*(2), 720–726.

137. Zhang, L., Shi, L., Huang, L., Zhang, J., Gao, R. & Zhang, D., 2014. Rational design of high-performance DeNO×catalysts based on Mn×Co3-x O4 nanocages derived from metal–organic frameworks. *ACS Catalysis*, *4*(6), 1753–1763.

138. Sennu, P., Christy, M., Aravindan, V., Lee, Y. G., Nahm, K. S. & Lee, Y. S., 2015. Two-dimensional mesoporous cobalt sulfide nanosheets as a superior anode for a

Li-ion battery and a bifunctional electrocatalyst for the Li–O2 system. *Chemistry of Materials*, *27*(16), 5726–5735.

139. Xu, G., Ding, B., Pan, J., Nie, P., Shen, L. & Zhang, X., 2014. High performance lithium–sulfur batteries: Advances and challenges. *Journal of Materials Chemistry A*, *2*(32), 12662–12676.

140. Demir-Cakan, R., Morcrette, M., Nouar, F., Davoisne, C., Devic, T., Gonbeau, D., Dominko, R., Serre, C., Férey, G. & Tarascon, J. M., 2011. Cathode composites for Li–S batteries via the use of oxygenated porous architectures. *Journal of the American Chemical Society*, *133*(40), 16154–16160.

141. Wang, Z., Li, X., Cui, Y., Yang, Y., Pan, H., Wang, Z., Wu, C., Chen, B. & Qian, G., 2013. A metal–organic framework with open metal sites for enhanced confinement of sulfur and lithium–sulfur battery of long cycling life. *Crystal Growth & Design*, *13*(11), 5116–5120.

142. Wang, Z., Dou, Z., Cui, Y., Yang, Y., Wang, Z. & Qian, G., 2014. Sulfur encapsulated ZIF-8 as cathode material for lithium–sulfur battery with improved cyclability. *Microporous and Mesoporous Materials*, *185*, 92–96.

143. Zheng, J., Tian, J., Wu, D., Gu, M., Xu, W., Wang, C., Gao, F., Engelhard, M. H., Zhang, J. G., Liu, J. & Xiao, J., 2014. Lewis acid–base interactions between polysulfides and metal organic framework in lithium sulfur batteries. *Nano Letters*, *14*(5), 2345–2352.

144. Zhou, J., Li, R., Fan, X., Chen, Y., Han, R., Li, W., Zheng, J., Wang, B. & Li, X., 2014. Rational design of a metal–organic framework host for sulfur storage in fast, long-cycle Li–S batteries. *Energy & Environmental Science*, *7*(8), 2715–2724.

145. Zheng, J., Tian, J., Wu, D., Gu, M., Xu, W., Wang, C., Gao, F., Engelhard, M. H., Zhang, J. G., Liu, J. & Xiao, J., 2014. Lewis acid–base interactions between polysulfides and metal organic framework in lithium sulfur batteries. *Nano Letters*, *14*(5), 2345–2352.

146. Zhou, J., Yu, X., Fan, X., Wang, X., Li, H., Zhang, Y., Li, W., Zheng, J., Wang, B. & Li, X., 2015. The impact of the particle size of a metal–organic framework for sulfur storage in Li–S batteries. *Journal of Materials Chemistry A*, *3*(16), 8272–8275.

147. Xu, G., Ding, B., Shen, L., Nie, P., Han, J. & Zhang, X., 2013. Sulfur embedded in metal organic framework-derived hierarchically porous carbon nanoplates for high performance lithium–sulfur battery. *Journal of Materials Chemistry A*, *1*(14), 4490–4496.

148. Xi, K., Cao, S., Peng, X., Ducati, C., Kumar, R. V. & Cheetham, A. K., 2013. Carbon with hierarchical pores from carbonized metal–organic frameworks for lithium sulphur batteries. *Chemical Communications*, *49*(22), 2192–2194.

149. Wu, H. B., Wei, S., Zhang, L., Xu, R., Hng, H. H. & Lou, X. W., 2013. Embedding sulfur in MOF-derived microporous carbon polyhedrons for lithium–sulfur batteries. *Chemistry–A European Journal*, *19*(33), 10804–10808.

10 Electrospinning as an Efficient Strategy for the Designing and Fabrication of Architectures for Energy Production, Conversion, and Storage Devices

Ailing Song, Hao Tian, and Guoxiu Wang

10.1 INTRODUCTION

Electrospinning is a special kind of electrostatic atomization, which is the process of spinning with a polymer solution under high electrostatic pressure. Electrospinning technology can manipulate materials on the nanoscale to prepare fibers with a diameter of tens to hundreds of nanometers, which makes products with high porosity, large specific surface area, diverse components and uniform diameter distribution. Electrospinning originated over 200 years ago with research on electrostatic atomization processes. In 1745, Bose et al. applied a high potential to the water surface at the end of the capillary and discovered that micro-jets would be ejected from its surface, thereby forming a highly dispersed aerogel (Bose 1745). This phenomenon is caused by an imbalance between the mechanical pressure and the electric field on the surface of the liquid. Subsequently, Cooley and Morton (1902), Zeleny (1914) et al. carried out in-depth research and phenomenon analysis and summary of the process properties respectively, such as the Taylor cone shape formed in the initial stage and the critical voltage. By 1934, the fabrication of fibers by electrospinning began when Formhals filed a patent technology for a device designed to produce polymer fibers using electrostatic repulsion forces, which for the first time detailed the reason why polymers formed jets under the action of high-voltage electrostatic fields (Formhals

DOI: 10.1201/9781003318859-10

1934). However, in the following 60 years, electrospinning has not attracted wide attention, and the research on it is mainly focused on the improvement of the device. Until the 1990s, the Reneker research group of Akron University in the US carried out a comprehensive study on the process and application of electrostatic spinning, which aroused the great interest of scientific researchers. Since then, electrospinning has been rapidly developing as a practical and feasible method to prepare polymers, composites, ceramics and metal microfibers in a wide variety of applications like bio-medicine, filtration, self-cleaning, and so on.

With the need for global renewable energy technology development, researchers are focused on developing and accelerating the implementation of holistic future energy systems that employ goal-driven interconnected technologies to maximize renewable energy generation, storage and conversion. In recent decades, researchers have pioneered innovative, interdisciplinary and integrated research and development using electrospinning for technological advances in molecular, optoelectronic and other mechanical capacities, conversion and energy storage systems. It is found that the properties of polymer solution such as concentration, viscosity and conductivity, process parameters such as voltage, perfusion speed and nozzle diameter, and environmental parameters such as temperature and humidity in the process of electrostatic spinning have a great influence on the morphology and characteristic feature of nanofibers. Based on these, a variety of architectures with diameters from nano- to micro-meters, as well as porous, beaded, ribbon, core-shell and three-dimensional hierarchical nanofibers and composites designed and fabricated from them with different characteristics, have been extensively developed to meet the needs of energy applications (Figure 10.1).

Functionalized electrospun nanofibers have great potential in the field of energy generation, storage and conversion applications including solar cells (López-Covarrubias et al. 2019), water splitting (Zhang et al. 2022), metal-ion batteries (Jung et al. 2016), metal-air batteries (Li et al. 2022), fuel cells (Banitaba et al. 2021), and solid-state batteries (Sharma et al. 2022) for electrodes, membrane separators, and electrolytes. Inimitable characteristics of high porosity, interconnectivity, and a large surface-to-volume ratio of nanofibers qualify their claims in the efficiency, capacity, specific energy, and cycle life required for the above applications. Moreover, simple, low-cost, and controllable processing routes make electrospinning an effective nanofabrication technology with great potential. This chapter summarizes the electrospinning procedure and its development in the preparation of nanofibers, with a discussion about the abovementioned energy applications.

10.2 THE DESIGNING AND FABRICATION OF ARCHITECTURES FROM ELECTROSPINNING

To meet the requirements of various energy production, conversion and storage devices and other applications, more and more electrospun polymer fibers and mats with various adjustable architectures have been manufactured by adjusting the composition of electrospinning solution, electric field intensity and other parameters, and changing the type of catcher in electrospinning device, etc., the resulting structures,

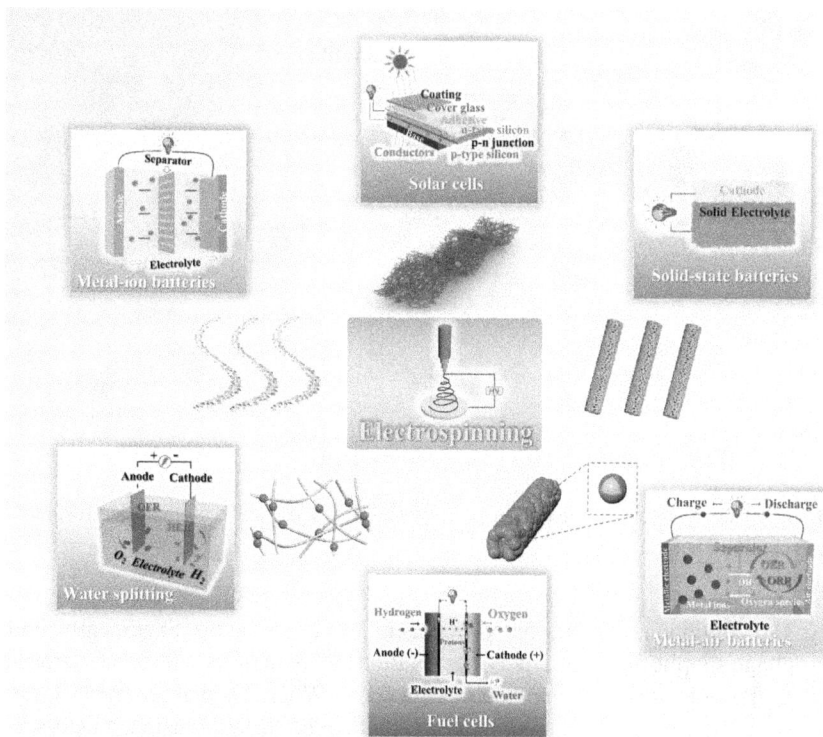

FIGURE 10.1 Pictures showing the schematic illustration of electrospinning technology and applications in energy production, conversion and storage devices. Picture credits: Lei Bai and Shenglu Song.

arrangements and morphology show great diversity. In the following, we will provide an insight into the nanofibrous architectures designed and fabricated from electrospinning, involving individual fibers (porous fibers, beaded fibers, ribbon fibers, core-shell fibers, and others) and fiber mats (random fibers, aligned fibers, patterned fibers, 3-D fibrous and other miscellaneous architectures).

10.2.1 SET-UP AND PROCEDURE OF ELECTROSPINNING

Electrostatic atomization is mainly used for paint atomization, which is the phenomenon of paint droplets splitting into fine particles under the action of the high-voltage electrostatic field. Electrospinning is a special case of electrostatic atomization. After entering the 21st century, electrospinning technology has entered a stage of high-speed development, and its mechanisms, laws and explanations have gradually formed and been revealed. The setup usually consists of four major components: a high-voltage power supply, a syringe pump, a spinneret, and a grounded conductive collector (Figure 10.2). Nanofiber production relies on electrostatic repulsion between surface

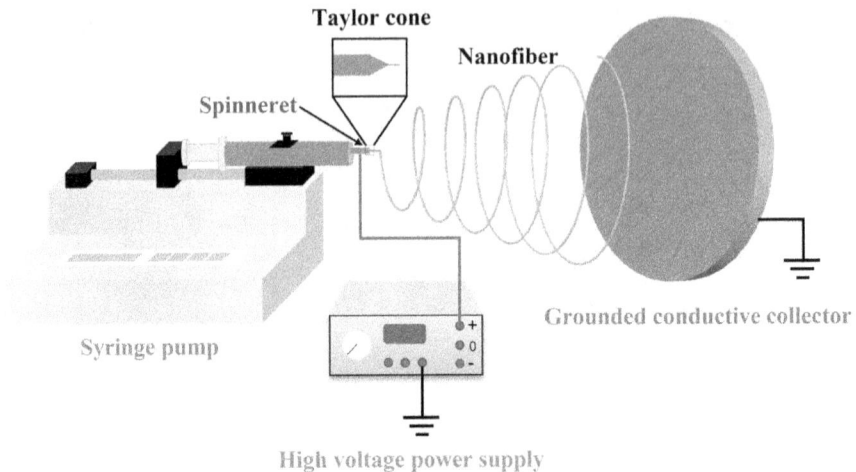

FIGURE 10.2 Schematic illustration of a typical setup for conventional electrospinning. Picture credits: Shenglu Song.

charges to continuously pump the viscoelastic jet out of the tip. Specifically, under high-voltage supply, droplets of polymer solution at the tip of the needle are deformed into a Taylor cone under strong electrostatic force. When the charge-repulsive force on the surface of the polymer solution is greater than its surface tension, a high-speed polymer jet will be formed on the surface of the Taylor cone at the tip of the conical drop. The jet will undergo deformation, volatilization and solidification under the action of electric field force, and finally deposit on the collector to form nanofibers.

Based on the above explanation of electrospinning, Table 10.1 illustrates the set-up parameters and procedure conditions in the electrospinning, which can influence fiber generation and nanostructure. As listed in the table, the diameter, morphology and structure of nanofibers can be adjusted by changing these parameters and conditions, especially applied voltage, solution flow rate and tip-to-collector distance.

In addition, modifications on the set-up like spinneret (including coaxial spinneret, triaxial spinneret and multichannel spinneret), by which polymer solution can be dispensed, will also affect some of the above parameters and conditions to prepare nanofibers with various structures (Yang et al. 2018). On this basis, after processing through the component adjustment of the polymer solution, manipulating the stacking pattern, and the combination of other methods, the nanofiber with secondary architectures can be obtained. We will briefly introduce polymer nanofibers and their secondary architectures in the following chapters.

10.2.2 ELECTROSPUN NANOFIBERS

Here, electrospun nanofibers refer to the nanofibers directly prepared by the electrospinning process, including organic nanofibers, inorganic nanofibers and their composite nanofibers. Organic nanofibers are mainly prepared in the early stages of

TABLE 10.1
The Set-up Parameters and Procedure Conditions in the Electrospinning

Events	Parameters	Effects
Set-up parameters	Applied voltage, solution pump rate, tip-to-collector distance, collector substrate, tip diameter, etc.	The voltage affects the surface charge density of the jet, the solution pump rate affects the stability of the Taylor cone and the diameter and solution amount of the jet, the tip-to-collector distance affects the jet path and the traveling time in the electric field, the collector substrate affects the structure and arrangement of the fiber, the tip diameter affects the morphology and diameter of the fiber, and then affects the morphology, structure, yield and quality of the nanofibers.
Polymer solution properties	Polymer molecular weight, solvent permittivity, conductivity, and volatility, solution concentration, viscosity, surface tension and conductivity, etc.	Polymer molecular weight affects the physical and chemical properties of the solution, solvent permittivity affects the thickness distribution of fiber, the conductivity affects the diameter, the volatility affects the solidification, solution concentration and viscosity affect the molecular entanglement, surface tension affects the shape of the jet, and conductivity affects the stretching effect of the jet, and then affects the morphology and structure of the nanofibers.
Environmental parameters	Humidity, temperature, etc.	Temperature affects the conductivity, viscosity and surface tension of the solution, and humidity affects the compatibility of the solvent, thus affecting the formation and morphology of the nanofibers.

the rise of electrospinning, which are divided into natural polymeric nanofibers and synthetic polymeric nanofibers. Inorganic nanofibers include metal nanofibers, oxide nanofibers, carbon and carbide nanofibers, etc. Based on massive works of literature put into the development of electrospun nanofibers, we have carried on the collation in this writing.

Natural polymeric nanofibers are mostly prepared from polysaccharides, proteins, polyesters, etc., due to good biocompatibility, degradability and non-toxicity, they have a wide range of applications in biology and medical fields, not detailed here. However, owing to the strong solubility of the synthetic polymers, a wide range of adjustable polymer solutions facilitates electrospun nanofibers with outstanding physical and chemical properties. This makes a large variety of synthetic polymeric nanofibers widely used in the energy field. According to different preparation methods, synthetic polymeric nanofibers can be divided into two types via solution spinning and melt spinning, respectively. According to the component distribution, it can be divided into single-component synthetic polymeric nanofibers and multicomponent synthetic polymeric nanofibers. Based on the properties of energy applications and the structure-activity relationships of different components for energy production, storage and conversion, we focus on this category.

A single variety of polymers, such as organic soluble polymers and bio-degradable polymers, are well developed to prepare single-component synthetic polymeric nanofibers. Organic soluble polymers, like polyacrylonitrile (PAN), polymethylmethacrylate (PMMA), and polyvinyl pyrrolidone (PVP), have been synthesized by mature spinning in dimethylformamide (DMF), tetrahydrofuran (THF) and other common organic solvents and water solution (Table 10.2) (Lin et al. 2010; Nitanan et al. 2012; Tao et al. 2017). There are several examples of organic soluble polymers. Biodegradable polymers (biopolymers) are a special class of "green" polymers with biocompatible, biodegradable and eco-friendly proper-ties that, after achieving their intended purpose, can be decomposed by the bac-terial decomposition process to produce natural by-products, such as gases (CO_2, N_2), water and inorganic salts. Due to the least effect on the rising environmental carbon footprint, research in the 21st century aims to develop advanced biopolymers to address environmental issues. With continuous efforts of researchers, biodegrad-able polymers will be applied to everything in commercial and industrial fields from food packaging to agriculture and healthcare (Agarwal, 2020). But because these polymers are usually synthesized through complex condensation reactions and have weak crystallinity, their applications in the energy field are still at a difficult stage of resolution. To obtain the outstanding performance of synthetic polymeric nanofibers in practical applications, multicomponent synthetic polymeric nanofibers are usu-ally adopted. When more than two synthetic polymers or components are used to dissolve in the same solvent, blends of synthetic polymer nanofibers can be produced

TABLE 10.2
Constants for Common Solvents

Solvent	Density (g/cm³)	Permittivity	Boiling point (°C)
Water (H_2O)	1.0	78.54~80.2	100
Ethanol (CH_3CH_2OH)	0.798	24.55	78.5
Methanol (CH_3OH)	0.792	32.6	64.7
Dimethylformamide ($HCON(CH_3)_2$)	0.945	36.7	153
Tetrahydrofuran (C_4H_8O)	0.889	7.6	66
Methylbenzene (C_7H_8)	0.867	2.38	110.6
Acetone (CH_3COCH_3)	0.790	20.7	56

by electrospinning. Multi-polymer, such as PS/PAN (Li et al. 2015), PAN/PANI (Qu et al. 2021), PLGA/PEO (Evrova et al. 2016), and single-polymer containing other components, such as PCL containing nanoparticles incorporating both calcium- and silica-based ceramics, can spin multicomponent synthetic polymeric nanofibers (Meka et al. 2019).

For metals, oxides, carbon and carbides, nitrides and other inorganic substances, electrospinning technology will enhance these materials' unique physicochemical properties and characteristics, especially specific surface area, size effects, and surface/interface effects for specific catalysis, fuel cells, and supercapacitor applications. As we all know, silicon-based anode materials have been devoted to batteries. In 2002, Shao et al. prepared silica (SiO_2) nanofibers with tetraethyl orthosilicate (TEOS) and PVA as polymers (Shao et al. 2002). This is the first time that the researchers have achieved the preparation of oxide nanofibers through electrospinning technology, which expands special advantages for the subsequent application of its derived materials in batteries. Typically, the synthesis process first goes through the preparation of a homogeneous polymer solution, then spinning to obtain inorganic/polymer hybrid nanofibers, and finally combined with calcination to remove organic components, and finally obtain inorganic nanofibers. Other common oxide nanofibers include titanium oxide (TiO_2), aluminum oxide (Al_2O_3), iron oxide (Fe_2O_3), zinc oxide (ZnO), etc. The metal oxide nanofibers produced by spinning have a high surface area, electron mobility, thermal stability, mechanical strength and surface defects. In the same period, metal nanofibers, carbon nanofibers, carbide nanofibers and nitride nanofibers also appeared. In addition, carbon nanofibers have excellent electrical conductivity, chemical stability and large specific surface area; metal nanofibers have good electrical conductivity and high-temperature resistance; carbide nanofibers possess high-temperature stability, corrosion resistance and mechanical strength; nitride nanofibers possess high melting point and hardness, chemical stability, corrosion resistance and excellent electrical and thermal conductivity. These respective characteristics make inorganic nanofibers suitable for exclusive applications in the energy fields. In addition, multicomponent inorganic nanofibers, such as ZnSe, CuO/NiO, MoO_2/C, $NiFe_2O_4$/C, Co/TiC/C, TiN/VN, etc., have also been reported to be used in catalysis and lithium-ion batteries. In this way, the composition controllability of electrospinning materials and their derivatives greatly increases their potential application in the energy field.

Furthermore, due to the energy and comprehensive requirements for the properties of various electrospun nanofibers and their derivatives, composite nanofibers from polymer and inorganic have been extensively developed. For example, in order to achieve improved electron transport capacity, but avoid oxidation and aggregation of metal nanoparticles, it can be combined with polymers to form polymer/metal composite nanofibers. Compared with the aforementioned synthesis process, the preparation of these nanofibers usually combines electrospinning technology with other methods, such as sol-gel method, hydrothermal method, deposition method (atomic layer deposition, liquid phase deposition), reduction method, gas-solid phase reaction, etc.

10.2.3 NANOFIBROUS ARCHITECTURES

From the perspective of physical properties, the nanofibers obtained by electrospinning are one-dimensional solid-state linear nanomaterials, with a unique three-dimensional network structure, excellent flexibility, tailored morphology, sub-micron and nano size, high surface area to volume ratio, adjustable pore size, high porosity, and directional strength. Variations in set-up parameters and procedure conditions, and other modifications on the device allow the production of a wide range of nanofibers with different compositions, morphologies and structures, including random orientation (Prabu et al. 2020; Faccini 2011), alignment (Pachón et al. 2014), patterning (Zhang et al. 2007), and porous (Xie et al. 2008), beaded (Li et al. 2014), ribbons (Stanishevsky et al. 2015), core-shell (Lin et al. 2018), hollow (Lee et al. 2014), multichannel tubular, tube-in-tube structures, etc. The structural diversity of the electrospun nanofibers is displayed in Figure 10.3.

The multi-characteristics and versatility are not only related to the properties of electrospun nanofibers and their derivatives but also extend to nanofibrous architectures with functionalization. More importantly, in the view of material synthesis, multifunctional complex architectures can be fabricated facilely and versatilely through electrospinning technology than the traditional "top-down" and "bottom-up" approaches. As a complex nanofibrous architecture, the cross-linked ceramic nanofibrous networks shown in Figure 10.4A were obtained through freeze-drying shaping and pyrolyzing flexible SiO_2 nanofibers with aluminoborosilicate (AlBSi) matrices, which as the basic structural building units are prepared by combining electrospinning technology with sol-gel strategy. The resulting lamellar-structured ceramic nanofibrous aerogels (CNFAs) inherited the individual characteristics and extended the comprehensive properties brought by original nanofibers and orientated nanofibrous architectures, and showed ultralow density (0.15 mg cm^{-3}), superelasticity, low thermal conductivity and fire resistance (Si et al. 2018).

Morphology engineering of constructing carbon-confined structure (encapsulation/coating structure) through electrospinning is an effective strategy to release and restrain the volume expansion in metal-based batteries. As shown in Figure 10.4B, Li et al. successfully synthesized a hierarchical nanodot-in-nanofiber structure. The active Sb_2Se_3 nanocrystallites are dual confined by 0-dimensional and 1-dimensional carbon layers from metal–organic-frameworks (MOFs) and polyacrylonitrile nanofibers (PAN NFs), respectively. Combined with the heat treatment process, the step of electrospinning leads to the uniform distribution of active Sb_2Se_3 nanocrystallites in the porous carbon matrix and dual-type carbon confinement, which can be expanded to accommodate and inhibit dramatic volume expansion during the charge/discharge process (Li et al. 2022). In addition, metal-organic frameworks (MOFs) structured into nanofibers by electrospinning is a typical nanofibrous architecture used for energy applications. For example, Deng et al. fabricated botryoid nanofibrous architectures with yolk-shell MnOx nanoparticles confined in carbonized nanofibers by electrospinning Mn-MOF, which demonstrate outstanding lithium storage performance (Figure 10.4C) (Yang et al. 2019). Furthermore, a novel N-doped MOF-based hierarchical carbon fiber (NPCF) was obtained through carbonizing ZIF-8/PAN electrospun nanofibers (Figure 10.4D). After pyrolysis, the structure

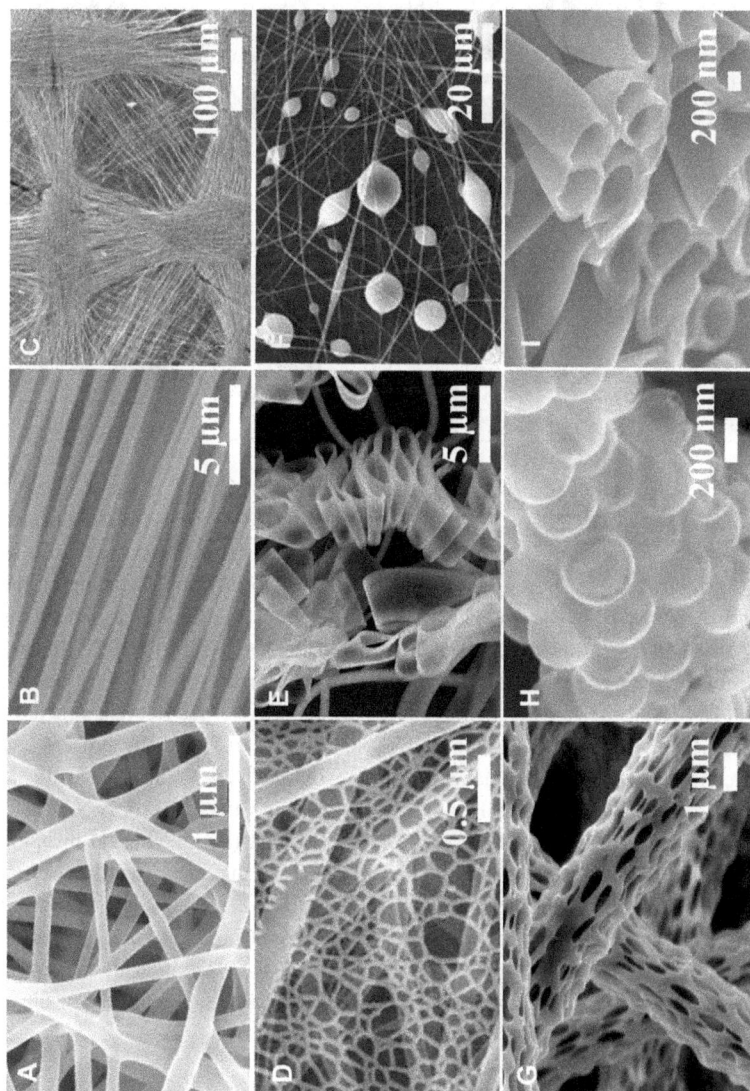

FIGURE 10.3 Structural diversity of the electrospun nanofibers. A. Random orientation from Prabu et al. (2020). B. Alignment structure from Pachón et al. (2014). C. Patterning structure from Zhang et al. (2007). D. Spider-web-like structure from Faccini et al. (2011). E. Ribbon structure from Stanishevsky et al. (2015). F. Beaded structure from Fawal et al. (2020). G. Porous structure from Xie et al. (2008). H. Core-shell structure from Lin et al. (2018). I. Hollow structure from Lee et al. (2014).

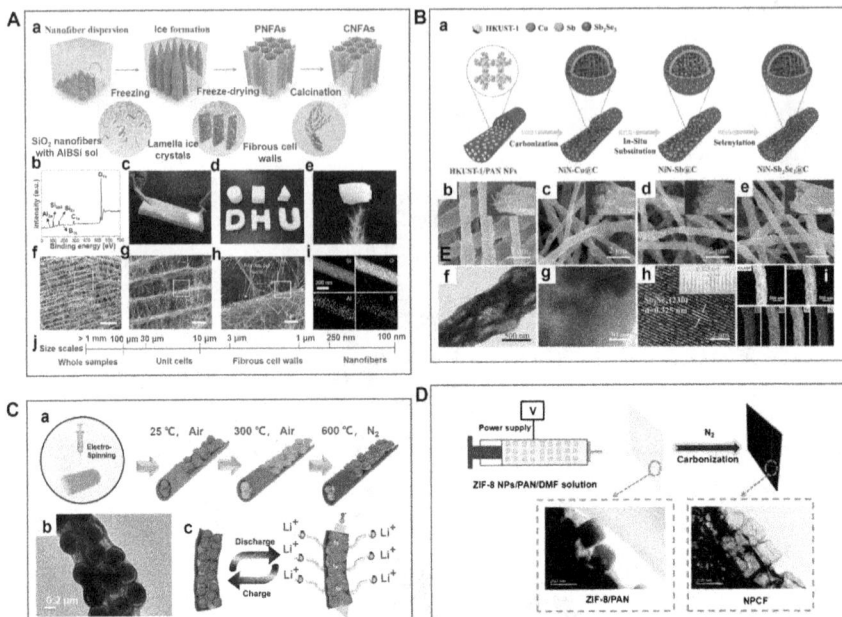

FIGURE 10.4 (A). Structure design and cellular architectures of CNFAs. (A-a) Schematic illustration of the fabrication of CNFAs. (A-b) XPS spectrum of CNFAs. (A-c) A CNFA heated by a butane blowtorch without any damage. (A-d) An optical image of CNFAs with diverse shapes. (A-e) An optical image showing a 20-cm³ CNFA (r=0.15 mg cm⁻³) standing on the tip of a feather. (A-f) to (A-h). Microscopic structure of CNFAs at different magnifications demonstrating the hierarchical nanofibrous cellular architecture. (A-i). STEM-EDS images of a single nanofiber with corresponding elemental mapping images of Si, O, Al, and B, respectively. (A-j). Schematic showing the three levels of hierarchy of the relevant structures (Si et al. 2018). (B). Schematic illustration of the NiN-Sb2Se3@C synthetic process and hierarchical architectures. (B-a). Schematic illustration of the NiN-Sb2Se3@C synthetic process. SEM images of (B-b). HKUST-1/PAN NFs, (B-c). NiN-Cu@C, (B-d). NiN-Sb@C, and (B-e). NiN-Sb2Se3@C. (B-f, g). TEM, (B-h). HRTEM, and (B-i). elemental mapping images of NiN-Sb2Se3@C. Insets in (b–e) are the SEM images of the fractured cross section. Inset in h is the intensity profiles of the d-spacing of Sb₂Se₃ (Li et al. 2022). (C). Schematic fabrication process of ysMnOx@NC and botryoid architectures. (C-a). Schematic fabrication process of ysMnOx@NC. (C-b).TEM images. (C-c). Schematic illustration of the cyclic "breathing" effect (Yang et al. 2019). (D). Preparation process of nanoporous carbon fibers (NCPFs) and hollow structure (Wang et al. 2017).

of end-product NPCF transformed from solid cubic into hollow cubic structure due to the confinement effect (the formation mechanism of the hollow structures is detailed in the original literature). After assembling NPCFS with meso-microporous hierarchical structure and high conductivity into supercapacitors (SCs), which exhibited high gravimetric and volumetric capacitance (Wang et al. 2017).

10.3 APPLICATIONS IN ENERGY PRODUCTION, CONVERSION, AND STORAGE DEVICES

To solve the global energy crisis and environmental pollution problems, the development of efficient and environmentally friendly renewable energy production, storage and conversion devices is one of the most urgent needs we face soon. The involved solar cells, water splitting, metal-ion batteries, metal-air batteries, fuel cells, solid-state batteries and other devices are of great value in solving this problem. With high porosity and specific surface area, good controllable fiber diameter and connected pore structure, electrospun nanofibers are good candidates for enhancing the conductivity of electrons and ions during the operating process of devices, thereby improving their electrochemical performance (Li et al. 2020; Zhang et al. 2022; Xia et al. 2021; Peng et al. 2020). This chapter provides a detailed overview of recent advances in electrospun nanofiber-based energy production, conversion and storage devices, describes the design and fabrication of anode, cathode, electrolyte, and separated nanofibers, and discusses their structural variables and devices' performance.

10.3.1 SOLAR CELLS

A solar cell (photovoltaic cell) is a device that can directly convert the energy of light into electricity through the photovoltaic effect. The development of solar cells has gone through stages from monocrystalline and polycrystalline silicon solar cells, amorphous silicon thin film solar cells to third-generation solar cells with high conversion efficiency (such as dye-sensitized solar cells and perovskite solar cells) and multi-junction solar cells. Dye-sensitized solar cells (DSSCs) have attracted much attention for their advantages of simple packaging, low cost and environmental friendliness. The counter electrode is an important part of DSSCs. Its function is to collect electrons from the external circuit and catalyze the redox reaction of the electrolyte. Composite materials with reduced interface resistance, rich catalytic active sites, and improved photovoltaic performance can be obtained by mixing and spinning active materials with poor conductivity and metal or carbon materials with excellent conductivity. For example, Qiu et al. successfully dispersed Ni and Mo_2C nanoparticles in carbon nanofibers (CNFs) by electrospinning technology and subsequence calcination and prepared composites with different mass ratios of Ni salt and Mo salt (Qiu et al. 2021). The test results show that when $Ni/Mo_2C(2:1)/CNFs$ composite is used as a counter electrode for DSSCs, the photoelectric conversion efficiency is 8.90%, which is higher than that of other $Ni/Mo_2C/CNFs$ counter electrodes and traditional Pt electrodes (8.07%).

10.3.2 WATER SPLITTING

Exploring the earth-abundant, low-cost, and efficient electrocatalysts for water splitting including hydrogen evolution reaction (HER) and oxygen evolution reaction (OER) is of great significance for uninterrupted hydrogen energy production. Electrospinning nanofibers are considered promising catalysts for this application due to their flexible preparation methods, controllable composition, unique one-dimensional characteristics and three-dimensional network structure. In order to meet the comprehensive requirements of catalytic performance, multicomponent and structural construction, electrospinning technology is usually combined with other strategies to prepare catalysts with nanofibrous architectures.

A typical example is that of Hu et al. who successfully established a free-standing CoP@CF electrocatalyst by combining electrospinning with surface metal-organic framework (MOF) functionalization (Figure 10.5A). In the process of electrospinning, a high-voltage power supply was 20 kV, the solution was sprayed at a rate of 0.03 ml min^{-1}, the roller speed rate was 700 rpm, and the collecting distance was placed at 13 cm. By introducing Co-based ZIF-67 into the electrospinning process and changing the phosphorus content in the precursor of electrospinning, the heavily doped carbon nanofibers with different phases of cobalt phosphates displayed prominent functional catalytic properties of HER and OER (Hu et al. 2022). In addition, due to high configuration entropy, lattice distortion, sluggish diffusion, and cocktail effects, high-entropy alloys (HEA) with exceptional properties have attracted extensive attention in catalysis nowadays. By combining the electrospinning technology and graphitization process, Hao et al. developed single-phase FeCoNiXRu solid solution HEA NPs (X = Cr, Mn, and Cu) in carbon nanofibers (FeCoNiXRu/CNFs), which achieved lower overpotentials towards HER and OER, respectively (Hao et al. 2022). As revealed by Figure 10.5B, large amounts of FeCoNiMnRu NPs were densely and uniformly anchored in CNFs, intertwined CNFs exhibited porous three-dimensional (3D) networks, visible lattice fringes and homogeneous distribution of elements were distinctly displayed. In this process, DMF was used as a solvent and PAN was used as polymer, anode voltage was 20 kV, the injection rate was 0.3 ml h^{-1}, and the distance between the collector and needle was 18 cm. Then the CVD process was carried out on the obtained FeCoNiXRu/PAN nanofibers to obtain the final FeCoNiXRu/CNFs. Based on homogeneous structures and uniform composition distribution brought by electrospinning, the significant charge redistribution and optimized energy barriers effectively play a stabilizing effect on the reaction intermediates, which greatly improves the efficiency of water dissociation under alkaline conditions (Figure 10.5C).

10.3.3 Metal-ion Batteries

Metal-ion batteries (MIBs) including Li-ion batteries, Na-ion batteries and Zn-ion batteries have a liquid electrolyte solution sandwiched in between a cathode (positive side) and an anode (negative side). MIBs hold promise for the grid and massive energy storage because they can mitigate the inherent intermittency of renewable energy. In MIBs, most of the highly active electrode materials produce serious volume expansion during the charge/discharge process, which leads to electrode crushing and poor cycle stability, thus hindering their practical application. As stated before, constructing carbon-confined structures through electrospinning is an effective strategy to solve the above issues as well as improve electronic conductivity. In particular, one-dimensional nanofibers with hollow or core-shell architectures can buffer volume changes, inhibit aggregation and provide rapid charge transfer.

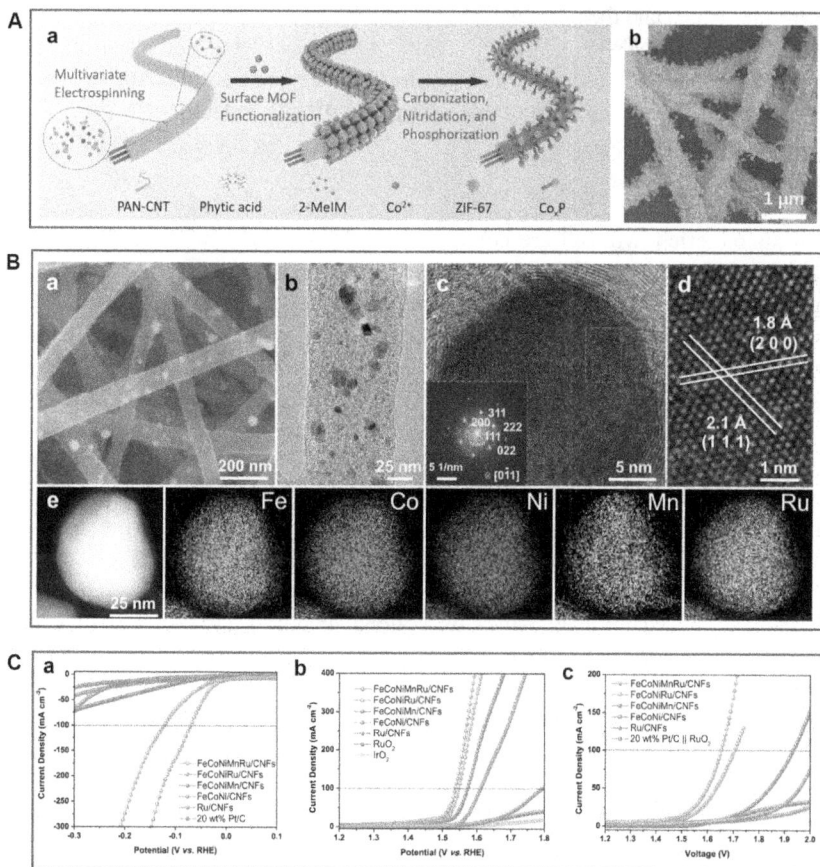

FIGURE 10.5 (A). Scheme for the synthesis of CoxP@CF-900 and SEM images after carbonization (Hu et al. 2022). (B). FE-SEM, TEM and HRTEM images of FeCoNiMnRu/ CNFs (The inset in c is the corresponding FFT pattern of a FeCoNiMnRu HEA NP), HAADF-STEM and the corresponding STEM-EDX mapping images of a FeCoNiMnRu HEA NP supported on CNFs. (C). Electrochemical performance. (C-a). HER polarization curves for the as-prepared electrocatalysts and commercial 20% Pt/C in 1.0 M KOH electrolyte. (C-b). OER polarization curves for the as-prepared electrocatalysts, commercial RuO₂ and IrO₂ in 1.0M KOH electrolyte. (C-c). Polarization curves for full water splitting by the as prepared electrocatalysts in a two-electrode configuration at a scan rate of 2 mV s⁻¹ (Hao et al. 2022).

A functionalized carbon-confined metal oxide nanofiber with a unique cube-in-tube structural construction is synthesized by Liu et al (2017). The multilevel hollow structure consists of poly(vinyl alcohol)-derived partly graphitized carbon nanotubes, randomly distributed manganese oxides in the carbon layer, and centrally distributed amorphous $CoSnO_3$ hollow cubes. The formation process consists of two steps, precursor-modified electrospinning, in which the modified power is $CoSn(OH)_6$

hollow nanocubes and the polymer is high-/middle-/low-molecular-weight poly(vinyl alcohol) (HMW-/MMW-/LMW-PVAs) with manganese acetate and pyrolysis process in air. When assembled as an anode in a Li-ion battery, this architecture delivered an outstanding reversible discharge capacity and a high rate capability.

10.3.4 METAL-AIR BATTERIES

Metal-air batteries (MABs) are composed of a base metal negative electrode and an air positive electrode. The active material of the positive electrode is the oxygen contained in the air, which makes the metal-air battery have a large specific energy density of more than 500 Wh kg kg^{-1}. The negative electrode is made of a metal such as zinc, lithium, aluminum, magnesium, and iron. In order to obtain a long-life service, the oxygen reduction reaction (ORR) and oxygen evolution reaction (OER) of air cathodes should be performed rapidly to improve energy conversion efficiency. Electrospinning technology contributes to high-performance and stable air catalysts. The electrospun nanofibers or nanofibrous architectures with further functionalization can be attached to the electrode substrate as an air catalyst or can be directly used as an independent integrated air cathode for metal-air batteries.

Single-atom catalysts (SACs) have been widely used in energy devices due to their maximum atom utilization efficiency, exceptional electronic structure and unsaturated coordination environment. When spinning them into nanofibers to make freestanding integrated air cathodes, it can not only avoid the use of binders to reduce production costs, but also maximize the utilization rate of metal atoms. Yang et al. (2022) used N, S doped porous carbon nanofiber membranes obtained by electrospinning to perform in situ engineering of highly exposed Fe-N$_4$/Cx sites as binder-free air electrode catalysts for zinc-air batteries, in which the ORR and OER processes are significantly promoted. It should be noted that, after pyrolysis, the introduction of ZnS particles in the electrospinning process will bring a hierarchical macro/meso/microporous structure to the final product. Therefore, combined with the enhanced properties of three phase reaction interfaces and external surface area by the structure construction, the Fe/PCNFs-950-assembled rechargeable zinc-air battery delivered a high specific capacity of 699 mAh g^{-1} at 5 mA cm^{-2}, a large peak power density of 228 mW cm^{-2} and a prolonged cycle life over 1000 h. Recently, Guo et al. (2023) also prepared a carbon nanofiber-supported Co$_2$O$_4$/CeO$_2$ heterostructure by electrospinning, in-situ growth and high-temperature annealing. It was applied to the air electrode of lithium-air batteries, showing excellent specific capacity and long-cycle stability.

In addition, electrospinning technology is also applied to metal anodes and separators. The main factors affecting the properties of the metal anode are hydrogen evolution corrosion, dendrite growth, passivation and morphological change. It is worth noting that electrospun nanofibers have ideal material compatibility, which can be adopted to construct integrated metal anodes to inhibit the corrosion and hydrogen evolution reaction. Lee et al. (2018) prepared a heteronanomat structure of zinc anode with polyetherimide (PEI) electrospun nanofibers and zinc powder/single-walled carbon nanotubes (SWCNTs) hybrid, which exhibited an outstanding performance for inhibiting hydrogen evolution, preventing the zinc anode from corrosion and morphological change. Remarkably, the good mechanical properties of the electrospun nanofibers can be perfectly inherited into the hybrid separators, which were demonstrated in the electrospun nanofiber mat-reinforced permselective composite membranes prepared by Lee et al. (2016).

10.3.5 Fuel Cells

In contrast to metal-air batteries, fuel cells use the chemical energy of hydrogen (H_2) or other fuels to cleanly and efficiently produce electricity. For the proton-exchange-membrane fuel cells (PEMFCs), a continuously supplied fuel (usually H_2) is fed to the anode and an oxidizing agent (usually air, O_2) is fed to the cathode, proton-exchange-membrane is used to separate the anode and the cathode. Liu et al. (2022) designed two self-supporting catalyst layers with Pt–M (M = Ni, Co) ultrathin and holey nanotubes by electrospinning and thermal treatment, which showed high-mass-activity and high-durability towards ORR in PEMFCs. This is mainly due to the high efficiency of mass transfer and high platinum exposure brought by the construction and functionalization of nanofibers. More notably, this work, for the first time in a single catalyst, achieved the US Department of Energy's mass activity goals and dual durability goals for PGM catalysts (Figure 10.6A). In addition to its application in the preparation of electrode catalysts, electrospinning techniques have also been used for the synthesis of PEM. The integration of a pair of insulating blocks combined with collector rotation in an electrospinning device can realize complex arrangements of nanofibers to solve the immiscible interactions between hydrophilic sulfonate groups in perfluorosulfonic acid (PFSA) and the hydrophobic nanoporous polytetrafluoroethylene (PTFE) matrix. Hwang et al. (2022) configured a reinforced composite membrane (RCM) with a cross-aligned PTFE (CA-PTFE) framework, in which the porous PTFE was introduced as a mechanical reinforcement and the PFSA was introduced as a proton conductive polymer. In the detailed synthetic process, the electric field was used to control the collector rotation to realize a cross-aligned PTFE matrix. The single-cell testing results demonstrate a long-lasting operation of CA-PTFE RCM (low hydrogen crossover less than 5 mA cm^{-2} at 0.4 V after 21,000 wet/dry cycles). After heat treatment, CA-PTFE RCM impregnated with PFSA can be achieved. As shown in Figure 10.6B, the grid-type CA-PTFE RCM impregnated with PFSA makes the membrane have adequate hydration, which can minimize swelling and promote the diffusion of protons by concentrated sulfonate groups.

10.3.6 Solid-state Batteries and Others

Recent progress and well-developed energy storage devices in the next generation have focused on flexible solid-state batteries (SSBs) due to their enhanced safety features and higher performance in energy density and cycle life. Different from metal-ion batteries, a solid-state battery uses a solid-state electrolyte (SSE) to transport the metal ions between the cathode and anode electrodes. Among a variety of technologies for the construction of solid-state electrolytes and flexible electrodes,

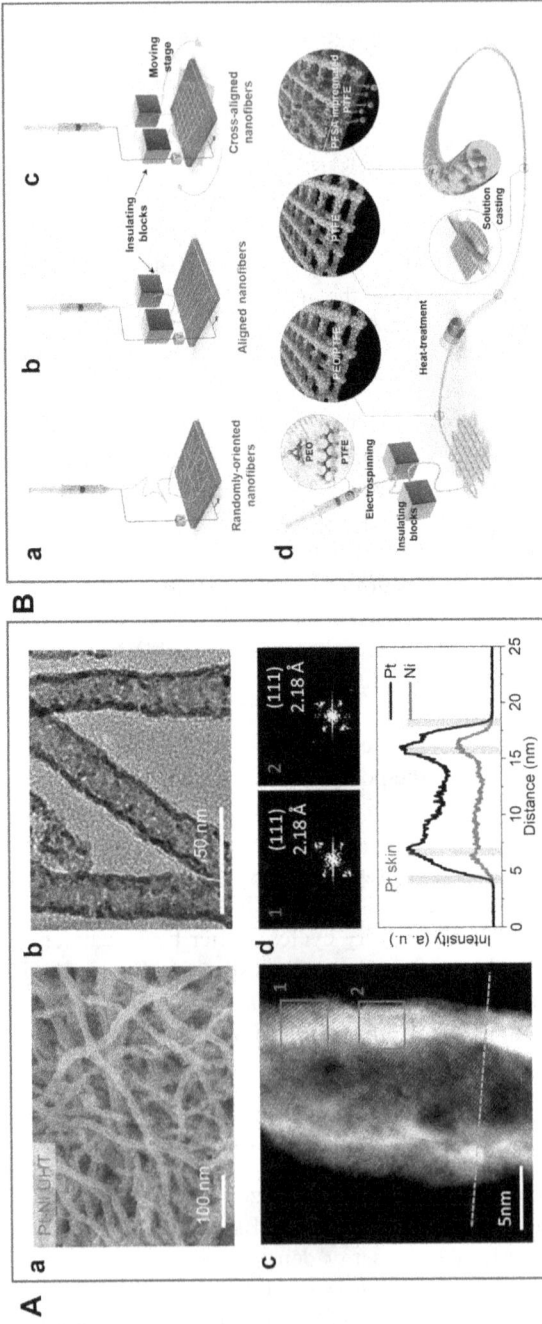

FIGURE 10.6 (A). Morphology, structure, and composition of Pt–Ni UHT. (A-a). SEM image, (A-b). TEM image, and (A-c). HAADF-STEM image accompanied by fast Fourier transformation (FFT) patterns of the marked spots (A-d) (Liu et al. 2022). (B). Schematic diagrams and images of collected nanofiber pattern by conventional and E-field guided electrospinning without and with collector rotation, and the schematic illustration of the synthesis of CA-PTFE and the preparation of CA-PTFE RCM (Hwang et al. 2022).

electrospinning is a straightforward and versatile technique to design and fabricate polymeric nanofibers into different morphologies.

In solid-state batteries, electrospun nanofibers used as polymer SSE is of great significance for improving ionic conductivity, interfacial contact, and mechanical integrity of the batteries. For example, Hu et al. (2020) fabricated a three-dimensional composite solid-state electrolyte membrane. The electrospun polyimide nanofibers film with mechanically robust and porous structure was used as a host,

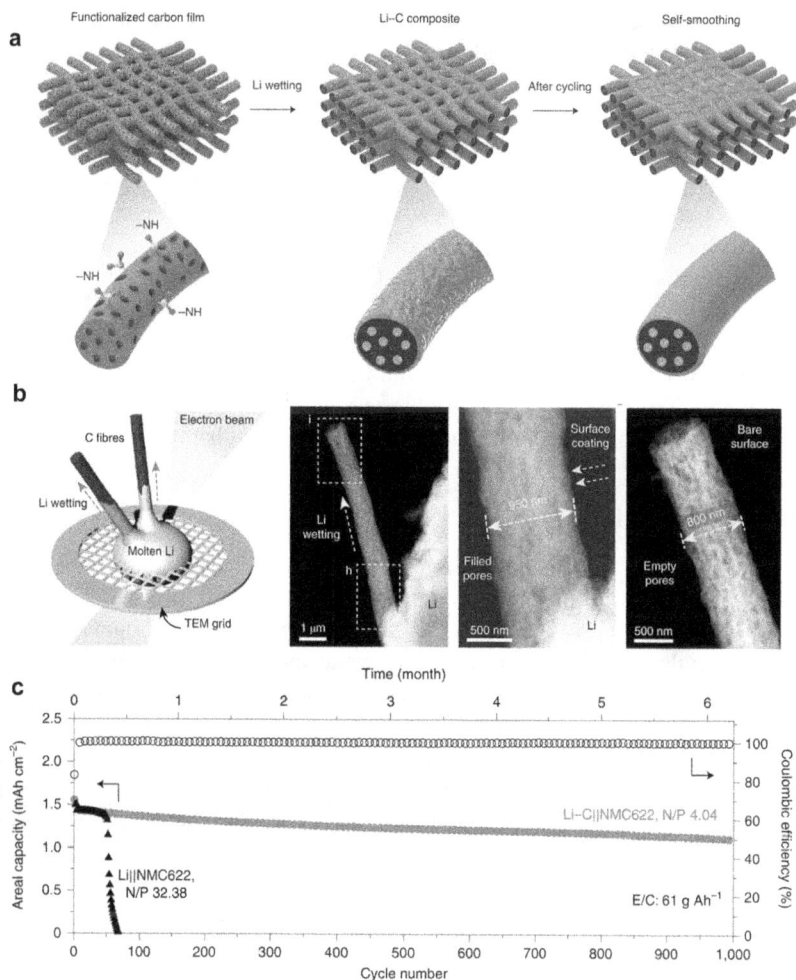

FIGURE 10.7 Illustration of self-smoothing behavior in the Li–C anode, the scheme shows the experimental set-up for the in-situ STEM measurement of Li infiltration into carbon fibers and STEM images of ongoing Li infiltration into individual carbon fiber, long-term cycling performance of Li-C‖NMC622 and Li‖NMC622 cells at a C/3 charge and 1 C discharge (Niu et al. 2019).

and the $Li_{6.75}La_3Zr_{1.75}Ta_{0.25}O_{12}$ (LLZTO) nanoparticles and PVDF polymer with lithium bis (trifluoromethane)sulfonimide (LiTFSI) were used as electrolyte filler. This composite solid-state electrolyte membrane had high ion conductivity of up to 1.23×10^{-4} S cm^{-1} and effectively inhibited Li dendrite growth during lithium plating/stripping processes. Electrospun carbon nanofibers (ECNFs) confined active materials have also been adopted as the anodes for SSBs to solve the large volume expansion. Tunable Sn structures spined in porosity-controlled carbon nanofibers were successfully prepared by Nam et al. (2015) for all-solid-state lithium-ion battery anodes, in which the carbon matrix acts to protect active materials and minimize anisotropic stress during discharge/charge processes. Batteries assembled with glass SSEs (77.5Li$_2$S-22.5P$_2$S$_5$) had excellent cycling performance.

Safety issues in practical applications related to uncontrolled dendrite formation of metals and volume fluctuation problems of the "hostless" metals have also become great challenges for high-energy batteries. With a large specific surface area, abundant porous structure, easy modification, high electronic conductivity and excellent mechanical properties, electrospun nanofibers-derived carbon fiber skeletons for metal anodes are a promising strategy that can solve the above problems. Niu et al. (2019) reported self-smoothing lithium–carbon (Li-C) anode by Li-wetting mesoporous and amine-functionalized carbon nanofibers. When coupled with a $LiNi_{0.6}Mn_{0.2}Co_{0.2}O_2$ cathode (Li-C||NMC622), the full cells delivered an excellent rate capability and long-term stable cycling. Especially, the full cells also displayed a cell-level energy density of 350–380 Wh kg kg^{-1} and a stable service life of up to 200 cycles under the realistic conditions for practical rechargeable lithium metal batteries (Figure 10.7).

10.4 CONCLUSION

Considering that our current energy, power supply and transportation infrastructure are plagued by many problems, more effective projects need to be implemented to solve the trouble faced by energy production, storage and transformation. In contrast to materials prepared by most conventional methods, electrospun nanofibers and nanofibrous architecture with remarkable characteristics especially large specific surface area, high porosity and interconnected structure, easy functionalization, mitigated charge transfer resistance and enhanced ionic conductivities, and mechanical strengths can effectively improve the performance of various materials of electrodes, electrolytes and separators involves in the abovementioned fields. Moreover, the simplicity of electrospinning technology and the tailored operation of adjustable parameters give nanofibers a high consistency of structure-activity correlation. The electrospinning technique is now popular for preparing one-dimensional tubular/fibrous nanomaterials assembled into two-dimensional/three-dimensional functionalized architectures, and great efforts devoted to continuing laboratory studies that are moving closer to practical applications should shed light on forthcoming electrospinning products for energy production, storage and conversion.

REFERENCES

Agarwal S. 2020. Biodegradable polymers: Present opportunities and challenges in providing a microplastic-free environment. *Macromolecular Chemistry and Physics* 221: 2000017.

Banitaba S. N., A. Ehrmann. 2021. Application of electrospun nanofibers for fabrication of versatile and highly efficient electrochemical devices: A review. *Polymers (Basel)* 13: 1741.

Bose G. M. 1745. *Researchers sur la cause et sur la veritable teorie de l'electricite publies par George Mathias Bose prof. en physique*. France: de l'imprimerie de Jean Fred. Slomac.

Cooley J. F. 1902. Apparatus for electrically dispersing fluids. *Compiler*. US: 692631.

Evrova O., V. Hosseini, V. Milleret, G. Palazzolo, D. Eberli. 2016. Hybrid randomly electrospun poly(lactic-co-glycolic acid):poly(ethylene oxide) (PLGA:PEO) fibrous scaffolds enhancing myoblast differentiation and alignment. *ACS Applied Materials Interfaces* 8(46): 31574–31586.

Faccini M., D. Amantia, S. V. Campos, C. Vaquero, J. M. L. Ipiña, L. Aubouy. 2011. Nanofiber-based filters as novel barrier systems for nanomaterial exposure scenarios. *Journal of Physics: Conference Series* 304: 012067.

Formhals A. 1934. Process and apparatus for preparing artificial threads. *Patents Specification* 1975: 504.

Guo S. Q., J. N. Wang, Y. X. Sun, L. C. Peng, C. J. Li. 2023. Interface engineering of Co_3O_4/CeO_2 heterostructure in-situ embedded in Co/N-doped carbon nanofibers integrating oxygen vacancies as effective oxygen cathode catalyst for Li-O_2 battery. *Chemical Engineering Journal* 452: 139317.

Hao J. C., Z. C. Zhuang, K. C. Cao, G. H. Gao, C. Wang, F. L. Lai, S. L. Lu, P. M. Ma, W. F. Dong, T. X. Liu, M. L. Du, H. Zhu. 2022. Unraveling the electronegativity-dominated intermediate adsorption on high-entropy alloy electrocatalysts. *Nature Communications* 13: 2662.

Hu J. K., P. G. He, B. C. Zhang, B. Y. Wang, L. Z. Fan. 2020. Porous film host-derived 3D composite polymer electrolyte for high-voltage solid state lithium batteries. *Energy Storage Materials* 26: 283–289.

Hu J. P., Y. G. Qin, H. Sun, Y. Ma, L. Lin, Y. Peng, J. Zhong, M. Z. Chen, X. H. Zhao, Z. Deng. 2022. Combining Multivariate Electrospinning with surface MOF functionalization to construct tunable active sites toward trifunctional electrocatalysis. *Small* 18: 2106260.

Hwang C. K., K. A. Lee, J. Lee, Y. Kim, H. Ahn, W. Hwang, B. K. Ju, J. Y. Kim, S. Y. Yeo, J. Choi, Y. E. Sung, I. D. Kim, K. R. Yoon. 2022. Perpendicularly stacked array of PTFE nanofibers as a reinforcement for highly durable composite membrane in proton exchange membrane fuel cells. *Nano Energy* 101: 107581.

Jung J. W., C. L. Lee, S. Yua, I. D. Kim. 2016. Electrospun nanofibers as a platform for advanced secondary batteries: a comprehensive review. *Journal of Materials Chemistry A* 4: 703–750.

Lee B. S., S. Y. Jeon, H. Park, G. Lee, H. S. Yang, W. R. Yu. 2014. New electrospinning nozzle to reduce jet instability and its application to manufacture of multi-layered nanofibers. *Scientific Reports* 4: 6758.

Lee D. G., H. W. Kim, J. M. Kim, K. H. Kim, S. Y. Lee. 2018. Flexible/rechargeable Zn-air batteries based on multifunctional heteronanomat architecture. *ACS Applied Materials Interfaces* 10: 22210–22217.

Lee H. J., J. M. Lim, H. W. Kim, S. H. Jeong, S. W. Eom, Y. T. Hong, S. Y. Lee. 2016. Electrospun polyetherimide nanofiber mat-reinforced, permselective polyvinyl alcohol composite separator membranes: A membrane-driven step closer toward rechargeable zinc–air batteries. *Journal of Membrane Science* 499: 526–537.

Li L., Z. Jiang, M. M. Li, R. S. Lia, T. Fang. 2014. Hierarchically structured PMMA fibers fabricated by electrospinning. *RSC Advances* 4: 52973–52985.

Li S. J., H. T. Guo, S. J. He, H. Q. Yang, K. M. Liu, G. G. Duan, S. H. Jiang. 2022. Advanced electrospun nanofibers as bifunctional electrocatalysts for flexible metal-air (O_2) batteries: Opportunities and challenges. *Materials & Design* 214: 110406.

Li P., Y. Qiao, L. L. Zhao, D. H. Yao, H. X. Sun, Y. F. Hou, S. Li, Q. Li. 2015. Electrospun PS/ PAN fibers with improved mechanical property for removal of oil from water. *Marine Pollution Bulletin* 93(1–2): 75–80.

Li Q. H., W. Zhang, J. Peng, D. D. Yu, Z. X. Liang, W. Zhang, J. W. Wu, G. Y. Wang, H. X. Li, S. M. Huang. 2022. Nanodot-in-nanofiber structured carbon-confined Sb_2Se_3 crystallites for fast and durable sodium storage. *Advanced Functional Materials* 32: 2112776.

Li X. Y., W. C. Chen, Q. R. Qian, H. T. Huang, Y. M. Chen, Z. Q. Wang, Q. H. Chen, J. Yang, J. Li, Y. W. Mai. 2020. Electrospinning-based strategies for battery materials. *Advanced Energy Materials*: 2000845.

Lin J., B. Ding, J. Yu. 2010. Direct fabrication of highly nanoporous polystyrene fibers via electrospinning. *ACS Applied Materials & Interfaces* 2(2): 521.

Lin L. L., F. Pei, J. Peng, A. Fu, J. Q. Cui, X. L. Fang, N. F. Zheng. 2018. Fiber network composed of interconnected yolk-shell carbon nanospheres for high-performance lithium-sulfur batteries. *Nano Energy* 54: 50–58.

Liu J. Y., S. Y. Liu, F. Z. Yan, Z. S. Wen, W. W. Chen, X. F. Liu, Q. T. Liu, J. X. Shang, R. H. Yu, D. Su, J. L. Shui. 2022. Ultrathin nanotube structure for mass-efficient and durable oxygen reduction reaction catalysts in PEM fuel cells. *Journal of the American Chemical Society* 144: 19106–19114.

Liu Z., R. T. Guo, J. S. Meng, X. Liu, X. P. Wang, Q. Li, L. Q. Mai. 2017. Facile electrospinning formation of carbon-confined metal oxide cube-in-tube nanostructures for stable lithium storage. *Chemical Communications* 53: 8284–8287.

López-Covarrubias J. G., L. Soto-Muñoz, A. L. Iglesias, L. J. Villarreal-Gómez. 2019. Electrospun nanofibers applied to dye solar sensitive cells: A review. *Materials* 12: 3190.

Meka S. R. K., V. Agarwal, K. Chatterjee. 2019. In situ preparation of multicomponent polymer composite nanofibrous scaffolds with enhanced osteogenic and angiogenic activities. *Materials Science & Engineering C* 94: 565–579.

Nam D. H., J. W. Kim, J. H. Lee, S. Y. Lee, H. A. S. Shin, S. H. Lee, Y. C. Joo. 2015. Tunable Sn structures in porosity-controlled carbon nanofibers for all-solid-state lithium-ion battery anodes. *Journal of Materials Chemistry A* 3: 11021–11030.

Nitanan T., P. Opanasopit, P. Akkaramongkolporn, T. Rojanarata, N. P. Supaphol. 2012. Effects of processing parameters on morphology of electrospun polystyrene nanofibers. *Korean Journal of Chemical Engineering* 29: 173–181.

Niu C. J., H. L. Pan, W. Xu, J. Xiao, J. G. Zhang, L. L. Luo, C. M. Wang, D. H. Mei, J. S. Meng, X. P. Wang, Z. Liu, L. Q. Mai, J. Liu. 2019. Self-smoothing anode for achieving high-energy lithium metal batteries under realistic conditions. *Nature Nanotechnology* 14: 594–601.

Prabu G. T. V., B. Dhurai. 2020. A novel profiled multi-pin electrospinning system for nanofiber production and encapsulation of nanoparticles into nanofibers. *Scientific Reports* 10: 4302.

Pachón E Y G., R. V. Graziano, R. M. Campos. 2014. Structure of poly(lactic-acid) PLA nanofibers scaffolds prepared by electrospinning. *IOP Conference Series: Materials Science and Engineering* 59: 012003.

Peng S. H., P. R. Ilango. 2020. *Electrospinning of Nanofibers for Zn-Air Battery*. Electrospinning of Nanofibers for Battery Applications, 121–139. Springer.

Qiu J., H. Wang, J. Wang, C. Wang. 2021. Synthesis of ternary Ni/Mo2C/carbon nanofibers as low-cost counter electrode for efficient dye-sensitized solar cells. *Chemical Research in Chinese Universities* 37: 480–487.

Qu C., P. Zhao, C. D. Wu, Y. Zhuang, J. M. Liu, W. H. Li, Z. Liu, J. H. Liu. 2021. Electrospun PAN/PANI fiber film with abundant active sites for ultrasensitive trimethylamine detection. *Sensors and Actuators B: Chemical* 338(1): 129822.

Shao C. L., H. Kim, J. Gong, 2002. A novel method for making silica nanofibers by using electrospun fibers of polyvinylalcohol/silica composite as precursor. *Nanotechnology* 13(5): 635–637.

Sharma J., G. Polizos, C. J. Jafta, D. L. Wood III, J. L. Li. 2022. Review—Electrospun inorganic solid-state electrolyte fibers for battery applications. *Journal of The Electrochemical Society* 169: 050527.

Si Y., X. Q. Wang, L. Y. Dou, J. Y. Yu, B. Ding. 2018. Ultralight and fire-resistant ceramic nanofibrous aerogels with temperature-invariant superelasticity. *Science Advances* 4: eaas8925.

Stanishevsky A., J. Wetuski, M. Walock, I. Stanishevskaya, H. Y. Lelièvre, E. Košťáková, D. Lukáš. 2015. Ribbon-like and spontaneously folded structures of tungsten oxide nanofibers fabricated via electrospinning. *RSC Advances* 5: 69534–69542.

Tao X. Y., S. X. Zhou, Z. M. Xiang, J. Ma, R. L. Hou, Y. B. Zhu, X. Y. Wei. 2017. Fabrication of continuous ZrB2 nanofibers derived from boron-containing polymeric precursors. *Journal of Alloys and Compounds* 697: 318–325.

Wang C. H., C. Liu, J. S. Li, X. Y. Sun, J. Y. Shen, W. Q. Han, L. J. Wang. 2017. Electrospun metal-organic framework derived hierarchical carbon nanofibers with high performance for supercapacitors. *Chemical Communications* 53: 1751–1754.

Xia C. F., Y. S. Zhou, C. H. He, A. I. Douka, W. Guo, K. Qi, B. Y. Xia. 2021. Recent advances on electrospun nanomaterials for zinc-air batteries. *Small Science* 1: 2100010.

Xie J. W., X. R. Li, Y. N. Xia. 2008. Putting electrospun nanofibers to work for biomedical research. *Macromolecular Rapid Communications* 29: 1775–1792.

Yang C., Y. Yao, Y. B. Lian, Y. J. Chen, R. Shah, X. H. Zhao, M. Z. Chen, Y. Peng, Z. Deng. 2019. A double-buffering strategy to boost the lithium storage of botryoid MnOx/C anodes. *Small* 15: 1900015.

Yang G., X. L. Li, Y. He, J. K. Ma, G. L. Ni, S. B. Zhou. 2018. From nano to micro to macro: Electrospun hierarchically structured polymeric fibers for biomedical applications. *Progress in Polymer Science* 81: 80–113.

Yang L. P., L. X. Yu, Z. H. Huang, F. Y. Kang, R. T. Lv. 2022. ZnS-assisted evolution of N,S-doped hierarchical porous carbon nanofiber membrane with highly exposed Fe-N₄/Cx sites for rechargeable Zn-air battery. *Journal of Energy Chemistry* 75: 430–440.

Zeleny J. 1914. The electrical discharge from liquid points, and a hydrostatic method of measuring the electric intensity at their surfaces. *Physical Review* 3(2): 69–91.

Zhang D. M., J. Chang. 2007. Patterning of electrospun fibers using electroconductive templates. *Advanced Materials* 19: 3664–3667.

Zhang Z. Y., X. Wu, Z. K. Kou, N. Song, G.D. Nie, C. Wang, F. Verpoort, S. C. Mu. 2022. Rational design of electrospun nanofiber-typed electrocatalysts for water splitting: A review. *Chemical Engineering Journal* 428: 131133.

11 3D Printed Materials for High Energy Production, Conversion, and Storage Devices

Manchu Mohan Krishna Sai, Deepti Ranjan Sahu, Amitava Mandal, and Jinoop AN

11.1 INTRODUCTION

Contrary to conventional subtractive and formative manufacturing, additive manufacturing (AM) introduced in the 1980s can be used for building components with complex designs using layer-by-layer manufacturing as per the sliced 3D model fed to the AM machine[1]. The dominance of AM technology is primarily due to the freedom of design with flexibility, waste minimization, on-demand manufacturing, and the ability to build parts using a wide range of materials including polymers, metals, and ceramics. In addition, AM allows the fabrication of final or near-final components with minimal post-processing. Due to these dominances, AM entered into several sectors such as aerospace, automobile, bio-medical, construction, defense, energy, internet-of-things (IoT), etc., and initiated design for additive manufacturing (DFAM) method to maximize the benefits of AM. According to ASTM/ISO standards, AM technology is classified into seven categories as vat photo-polymerization/stereo lithography (SLA), material extrusion, binder jetting (BJ), material jetting, (MJ), powder bed fusion (PBF), directed energy deposition (DED), and laminated object manufacturing (LOM)[1,2]. The detailed selection of each printing method, materials, and their capabilities lies beyond the scope of this article.

Over the past decade, AM, most commonly known as 3D Printing (3DP), influenced and caused breakthrough developments in various fields, bringing life to imaginable and impossible things. 3DP of wearable energy storage devices (ESDs), high-resolution electronics[3], affordable rocket engines[4], bio-printing tissues and organoids using bio-ink, and custom-made prosthetics[5,6], houses using novel construction materials, and industrial wastage are a few examples[7]. There is a demand for micro-robots in the medical field, for better target drug delivery systems,

DOI: 10.1201/9781003318859-11

conducting biopsies, and fertilization assistance. University of Toronto engineers developed a magnetized medical robot using 3D printing of a millimeter scale and use piezoelectric powered with microscopic needles over a flexible material, squirms through fluid channels, and collects tissue samples with its tiny mechanical jaws under magnetic fields[8,9]. 3DP of lithium-ion micro-batteries to power miniature drones, robots, and implanted smart electronics[10,11] were expanding the applications of the technology to the energy sector.

ESDs play an indispensable role in building sustainable energy storage systems that can efficiently generate, convert and store energy from one form to another. In recent years, the emergence of microelectronic systems such as micro-electromechanical systems (MEMS), micro-sensors, implantable miniaturized medical devices, micro-robots, and smart portable/wearable microsystems demands the development of on-chip integration of miniaturized energy storage (MESD) devices[12–15]. Unlike conventional ESDs, MESDs are compact, compatible, and with customizable power requirements according to microelectronic systems and satisfy the architectures with high aspect ratio, enhanced areal energy densities, power densities, and extremely lightweight. Micro-batteries (MBs), and micro-supercapacitors (MSCs) are promising MESDs to satisfy the requirements of high aspect ratio, enhanced areal energy densities, power densities, and rapid charging-discharging rates. Especially batteries take longer charging/discharging and possess relatively high energy densities (30–170 Wh kg^{-1}). Supercapacitors are widely used in electronic systems where fast and frequent discharging/charging (power densities) are essential but with poor energy densities (less than 10 Wh kg^{-1})[16].

In the past decade, extensive research and development happened in MBs and MSCs by exploring new electrode materials enhancing electrochemical performance with ease of manufacturing[17–20]. There are versatile ways of producing ESDs like atomic layer depositions, electrochemical depositions, chemical vapor depositions, electrophoretic depositions etc., which have limitations for integration, customization, and large production. Conventional fabrication techniques have limitations in control over the geometry and architecture of electrodes and solid-state electrolytes. 2D designs of traditional ESDs require a larger area footprint to generate high capacity, which is not preferable for portable ESDs. Building thicker electrodes is a performance optimization strategy to enhance areal capacitance and energy density by increasing active material and not compromising the ion diffusion rate. The 3D structural electrode provides shorter diffusion paths and enhances energy density by creating a porous structure. 3DP of ESD aids customizable designs (2D and 3D), on-chip integration, tailored material properties, almost negligible waste, economical, less areal footprint, and energy conservation.

Thus, the present chapter describes the 3DP systems and material considerations for ESD application; design, and architecture; printing processes; and performance optimization strategies followed by the challenges involved in the development. Further, the influence of 3DP fabrication techniques on the design of printing materials and the electrochemical performance of printed devices is discussed.

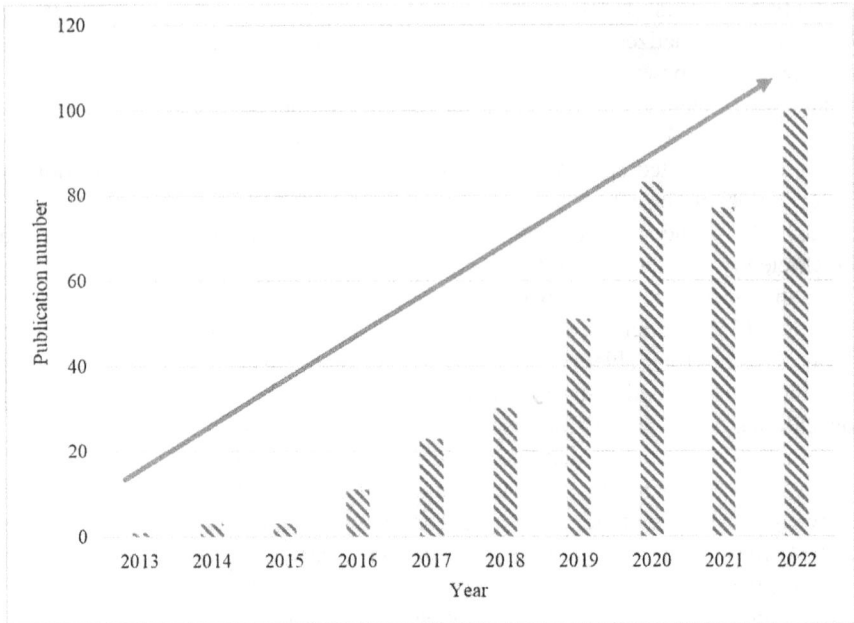

FIGURE 11.1 Statistical analysis regarding a number of research publications on 3D-printing of ESDs since 2013 (Web of Science using the keywords energy storage devices, 3D printing).

11.1.1 BACKGROUND OF 3DP TECHNOLOGIES IN FABRICATING ESD

Presently, the world is moving in the era of Industry 4.0, bringing life to many fictional technologies and molding various fields according to the industry's requirements. Especially, manufacturing and sustainable energy systems have undergone many changes, enabling 3DP of miniaturized ESDs. 3DP has a catching research trend of producing ESDs by exploring various design architectures, materials, and performance optimization strategies (refer Figure 11.1)[21].

Researchers are investigating standardizing 3DP technology in fabricating electrochemical ESDs and exploring AM/3DP process optimization using various materials based on carbon, lithium-ion, sodium-ion, magnesium-ion, aluminium-ion, conducting polymers, and metal oxides [22–27]. In 2007, 2D-integrated Li-ion batteries were fabricated successfully, which opened the door to wearable ESDs[28]. In 2013, Sun *et al.* successfully standardized a 3DP interdigitated battery, which outperformed most traditionally manufactured Li-ion batteries with high areal and power densities[11]. Subsequently, many design architectures, and novel materials (graphene, carbon nanotubes, etc.) were introduced to enhance the electrochemical performance of ESDs[29–35].

11.2 MATERIAL CONSIDERATIONS AND CLASSIFICATION

11.2.1 MATERIAL CONSIDERATIONS

ESDs are made of many components such as the electrolyte, the active electrodes, and the current collector. Among them, MESDs are expected to exhibit phenomenal electrochemical performance to be integrated with various energy supply/harvesting systems like solar cells, piezoelectric nanogenerators (PENG), triboelectric nanogenerators

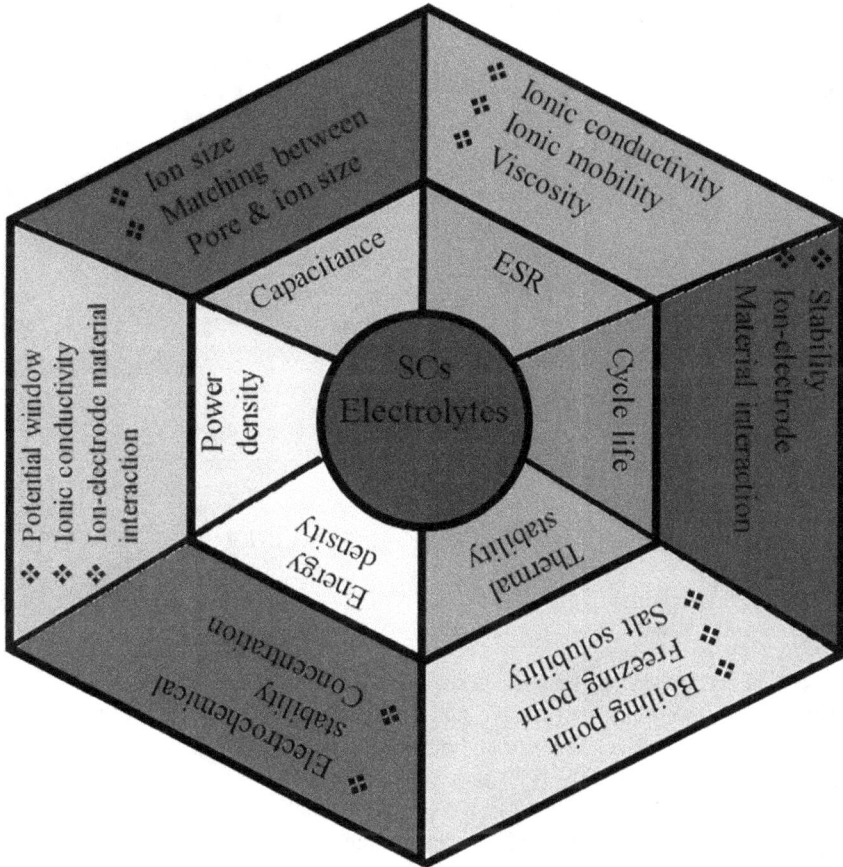

FIGURE 11.2 (a) Effects of electrolyte on supercapacitors and (b) electrochemical characteristics of various ESDs.

Source: [36].

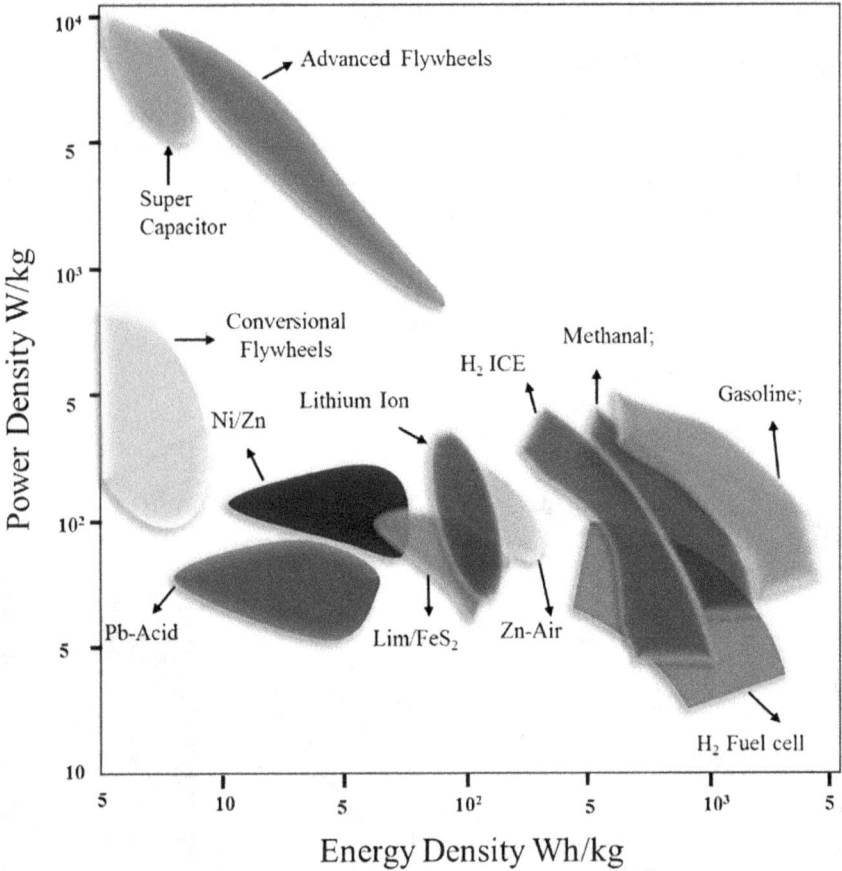

FIGURE 11.2 (Continued)

(TENG), etc. Carbon-based materials such as CNTs, graphene, and activated carbon are the most commonly used active electrode materials in ESDs because of their high energy density, low areal/volumetric capacitance, and printability. Flexible substances such as polyimide (PI), paper, polyethylene (PE), conductive fabric tape, polyethylene terephthalate (PET), PVA-H3PO4, polydimethylsiloxane (PDMS), polyethersulfone (PES), polyethylene naphthalate (PEN), etc. as electrolytes exhibit excellent flexibility, resistance to high temperatures and remain stable in alkaline solutions. The critical issue of the ESD's working potential depends on electrolyte characteristics like operating temperature, safety, ionic conductivity, and potential range (refer to Figure 11.2a). In general, electrolytes can be classified as aqueous, organic, and ionic-based systems[14].

Table 11.1 summarizes the solid-state electrolytes developed over the past decade corresponding to the class of voltages. Aqueous electrolytes have high

TABLE 11.1

Developed Solid-state Electrolytes for ESDs and Corresponding Working Voltages

Type	Electrolyte	Voltage (V)	Ref
Aqueous-based	GO-PAA	0.08	[22]
	PVA/H$_3$PO$_4$	0-0.8	[23]
	PVA/LiCl	0-1.4	[24]
	CMC/Na$_2$SO$_4$	0-1.5	[25]
	SiO$_2$-LiTFSI	0-2	[37]
Organic-based	PC/Et$_4$NBF$_4$	0-1	[26]
	PC-PMMA-LiClO$_4$	0-1.6	[38]
	ADN/SN/LiTFSI/PMMA	0-2	[39]
Ionic liquid-base	LiTFSI-P14TFSI-PVDF-HFP	0-3	[27]
	P(VDF-HFP)/[EMIMBF4]	0-3	[23]
	EMImNTF2	0-3	[40]
	P(VDF-HFP)/[EMIM][TFI]	0-1.5	[41]

ionic conductivity, good safety, and low cost. But, the voltage is limited due to the decomposition of water. The organic electrolyte can supply much higher voltages than aqueous electrolytes, but the toxic, flammable, and volatile properties of these electrolytes provide safety issues in wearable electronics. The ionic liquid-based electrolyte has a high viscosity, exhibits wide voltage ranges, and excellent stability, flexible for high temperature applications.

11.3 CLASSIFICATION OF ESDS

ESDs can be classified based on many parameters such as the kind of energy storage system (mechanical, chemical, thermal, electrochemical, fuel, and electrical systems), size of the system depending on the demand for energy/powder (miniaturized/ absolute), and portable/ fixed ESDs. The miniaturized ESDs are mainly classified into micro-batteries (MBs) and micro-supercapacitors (MSCs) as shown in Figure 11.3 to address the energy demands in current world applications.

MBs possess intrinsic characteristics of high energy density, durability, and stable operating life of more than 2000 cycles and are considered worthy for miniaturized energy sources to integrate into microsystems. MBs consist of a cathode, anode, electrolyte, and current collectors for discharging and charging cycles. Recently, extensive research has been carried out on designing MB electrodes (cathode and anode) to enhance their energy density, power density, life span, and difficulties entangled in battery on-chip integration[42]. Based on different valances of working cations, MBs can be divided into alkali-ion MBs (e.g. lithium/potassium/sodium-ion) and multivalent cation MBs (e.g. zinc/magnesium/aluminum-ion)[43,44] (refer to Figure 11.2b).

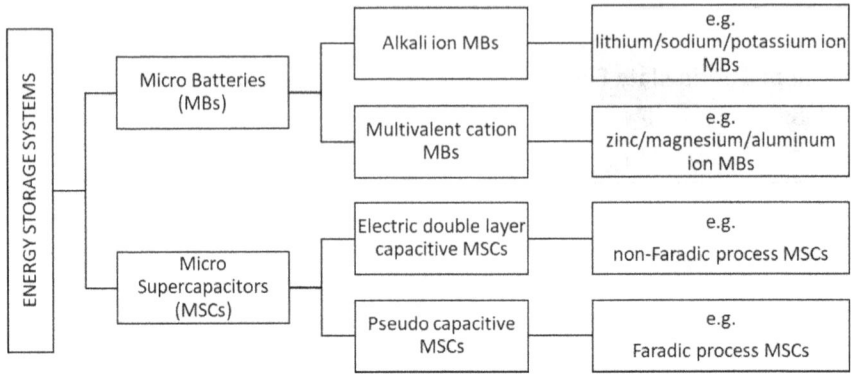

FIGURE 11.3   Classification of ESDs.

11.3.1 ALKALI-ION MICRO-BATTERIES (AIMBS)

During the past decade, several advanced configuration designs have been reported for alkali-ion MBs. In 1983, the world's first thin-film lithium-ion micro-battery (LIMBs) was developed by Kanehori et al. using chemical vapor deposition and magnetron sputtering by considering lithium metal as an anode, titanium disulfide (TiS_2) as cathode, and $Li_3Si_{0.6}P_{0.4}O_4$ as the glassy solid electrolyte without any binder or additive[45]. The LIMBs delivered a stable life of 2000 cycles due to a relatively short Li-ion diffusion pathway. Later on, many micro-manufacturing techniques such as atomic layer deposition, physical vapor deposition, sol-gel solution, and thermal evaporation were exploited in thin-film MBs fabrication. 3D microelectrodes with thicker layers show a promising performance with high energy density as well as superior power density on a defined footprint. In consideration of the great advantages of 3D electrodes and controllable fabrication process, 3DP has proven its potential for designing unique architectures of ESDs with an exponential enhancement of areal and power densities due to its precise fabrication, efficiency, economy, design freedom, and rapid printability. Pikul et al. reported the fabrication of 3D interdigitated micro-battery architectures (3D IMBA) of high-powered LIMBs consisting of 3D bi-continuous electrodes of 30 μm size with a nanoporous structure; lithium-ion manganese oxide ($LiMnO_2$) as cathode and nickel tin (NiSi) as an anode, which yielded an energy density of 15.7 μWh cm^{-2} μm^{-1} and high power density of 7.4 mW cm^{-2} μm^{-1}[46]. Even though LIMBs exhibit high energy and power densities, the safety hazards associated with instability, and flammable and poisonous electrolytes seriously obstruct their use in MESDs. In an exploration of better materials, potassium-ion batteries (PIB) and sodium-ion batteries (SIB) are promising and considered as the next-generation batteries AIMBs. Presently, potassium-ion micro-batteries (PIMBs) and sodium-ion micro-batteries (SIMBs) are in a nascent stage. Aqueous PIMBs were found to have great potential as highly safe ESDs with standard energy density, and fast charging/ discharging rates[42,47]. In SIMBs, creating an effectual conductive network for ion diffusion and electron transport is challenging. Zheng et al fabricated a non-aqueous SIMB in an interdigital structure, where sodium vanadate phosphate is a cathode and sodium titanate an anode with sodium tetrafluoroborate-based ionogel electrolyte[42,48]. SIMB exhibits long-term cyclic stability of 10 mA h cm^{-3} and remained still after 3000 cycles at 20°C. The proposed SIMB is considered to be the ideal design for basic research of charging/discharging cycles based on in-situ characterizations.

11.3.2 Multivalent Cation Micro-batteries (MVCMBs)

In MVCMBs, multivalent ions like Ca^{2+}, Mg^{2+}, Al^{3+}, and Zn^{2+} are used as the active charge carrier in electrolytes. Since MVCMBs (e.g. zinc/magnesium/aluminium-ion batteries) exhibit high energy density, high capacity density, fast charge transfer dynamics, environmental protection, and less sensitivity to water and air conditions, they are preferred for various applications under outdoor environment conditions. Typically, in zinc-ion batteries, the anode is made of zinc metal, which contains Zn^{2+} ions in an aqueous electrolyte, and a layered structured cathode for Zn^{2+} intercalation/ deintercalation. Researchers have developed a variety of Zn-based batteries like zinc-vanadium (Zn-V), zinc-manganese (Zn-Mn), and zinc-nickel (Zn-Ni) batteries for various applications due to their stability in the outdoor environment and high abundance in the earth. The main disadvantage of zinc-based batteries is their cycling instability. The development of miniaturized zinc-ion ESDs is in its deficient stage. Most investigations on MVCMBs result in achieving higher areal/ volumetric energy densities by increasing the thickness of the microelectrodes, due to an increase in mass loading of the active material of ESDs. However, the power density declined due to longer electron and ion diffusion distances. 3D microstructured microelectrode exhibits promising electrochemical performance because of high surface area, high-efficiency electron diffusion paths, and short ion transport distances[42].

11.3.3 Micro-supercapacitors (MSCs)

MSCs are specifically considered ESDs because of their fast charging capabilities, high power densities, high specific capacitance, long cyclic stability, and eco-friendly nature. MSCs are extensively used in airbags, portable flexible electronics, power backup devices for personal computers and mobiles, and electrical vehicles. MSCs are classified into two types, (i) electrochemical double-layer capacitors (EDLC), and (ii) pseudocapacitance (PC).

EDLC MSCs work on the non-faradaic process (no chemical reaction), which is used in applications where high demand is there for energy and power densities. EDLSs store electrical energy directly in the charge separation method. EDLC carbon-based materials possess superior electrical properties like high electrical conductivity, high electrochemical stability, and high specific surface area. Carbon nanotubes

(CNT), graphene, active carbon, and carbon-based matrix composites are commonly used as the EDLC's electrode material as a result of their interesting properties. Even though carbonaceous electrode material possesses superior electrical properties, it exhibits poor cyclability and low energy density[49–52]. Researchers have developed conducting polymer-based electrodes with high specific capacitance, tunable electrical conductivity, and flexibility, but with poor electrochemical stability[40,53]. To improve the energy density, Pseudocapacitive MSCs were developed which store energy by reversible redox reactions with transition metal oxides/hydroxides/nitrides/carbides/conducting polymers(CP)[14,54–56] (refer Figure 11.2(a)).

11.3.4 PSEUDOCAPACITIVE MICRO-SUPERCAPACITORS (PC-MSCs)

In comparison to EDLC MSCs, PC-MSCs possess relatively high specific capacitance and superior energy density, but inferior cycling stability. Conductive polymer-based PC-MSCs based on polyaniline, polypyrrole[57], disulfides[58], hydroxides[59], and transition metal oxides or nitrides[60] often exhibit higher specific capacitance because of reversible and rapid Faradaic reactions at PC-MSCs electrodes. In PC-MSC, three types of chemical reactions occur at the electrodes: (i) reversible adsorption, (ii) redox reactions (occurs in transition metal oxides), (iii) reversible electrochemical doping (occurs in conducting polymer-based electrodes). Ruthenium dioxide (RuO_2) electrode material (nanoparticles) recorded an ultrahigh capacitance value of 1200 F/g using surface oxidation/reduction methods of electric charge storage[42,61–64]. Under various bending tests, all-solid-state planar MSCs exhibit excellent mechanical stability and significant potential as next-generation integrated MESDs for microsystem applications.

11.4 POTENTIAL DESIGNS FOR 3D PRINTED ESDS

11.4.1 DESIGN ARCHITECTURE

Overall electrochemical performance of ESDs is greatly influenced by the design architecture and results in diverse applications depending on the demanding configuration. The demand-associated application requirements and configurations further impact and cause the development of various morphologies and manufacturing techniques for fabrication. The configuration design based on the recent studies of MESDs (both MBs and MSCs) can be classified into five types: (i) two-dimensional (2D) stacked micro-devices, (ii) two-dimensional planar interdigital micro-devices, (iii) two-dimensional (2D) arbitrary-shaped micro-devices, (iv) three-dimensional (3D) planar micro-devices, (v) wire-shaped (1D) micro-devices (refer Figure 11.4).

2D stacked micro-devices consist of a stack of multilayer structures with an actual battery structure (planar substrate, a bottom current collector, cathode, separator, electrolyte, anode, and top current collector) fabricated vertically through a layer-by-layer manufacturing process. This design is most suitable for flexible thin-film batteries integration with flexible electronic systems, [42,65]. 2D planar interdigital micro-devices are the typical in-plane structure consisting of two neighbor electrodes separated by millimeter/nanometer interspacing in parallel arrays and arranged on a substrate interdigitally. This design provides rapid and expedient ionic transport between the

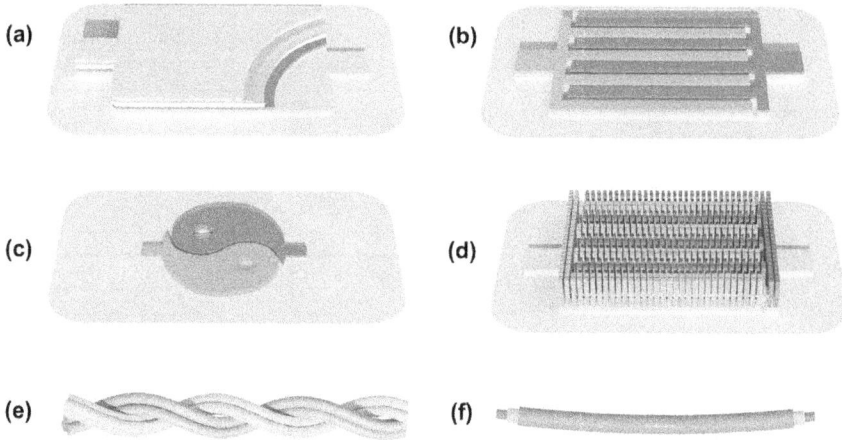

FIGURE 11.4 Design architectures of 3D printed ESDs (a) 2D stacked, (b) 2D planar interdigital, (c) 2D arbitrary-shaped, (d) 3D planar, (e) cable-shaped (wire-shaped), and (f) coaxial-fiber-shaped (wire-shaped) micro-devices.

Source: [42].

cathode and anode, exhibiting higher power density than the 2D stacked micro-devices design. Fabrication techniques like screen printing, mask-assisted filtration, spray coating, plasma etching, photolithography, laser scribing, inkjet printing, and electro-chemical deposition are devised to produce functional 2D planar interdigital design ESDs in various applications [42][66]. 2D arbitrary-shaped micro-devices are similar to 2D planar interdigital micro-devices with customizable electrode shapes according to the application requirement. But, it is typically fabricated using a mask-assisted fil-tration technique. 3D planar micro-devices consist of interdigitally arranged cathodes and anodes as vertically aligned pillars on the current collectors. Developing 3D pillars follows the concept of 3D MBs to enlarge areal capacitance and energy densities. Due to the structural superiority of 3D architectures, the 3D microelectrodes are anticipated to raise the energy density per unit area with a greatly increased loading amount of active substances[33]. Wire-shaped micro-devices use wires that are innately flexible in all directions and have the potential for using in extreme applications and catching attention in recent research. It is a new approach to developing fiber-shaped capacitors, and high-performance power sources for portable and wearable electronics[67–69]. FDM and SLA-based 3DP processes are used to print sandwich/stacked design batteries, whereas DIW and IJP techniques are mostly preferred to print in-plane designs.

11.4.2 3DP of ESDs

Fabrication of ESDs has its challenges, mainly systematic hierarchical assembly of each component, which are current collectors, electrodes, separators, and electrolytes. In 3DP, each printing technique has a unique printing method using specific feed material (refer Figure 11.5). Therefore, understanding each printing method, material, and its capabilities is essential for the design and fabrication of ESDs. Inkjet printing

```
                    ┌── Stereolithography (SLA) ──── Photo polymers
              ┌─Wet─┼──── Inkjet Printing (IJP) ──── Newtonion fluid
              │     └─── Direct Ink Writing (DIW) ── Shear-thinnig paste
  Materials ──┤      ┌─ Fused Deposition
              │      │   Modeling (FDM) ─────────── Filament
              └─Dry──┤   Laminated Object
                     │   Manufacturing (LOM) ────── Laminate
                     ├─ Powder Bed Fusion (PBF) ─── Powder
                     └─ Direct Energy Deposition
                        (DED) ───────────────────── Powder/ Wire
```

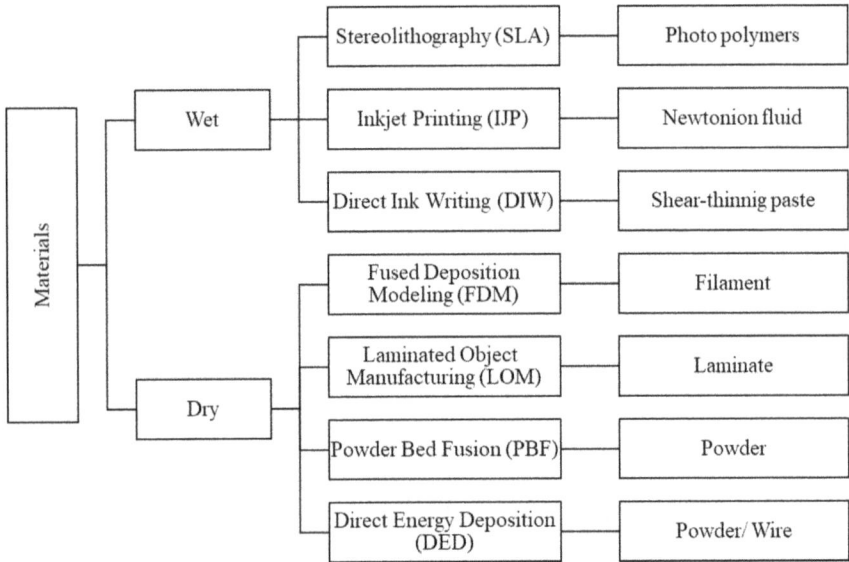

FIGURE 11.5   ESDs 3DP methods with associated materials.

(IJP), direct ink writing (DIW), selective laser sintering (SLS), and fusion deposition modeling (FDM) are the most commonly used 3DP techniques for ESDs.

Inkjet printing (IJP) follows droplet-based material deposition technology with multi-material printing capabilities. IJP is a highly suitable process for conducting polymer-based electrodes for EDLC MSCs. There are two different mechanisms for generating liquid droplets: continuous IJP and drop-on-demand IJP. A Newtonian fluid material is required to prepare ink and the active material is dispersed into the solvent. The challenges involved in the process are:

  i. *Formation of liquid material:* The active material may cause suspended particle aggregation in the carrier liquid if the active medium is not in liquid form (e.g. carbon, graphene, metal nanoparticles, and carbon nanotubes)[70]. In order to attain acceptable characteristics, other substances like surfactants are added to the liquid.

 ii. *Droplet formation:* Conversion of the continuous volume of liquid ink into small discrete droplets is challenging. The addition of tiny particles can change droplet behavior.

iii. *Control over droplet deposition:* In the printing process, relative motion between the substrate and the print head is established. To overcome this, the calculation of droplet size, trajectory, velocity, location, and quality of the impact must be considered. If the droplet creates smaller droplets, satellites are formed due to breaking off from the main droplet during flight, and if the droplet creates splashes on the impact, it causes a larger spread area than intended, which is called a crown.

 iv. *Conversion of liquid material into solid geometry:* IJP relies on a phase change to convert liquid material droplets into solid geometry. Evaporation of the liquid portion of the solution, solidification of melted material, curing of photopolymers, and chemical reactions are the phase change modes engaging in the IJP method[1].

A good quality print is possible only with optimal printable fluids, and there is a complex and precise physics and fluid mechanics involved in terms of Reynolds number (Re), Weber number (We), and Ohnesorge number (Oh) as seen in Eqs. 11.1–11.3.

$$Re = \frac{\rho v a}{\eta} \tag{11.1}$$

$$We = \frac{\rho a v^2}{\gamma} \tag{11.2}$$

$$Oh = \frac{\sqrt{We}}{Re} \tag{11.3}$$

Where ρ, η, and γ are the density, dynamic viscosity, and surface tension of the fluid, respectively, while a and v are the characteristic length and velocity, respectively. The minimum velocity of the drop from the nozzle for easy ejection is v_{min} for nozzle diameter d_n as shown in equation 11.4.

$$v_{min} = \sqrt{\left(\frac{4\gamma}{\rho d_n}\right)} \tag{11.4}$$

$$We = v_{min}\sqrt{\left(\frac{\rho d_n}{\lambda}\right)} > 4 \tag{11.5}$$

For stable drop generation 10>Z>1, where Z= 1/Oh.

Direct Ink Writing (DIW) is an extrusion-based direct material deposition with paste-like ink possessing shear-thinning behavior. The outstanding advantage of DIW is the availability of a significant number of printable ink types such as sol-gel inks, polyelectrolyte inks, nanoparticle-filled inks, fugitive organic inks, and high shear-thinning colloidal suspension inks. These inks solidify due to solvent-driven reactions, gelation, evaporation, thermal energy, or solvent-induced phase changes. Comparatively, DIW is more promising than IJW because of its simplicity, good rheological properties, high loading of a versatile material, less cost, and more control over material deposition. These merits of DIW make it the more desirable/suitable 3DP process to fabricate MBs and supercapacitors[11,32,71]. In recent years, many breakthroughs in ESD electrode fabrication using DIW have been reported[16,72]. Based on printing materials (e.g. $LiMnO_2$, $Li_4Ti_5O_{12}$, $LiFePO_4$, CNT, and graphene oxide/GO), a few interesting studies on DIW ESDs fabrications and their electrochemical performances were summarized in Tables 11.2 and 11.3.

TABLE 11.2
3D Printed Micro-batteries and Electrochemical Performance

AM technique	Cathode	Anode	Electrolyte	Energy/Power density	Specific capacity	Ref
DIW (Interdigitated)	$LiFePO_4$ (LFP)	$Li_4Ti_5O_{12}$ (LTO)	$LiClO_4$ non-aqueous-liquid	9.7 J cm^{-2} at 2.7 mW cm^{-2}	1.2 mA h cm^{-2}	[34]
DIW (3D-planner)	$LiMnO_2$ (LMO)	LTO	$LiPF_6$	64.6 J cm^{-2} at 2.3 mW cm^{-2}	117 mA h cm^{-2}	[29]
DIW (Interdigitated)	LFP	LTO	LiTFSI/PC	20 mW h cm^{-2} at 1 mW cm^{-2}	14.5 mA h cm^{-2}	[73]
SLA (3D-scaffolds)	Li	Li	$Li_{1.4}Al_{0.4}Ge_{1.6}$-$(PO_4)_3$ (LAGP)			[74]
SLA (Interdigitated)	LFP	LTO	$LiClO_4$			[75]
IJP (Stacked)	$LiCoO_2$ (LCO)	SiO_2	$LiPF_6$		812.7 mA h g^{-1} at 33 μA cm^{-2}	[76]
IJP (Pillar)	LFP	Li-metal	$LiPF_6$		151.3 mA h g^{-1} at 0.1C	[77]
FDM (stacked)	LTO	LFP	$LiPF_6$		40 mA h g^{-1} at 10 mA g^{-1}	[78]
FDM (Framework)	$Na_3V_3PO_4$	GO	$LiClO_4$		1.26 mA h cm^{-2} at 0.2C	[79]

TABLE 11.3
3D Printed Micro-supercapacitors and Electrochemical Performance

AM technique	Electrodes	Electrolyte	Energy/Power density	Specific capacity	Ref
DIW (1D-scaffold)	GO/Ag; GO/MWCNT	KOH	0.021 mW h cm^{-2} at 11.4 mW cm^{-2}	639.6 mFcm^{-2} at 4 mA h cm^{-2}	[86]
DIW (3D-structural)	Pristine GO/MWNTs	H_2SO_4	26.32 KW kg^{-1} and 14.73 W h kg^{-1}	100 F g^{-1} at 3 A g^{-1}	[87]
IJP (Interdigitated)	Graphene	Na_2SO_4		0.82 mF cm^{-2} at 10V	[88]
IJP (Interdigitated)	GO	H_2SO_4	6.74 W h kg^{-1} at 0.19 kW kg^{-1}	132 F g^{-1} at 0.01V s^{-1}	[89]
SLS (Interdigitated)	Ti_6Al_4V	PVA-H_3PO_4		1920 mFcm^{-2} at 3.74 mA h cm^{-2}	[90]
SLS (Interdigitated)	polyamide	LiCl-PVA		42.6 mFcm^{-2} at 0.1 mA h cm^{-2}	[51]
FDM (3D disc)	92% PLA+8% Graphene	KOH		15.8 mA h g^{-1} at 10 mA g^{-1}	[91]
SLA (Interdigitated)	Silver-nitrate ($AgNO_3$) with polyethylene glycol di-acrylate-based composite resin	$LiClO_4$/PVA-gel		3.01 mF g^{-1} at 0.2V s^{-1}	[92]

Fused deposition modeling (FDM) is an extrusion-based AM technique, which prints solid feed materials (mostly thermoplastic polymers) by melting and solidifying by natural cooling through a CNC-controlled nozzle according to a predefined path of 3D design. The minimum feature that the FDM machine can print depends on the nozzle size and the actual shape produced is dependent on the acceleration, deceleration of the printing head/nozzle head, and viscoelastic behavior of the material as it solidifies on deposition. Solidification of deposited material is based on the crystallization of polymer and chain entanglement. In FDM, drawing sharp corners is impossible and the printing time can be reduced with the use of high-speed motors and multiple printing heads. In the FDM process, maintaining optimal temperatures of the build platform is essential to overcome the major issue of solidification of the deposited layer even before the next layer deposition, which causes wrapping. MB and MSC electrode materials are added as additives into the solid feed material like a polymer matrix to fabricate electrodes (e.g. PLA/graphene, PLA/carbon, ABS/graphene, and ABS/carbon). The research claims that a large percentage of PLA was preventing conducting agents from contacting active material. Researchers have developed strategies such as electrodeposition, stamping strategy, and 3D dual printing methods to enhance performance without compromising the FDM printing process[16,31,80].

Vat-polymerization/stereolithography (SLA) is the first developed AM technique, which uses radiation-curable resins in liquid form as printing material. These radiation-curable resins (photopolymers) are irradiated under ultraviolet (UV) in a vat layer by layer subsequently to build the 3D structure. SLA is a complicated process that needs an ultraclean working environment, long fabrication duration, the requirement of post-processing, high resolution, and precision. To address problems of the conventional SLA process, multiple process variants were developed which include digital light processing (DLP), mask projection SLA, two-photon polymerization (TPP), and projection micro-SLA. Researchers have synthesized a graphene-based photocurable resin to print high-surface macro-scale architecture, which exhibits enhanced mechanical and electrochemical performance[81]. Ultra-light, ultra-stiff metallic, and ceramic micro-lattices can be fabricated with a combination of nano-coating, SLA, and post-processing. Therefore, SLA is a promising high-resolution 3DP technique for printing ESDs.

Selective laser sintering (SLS) is a powder bed fusion technique of AM where it fuses various materials (ceramic, metal, and polymer) in powder form in a layer-by-layer fashion using a laser beam. Laser processing parameters, material powder particle size and its quality, and processing environment play a critical role in build quality. SLS is one of the efficient ways to fabricate carbon-based composite electrodes with high resolution. Tailored properties of ESDs can be achieved with binder incorporated into carbon structure during laser processing[82,83]. The technology has also been used to build helical-shaped stainless steel electrodes using full melting (Selective Laser Melting (SLM)) that can be used as platforms for various electrochemical devices such as pseudo-capacitors, oxygen evolution catalysts, and pH sensors. In conclusion, SLS offers a simple and affordable method for fabricating electrodes of ESDs with novel designs. The mechanical and electrochemical characteristics of such 3D-printed functional devices can be significantly influenced by the powder's physical characteristics, such as chemical composition, shape, and size. The high minimum attainable dimensions (in the millimeter range) and the roughness of the metal surface are the main drawbacks of the SLS process.

Laminated Object Manufacturing (LOM) uses the layer-by-layer joining of 2D planar sheets either by gluing/ welding together followed by laser/knife cutting (into dice) along the input design boundaries, and the process continues with the spreading of new sheets till the final structure is complete. The laminated sheet can be paper, plastic, or metal. LOM can be a more suitable process to produce sandwich architecture ESDs. Researchers developed a laser-induced graphene (LIG) process to synthesize porous graphene film structure by laser photo-thermal treatment of polyamide (PA) film in an ambient atmosphere[84,85].

In comparison, low cost, high resolution, rapid printing speed, efficiency, and broad material printability make DIW preferable over other AM techniques for fabricating ESDs.

11.5   FUTURE SCOPE

The world is looking for self-powered and sustainable devices. For modern gadgets, mechanically robust and high-flexibility ESDs are required. However, complications are entangled with integrating all the components in a single-step manufacturing process with tailored properties. The developed 3DP techniques have limitations such as process optimization, material performance, production time, and cost of the final product. Ultrafine microstructure of active material is essential to enhance the electrochemical performance of ESDs, and 3DP technology is a suitable process to produce such components even with complex designs. Energy harvesting/storage systems need very thin and highly flexible ESDs which can be printed efficiently using extrusion and material jetting processes by suspending active materials into the ink with optimized parameters and process control. Future research on 3DP for ESDs should be mainly directed at the integration of printed ESDs with energy harvesting devices and electronic systems. Demand for the development of stable and high electrochemical performances of ESDs is essential for high-temperature applications. Researchers are exploring various combinations of materials, design architectures, and fabrication techniques to further increase the ESD's lifecycle, charge/discharge rate, energy, and power densities at low cost.

11.6   CONCLUSION

A detailed study of ESDs and 3DP fabrication methods was summarized successfully in this chapter. The research trend and reversal of ESD fabrication towards 3DP were discussed. The capability of 3DP and the complications involved in printing were discussed in detail. DIW technique is highly accurate and is currently the best and preferable 3DP technique for ESD fabrication. DIW is mainly recommended for printing all fiber quasi-solid-state LIBs, CNT-based micro-supercapacitors, and hybrid aerogel-based supercapacitors. FDM has a low operating cost, large build size, and rapid printing capability, but limited extrusion temperature, and surface finishing treatment are needed. FDM is suitable for printing composite electrode printing, stamping of solid-state printing of micro-supercapacitors. SLS/SLM is used to fabricate helical-shaped IrO_2 capacitors and metal electrode matrix structures of supercapacitors. SLA printed parts have high resolution and good surface quality. But limited to printing photocurable materials, and suitable to fabricate 3D microstructured electrodes and LFP tri-layered structured cells. IJP can print larger areas, but a limited range of inks, and printable speeds can be used. The IJP technique is used to print graphene-based micro-supercapacitors and stacked/planar solid-state flexible supercapacitors. BJP technique eliminates the usage of solvents in fabrication ESDs, but the strength of the printed parts/electrodes is limited along with more surface roughness. Printing of thick graphene 3D electrodes of supercapacitors has been carried out. The LOM process has low resolution and more anisotropy, and printing of 3D laser-induced graphene-based Li-ion capacitors has been carried out.

3DP of ESDs is in the infancy stage of development. Design architecture, fabrication process, and material processing optimization are uncharted to enhance the electrochemical performance of fabricated ESDs using 3DP. It can be anticipated that with continuous development, brainstorming, and further research, 3DP of ESDs can unfold and play a critical role as the most promising fabrication technique for next-generation ESDs.

REFERENCES

[1] I. Gibson, D. Rosen, B. Stucker, 3D Printing, *Rapid Prototyping, and Direct Digital Manufacturing*, 2015. http://link.springer.com/10.1007/978-1-4939-2113-3

[2] C. Chen, X. Wang, Y. Wang, D. Yang, F. Yao, W. Zhang, B. Wang, G.A. Sewvandi, D. Yang, D. Hu, Additive manufacturing of piezoelectric materials, *Adv. Funct. Mater.* 30 (2020). https://doi.org/10.1002/adfm.202005141

[3] S. Zhang, Y. Liu, J. Hao, G.G. Wallace, S. Beirne, J. Chen, S. Zhang, G.G. Wallace, S. Beirne, J. Chen, Y. Liu, 3D-Printed wearable electrochemical energy devices, *Adv. Funct. Mater.* 32 (2022) 2103092. https://doi.org/10.1002/ADFM.202103092

[4] B. West, E. Robertson, R. Osborne, M. Calvert, *Additive Manufacturing for Affordable Rocket Engines*. (2016).

[5] A. Farazin, C. Zhang, A. Gheisizadeh, A. Shahbazi, 3D bio-printing for use as bone replacement tissues: A review of biomedical application, *Biomed. Eng. Adv.* 5 (2023) 100075. https://doi.org/10.1016/J.BEA.2023.100075

[6] H. Bhuskute, P. Shende, B. Prabhakar, 3D Printed personalized medicine for cancer: Applications for betterment of diagnosis, prognosis and treatment, *AAPS PharmSciTech.* 23 (2022) 1–12. https://doi.org/10.1208/S12249-021-02153-0/TABLES/2

[7] D. Dey, D. Srinivas, B. Panda, P. Suraneni, T.G. Sitharam, Use of industrial waste materials for 3D printing of sustainable concrete: A review, *J. Clean. Prod.* 340 (2022) 130749. https://doi.org/10.1016/J.JCLEPRO.2022.130749

[8] T. Xu, J. Zhang, M. Salehizadeh, O. Onaizah, E. Diller, Millimeter-scale flexible robots with programmable three-dimensional magnetization and motions, *Sci. Robot.* 4 (2019). https://doi.org/10.1126/SCIROBOTICS.AAV4494/SUPPL_FILE/AAV4494_SM.PDF

[9] No assembly required: U of T Engineering researchers automate microrobotic designs, (n.d.). https://news.engineering.utoronto.ca/automate-microrobotic-designs-eric-dil ler/ (accessed February 17, 2023).

[10] Z. Wang, Y. Chen, Y. Zhou, J. Ouyang, S. Xu, L. Wei, Miniaturized lithium-ion batteries for on-chip energy storage, (2022). https://doi.org/10.1039/d2na00566b

[11] K. Sun, T.S. Wei, B.Y. Ahn, J.Y. Seo, S.J. Dillon, J.A. Lewis, 3D printing of interdigitated Li-ion microbattery architectures, *Adv. Mater.* 25 (2013) 4539–4543. https://doi.org/10.1002/ADMA.201301036/ABSTRACT

[12] W. Liu, M.-S. Song, B. Kong, Y. Cui, W. Liu, M.-S. Song, B. Kong, Y. Cui, Flexible and stretchable energy storage: Recent advances and future perspectives, *Adv. Mater.* 29 (2017) 1603436. https://doi.org/10.1002/ADMA.201603436

[13] J. Kim, R. Kumar, A.J. Bandodkar, J. Wang, J. Kim, R. Kumar, A.J. Bandodkar, J. Wang, Advanced materials for printed wearable electrochemical devices: A review, *Adv. Electron. Mater.* 3 (2017) 1600260. https://doi.org/10.1002/AELM.201600260

[14] R. Jia, G. Shen, F. Qu, D. Chen, Flexible on-chip micro-supercapacitors: Efficient power units for wearable electronics, *Energy Storage Mater.* 27 (2020) 169–186. https://doi.org/10.1016/J.ENSM.2020.01.030

[15] Y. Zhang, W. Bai, X. Cheng, J. Ren, W. Weng, P. Chen, X. Fang, Z. Zhang, H. Peng, Y. Zhang, W. Bai, X. Cheng, J. Ren, W. Weng, P. Chen, X. Fang, Z. Zhang, H. Peng, Flexible and stretchable lithium-Ion batteries and supercapacitors based on electrically conducting carbon nanotube fiber springs, *Angew. Chemie Int. Ed.* 53 (2014) 14564– 14568. https://doi.org/10.1002/ANIE.201409366

[16] P. Chang, H. Mei, S. Zhou, K.G. Dassios, L. Cheng, 3D printed electrochemical energy storage devices, *J. Mater. Chem. A.* 7 (2019) 4230–4258. https://doi.org/10.1039/c8t a11860d

[17] G. Zhang, Y. Han, C. Shao, N. Chen, G. Sun, X. Jin, J. Gao, B. Ji, H. Yang, L. Qu, Processing and manufacturing of graphene-based microsupercapacitors, *Mater. Chem. Front.* 2 (2018) 1750–1764. https://doi.org/10.1039/C8QM00270C

[18] J. Qin, P. Das, S. Zheng, Z.S. Wu, A perspective on two-dimensional materials for planar micro-supercapacitors, *APL Mater.* 7 (2019) 090902. https://doi.org/10.1063/ 1.5113940

[19] H. Li, J. Liang, Recent development of printed micro-supercapacitors: Printable materials, printing technologies, and perspectives, *Adv. Mater.* 32 (2020) 1805864. https://doi.org/10.1002/ADMA.201805864

[20] W. Zuo, R. Li, C. Zhou, Y. Li, J. Xia, J. Liu, W.H. Zuo, J.L. Xia, J.P. Liu, R.Z. Li, C. Zhou, Y.Y. Li, Battery-supercapacitor hybrid devices: Recent progress and future prospects, *Adv. Sci.* 4 (2017) 1600539. https://doi.org/10.1002/ADVS.201600539

[21] 3D printing of energy storage devices, Analyze Results, (n.d.). www.webofscience. com/wos/woscc/analyze-results/d75d19dd-1fd5-400c-88cb-9ce583d6ec15-7380d e53?state=%7B%22backlink%22:false%7D (accessed February 26, 2023).

[22] X. Jin, G. Zhang, G. Sun, H. Yang, Y. Xiao, J. Gao, Z. Zhang, L. Jiang, L. Qu, Flexible and high-performance microsupercapacitors with wide temperature tolerance, *Nano Energy.* 64 (2019) 103938. https://doi.org/10.1016/J.NANOEN.2019.103938

[23] S. Zheng, C. Zhang, F. Zhou, Y. Dong, X. Shi, V. Nicolosi, Z.S. Wu, X. Bao, Ionic liquid pre-intercalated MXene films for ionogel-based flexible micro-supercapacitors with high volumetric energy density, *J. Mater. Chem. A.* 7 (2019) 9478–9485. https:// doi.org/10.1039/C9TA02190F

[24] L. Qin, Q. Tao, X. Liu, M. Fahlman, J. Halim, P.O.Å. Persson, J. Rosen, F. Zhang, Polymer-MXene composite films formed by MXene-facilitated electrochemical

polymerization for flexible solid-state microsupercapacitors, *Nano Energy*. 60 (2019) 734–742. https://doi.org/10.1016/J.NANOEN.2019.04.002

[25] F. Clerici, M. Fontana, S. Bianco, M. Serrapede, F. Perrucci, S. Ferrero, E. Tresso, A. Lamberti, In situ MoS₂ decoration of laser-induced graphene as flexible supercapacitor electrodes, *ACS Appl. Mater. Interfaces*. 8 (2016) 10459–10465. https://doi.org/10.1021/ACSAMI.6B00808/ASSET/IMAGES/LARGE/AM-2016-00808B_0007.JPEG

[26] H. Li, Q. Zhao, W. Wang, H. Dong, D. Xu, G. Zou, H. Duan, D. Yu, Novel planar-structure electrochemical devices for highly flexible semitransparent power generation/storage sources, *Nano Lett*. 13 (2013) 1271–1277. https://doi.org/10.1021/NL4000079/SUPPL_FILE/NL4000079_SI_001.PDF

[27] S. Zheng, J. Ma, Z.S. Wu, F. Zhou, Y.B. He, F. Kang, H.M. Cheng, X. Bao, All-solid-state flexible planar lithium ion micro-capacitors, *Energy Environ. Sci*. 11 (2018) 2001–2009. https://doi.org/10.1039/C8EE00855H

[28] S. Zheng, X. Shi, P. Das, Z.S. Wu, X. Bao, The road towards planar microbatteries and micro-supercapacitors: From 2D to 3D device geometries, *Adv. Mater*. 31 (2019) 1900583. https://doi.org/10.1002/ADMA.201900583

[29] J. Li, M.C. Leu, R. Panat, J. Park, A hybrid three-dimensionally structured electrode for lithium-ion batteries via 3D printing, *Mater. Des*. 119 (2017) 417–424. https://doi.org/10.1016/J.MATDES.2017.01.088

[30] C.W. Foster, M.P. Down, Y. Zhang, X. Ji, S.J. Rowley-Neale, G.C. Smith, P.J. Kelly, C.E. Banks, 3D printed graphene based energy storage devices, *Sci. Rep*. 7 (2017). https://doi.org/10.1038/SREP42233

[31] B.L. Ellis, P. Knauth, T. Djenizian, B.L. Ellis, T. Djenizian, P. Knauth, Three-dimensional self-supported metal oxides for advanced energy storage, *Adv. Mater*. 26 (2014) 3368–3397. https://doi.org/10.1002/ADMA.201306126

[32] J. Hu, Y. Jiang, S. Cui, Y. Duan, T. Liu, H. Guo, L. Lin, Y. Lin, J. Zheng, K. Amine, F. Pan, J. Hu, Y. Jiang, S. Cui, Y. Duan, T. Liu, H. Guo, L. Lin, Y. Lin, J. Zheng, K. Amine, F. Pan, 3D-Printed cathodes of LiMn1−xFexPO4 nanocrystals achieve both ultrahigh rate and high capacity for advanced lithium-Ion battery, *Adv. Energy Mater*. 6 (2016) 1600856. https://doi.org/10.1002/AENM.201600856

[33] J. Qin, P. Das, S. Zheng, Z.S. Wu, A perspective on two-dimensional materials for planar micro-supercapacitors, *APL Mater*. 7 (2019) 090902. https://doi.org/10.1063/1.5113940

[34] X. Li, H. Li, X. Fan, X. Shi, J. Liang, X. Li, H. Li, X. Fan, X. Shi, J. Liang, 3D-Printed stretchable micro-supercapacitor with remarkable areal performance, *Adv. Energy Mater*. 10 (2020) 1903794. https://doi.org/10.1002/AENM.201903794

[35] Z. Wang, Y. Chen, Y. Zhou, J. Ouyang, S. Xu, L. Wei, Miniaturized lithium-ion batteries for on-chip energy storage, *Nanoscale Adv*. 4 (2022) 4237–4257. https://doi.org/10.1039/d2na00566b

[36] E. Dhandapani, S. Thangarasu, S. Ramesh, K. Ramesh, R. Vasudevan, N. Duraisamy, Recent development and prospective of carbonaceous material, conducting polymer and their composite electrode materials for supercapacitor — A review, *J. Energy Storage*. 52 (2022) 104937. https://doi.org/10.1016/J.EST.2022.104937

[37] W.J. Hyun, E.B. Secor, C.H. Kim, M.C. Hersam, L.F. Francis, C.D. Frisbie, Scalable, Self-Aligned printing of flexible graphene micro-supercapacitors, *Adv. Energy Mater*. 7 (2017) 1700285. https://doi.org/10.1002/AENM.201700285

[38] J. Yun, Y. Lim, H. Lee, G. Lee, H. Park, S. Yeong Hong, S. Woo Jin, Y. Hui Lee, S.-S. Lee, J. Sook Ha, J. Yun, H. Lee, H. Park, S.Y. Hong, Y.H. Lee, J.S. Ha, Y. Lim, G. Lee, S.W. Jin, S. Lee, A patterned graphene/ZnO UV sensor driven by integrated

asymmetric micro-supercapacitors on a liquid metal patterned foldable paper, *Adv. Funct. Mater.* 27 (2017) 1700135. https://doi.org/10.1002/ADFM.201700135

[39] G. Lee, J.W. Kim, H. Park, J.Y. Lee, H. Lee, C. Song, S.W. Jin, K. Keum, C.H. Lee, J.S. Ha, Skin-Like, dynamically stretchable, planar supercapacitors with buckled carbon nanotube/mn-mo mixed oxide electrodes and air-stable organic electrolyte, *ACS Nano.* 13 (2019) 855–866. https://doi.org/10.1021/ACSNANO.8B08645/ASSET/IMAGES/MEDIUM/NN-2018-08645F_M005.GIF

[40] N. Kurra, M.K. Hota, H.N. Alshareef, Conducting polymer micro-supercapacitors for flexible energy storage and Ac line-filtering, *Nano Energy.* 13 (2015) 500–508. https://doi.org/10.1016/J.NANOEN.2015.03.018

[41] H. Kim, J. Yoon, G. Lee, S.H. Paik, G. Choi, D. Kim, B.M. Kim, G. Zi, J.S. Ha, Encapsulated, high-performance, stretchable array of stacked planar micro-supercapacitors as waterproof wearable energy storage devices, *ACS Appl. Mater. Interfaces.* 8 (2016) 16016–16025. https://doi.org/10.1021/ACSAMI.6B03504/ASSET/IMAGES/LARGE/AM-2016-03504Y_0007.JPEG

[42] H. Liu, G. Zhang, X. Zheng, F. Chen, H. Duan, Emerging miniaturized energy storage devices for microsystem applications: From design to integration, *Int. J. Extrem. Manuf.* 2 (2020). https://doi.org/10.1088/2631-7990/abba12

[43] F.N. Crespilho, G.C. Sedenho, D. De Porcellinis, E. Kerr, S. Granados-Focil, R.G. Gordon, M.J. Aziz, Non-corrosive, low-toxicity gel-based microbattery from organic and organometallic molecules, *J. Mater. Chem. A.* 7 (2019) 24784–24787. https://doi.org/10.1039/C9TA08685D

[44] W. Li, L.C. Bradley, J.J. Watkins, Copolymer solid-state electrolytes for 3d microbatteries via initiated chemical vapor deposition, *ACS Appl. Mater. Interfaces.* 11 (2019) 5674. https://doi.org/10.1021/ACSAMI.8B19689/ASSET/IMAGES/MEDIUM/AM-2018-19689K_M003.GIF

[45] K. Kanehori, K. Matsumoto, K. Miyauchi, T. Kudo, Thin film solid electrolyte and its application to secondary lithium cell, *Solid State Ionics.* 9–10 (1983) 1445–1448. https://doi.org/10.1016/0167-2738(83)90192-3

[46] C. Ke Sun, T.-S. Wei, B. Yeop Ahn, J. Yoon Seo, S.J. Dillon, J.A. Lewis, K. Sun, S.J. Dillon, T.S. Wei, B.Y. Ahn, J.A. Lewis, J.Y. Seo, 3D Printing of interdigitated Li-Ion microbattery architectures, *Adv. Mater.* 25 (2013) 4539–4543. https://doi.org/10.1002/ADMA.201301036

[47] Y.Q. Li, H. Shi, S.B. Wang, Y.T. Zhou, Z. Wen, X.Y. Lang, Q. Jiang, Dual-phase nanostructuring of layered metal oxides for high-performance aqueous rechargeable potassium ion microbatteries, *Nat. Commun.* 101(10) (2019) 1–9. https://doi.org/10.1038/s41467-019-12274-7

[48] S. Zheng, H. Huang, Y. Dong, S. Wang, F. Zhou, J. Qin, C. Sun, Y. Yu, Z.S. Wu, X. Bao, Ionogel-based sodium ion micro-batteries with a 3D Na-ion diffusion mechanism enable ultrahigh rate capability, *Energy Environ. Sci.* 13 (2020) 821–829. https://doi.org/10.1039/C9EE03219C

[49] T. Nathan-Walleser, I.M. Lazar, M. Fabritius, F.J. Tölle, Q. Xia, B. Bruchmann, S.S. Venkataraman, M.G. Schwab, R. Mülhaupt, 3D Micro-extrusion of graphene-based active electrodes: towards high-rate ac line filtering performance electrochemical capacitors, *Adv. Funct. Mater.* 24 (2014) 4706–4716. https://doi.org/10.1002/ADFM.201304151

[50] M.F. El-Kady, R.B. Kaner, Scalable fabrication of high-power graphene micro-supercapacitors for flexible and on-chip energy storage, *Nat. Commun.* 41(4) (2013) 1–9. https://doi.org/10.1038/ncomms2446

[51] S. Wang, Y. Yu, S. Luo, X. Cheng, G. Feng, Y. Zhang, Z. Wu, G. Compagnini, J. Pooran, A. Hu, All-solid-state supercapacitors from natural lignin-based composite film by laser direct writing, *Appl. Phys. Lett.* 115 (2019) 083904. https://doi.org/10.1063/1.5118340

[52] J. Zhao, Y. Zhang, J. Yan, X. Zhao, J. Xie, X. Luo, J. Peng, J. Wang, L. Meng, Z. Zeng, C. Lu, X. Xu, Y. Dai, Y. Yao, Fiber-shaped electrochemical capacitors based on plasma-engraved graphene fibers with oxygen vacancies for alternating current line filtering performance, *ACS Appl. Energy Mater.* 2 (2019) 993–999. https://doi.org/10.1021/ACSAEM.8B02060/ASSET/IMAGES/LARGE/AE-2018-02060T_0005.JPEG

[53] M. Criado-Gonzalez, A. Dominguez-Alfaro, N. Lopez-Larrea, N. Alegret, D. Mecerreyes, Additive manufacturing of conducting polymers: recent advances, challenges, and opportunities, *ACS Appl. Polym. Mater.* 3 (2021) 2865–2883. https://doi.org/10.1021/ACSAPM.1C00252/ASSET/IMAGES/LARGE/AP1C00252_0007.JPEG

[54] H. Bishwakarma, A.K. Das, Synthesis of zinc oxide nanoparticles through hybrid machining process and their application in supercapacitors, *J. Electron. Mater.* 49 (2020) 1541–1549. https://doi.org/10.1007/S11664-019-07835-X/METRICS

[55] H. Bishwakarma, R. Tyagi, N. Kumar, A.K. Das, Green synthesis of flower shape ZnO-GO nanocomposite through optimized discharge parameter and its efficiency in energy storage device, *Environ. Res.* 218 (2023) 115021. https://doi.org/10.1016/J.ENVRES.2022.115021

[56] H. Bishwakarma, N. Kumar, M. Anand, A.K. Das, P.K. Singh, S.K. Mishra, Magnetic assisted synthesis of ZnO nanoparticle by electrochemical discharge method, *Mater. Today Proc.* 56 (2022) 857–861. https://doi.org/10.1016/J.MATPR.2022.02.511

[57] Z.S. Wu, K. Parvez, S. Li, S. Yang, Z. Liu, S. Liu, X. Feng, K. Müllen, Alternating stacked graphene-conducting polymer compact films with ultrahigh areal and volumetric capacitances for high-energy micro-supercapacitors, *Adv. Mater.* 27 (2015) 4054–4061. https://doi.org/10.1002/ADMA.201501643

[58] Q. Jiang, N. Kurra, C. Xia, H.N. Alshareef, Q. Jiang, N. Kurra, C. Xia, H.N. Alshareef, Materials science, hybrid microsupercapacitors with vertically scaled 3d current collectors fabricated using a simple cut-and-transfer strategy, *Adv. Energy Mater.* 7 (2017) 1601257. https://doi.org/10.1002/AENM.201601257

[59] Z. Hu, X. Xiao, H. Jin, T. Li, M. Chen, Z. Liang, Z. Guo, J. Li, J. Wan, L. Huang, Y. Zhang, G. Feng, J. Zhou, Rapid mass production of two-dimensional metal oxides and hydroxides via the molten salts method, *Nat. Commun.* 81(8) (2017) 1–9. https://doi.org/10.1038/ncomms15630

[60] M. Boota, B. Anasori, C. Voigt, M.-Q. Zhao, M.W. Barsoum, Y. Gogotsi, M.B. Boota, C. Anasori, M.-Q. Voigt, M.W. Zhao, Y. Barsoum, A.J. Gogotsi, Pseudocapacitive electrodes produced by oxidant-free polymerization of pyrrole between the layers of 2d titanium carbide (MXene), *Adv. Mater.* 28 (2016) 1517–1522. https://doi.org/10.1002/ADMA.201504705

[61] C.-C. Hu, W.-C. Chen, K.-H. Chang, How to achieve maximum utilization of hydrous ruthenium oxide for supercapacitors, *J. Electrochem. Soc.* 151 (2004) A281. https://doi.org/10.1149/1.1639020/XML

[62] S.K. Mandal, S. Kumar, P.K. Singh, S.K. Mishra, H. Bishwakarma, N.P. Choudhry, R.K. Nayak, A.K. Das, Performance investigation of CuO-paraffin wax nanocomposite in solar water heater during night, *Thermochim. Acta.* 671 (2019) 36–42. https://doi.org/10.1016/J.TCA.2018.11.003

[63] P.K. Singh, H. Bishwakarma, Shubham, A.K. Das, Study of annealing effects on Ag_2O nanoparticles generated by electrochemical spark process, *J. Electron. Mater.* 46 (2017) 5715–5727. https://doi.org/10.1007/S11664-017-5614-6/METRICS

[64] P.K. Singh, Shubham, N.K. Singh, H. Bishwakarma, M. Hussain, A.K. Das, B.H. Prasad, Effect of annealing on silver oxide nano particle generated by electrochemical discharge machining, *Mater. Today Proc.* 5 (2018) 26804–26809. https://doi.org/10.1016/J.MATPR.2018.08.160

[65] M. Koo, K. Il Park, S.H. Lee, M. Suh, D.Y. Jeon, J.W. Choi, K. Kang, K.J. Lee, Bendable inorganic thin-film battery for fully flexible electronic systems, *Nano Lett.* 12 (2012) 4810–4816. https://doi.org/10.1021/NL302254V/SUPPL_FILE/NL302 254V_SI_002.AVI

[66] N.A. Kyeremateng, T. Brousse, D. Pech, Microsupercapacitors as miniaturized energy-storage components for on-chip electronics, *Nat. Nanotechnol.* 121(12) (2016) 7–15. https://doi.org/10.1038/nnano.2016.196

[67] Q. Zhang, C. Li, Q. Li, Z. Pan, J. Sun, Z. Zhou, B. He, P. Man, L. Xie, L. Kang, X. Wang, J. Yang, T. Zhang, P.P. Shum, Q. Li, Y. Yao, L. Wei, Flexible and high-voltage coaxial-fiber aqueous rechargeable Zinc-Ion battery, *Nano Lett.* 19 (2019) 4035–4042. https://doi.org/10.1021/ACS.NANOLETT.9B01403/ASSET/IMAGES/LARGE/NL-2019-014032_0006.JPEG

[68] K. Wang, X. Zhang, J. Han, X. Zhang, X. Sun, C. Li, W. Liu, Q. Li, Y. Ma, High-performance cable-type flexible rechargeable zn battery based on MnO2@CNT fiber microelectrode, *ACS Appl. Mater. Interfaces.* 10 (2018) 24573–24582. https://doi.org/10.1021/ACSAMI.8B07756/ASSET/IMAGES/LARGE/AM-2018-07756Q_0 002.JPEG

[69] E. Gibertini, L. Magagnin, E. Gibertini, L. Magagnin, PEDOTS:PSS@KNF wire-shaped electrodes for textile symmetrical capacitor, *Adv. Mater. Interfaces.* 9 (2022) 2200513. https://doi.org/10.1002/ADMI.202200513

[70] K.H. Choi, J.T. Yoo, C.K. Lee, S.Y. Lee, All-inkjet-printed, solid-state flexible supercapacitors on paper, *Energy Environ. Sci.* 9 (2016) 2812–2821. https://doi.org/10.1039/C6EE00966B

[71] K. Fu, Y. Wang, C. Yan, Y. Yao, Y. Chen, J. Dai, S. Lacey, Y. Wang, J. Wan, T. Li, Z. Wang, Y. Xu, L. Hu, Graphene Oxide-Based electrode inks for 3d-printed Lithium-Ion batteries, *Adv. Mater.* 28 (2016) 2587–2594. https://doi.org/10.1002/ADMA.201505391

[72] A. Ambrosi, M. Pumera, 3D-printing technologies for electrochemical applications, *Chem. Soc. Rev.* 45 (2016) 2740–2755. https://doi.org/10.1039/C5CS00714C

[73] K. Sun, T.-S. Wei, B.Y. Ahn, J.Y. Seo, S.J. Dillon, J.A. Lewis, 3D printing of interdigitated Li-Ion microbattery architectures, *Adv. Mater.* 25 (2013) 4539–43. https://doi.org/10.1002/ADMA.201301036

[74] S. Zekoll, C. Marriner-Edwards, A.K.O. Hekselman, J. Kasemchainan, C. Kuss, D.E.J. Armstrong, D. Cai, R.J. Wallace, F.H. Richter, J.H.J. Thijssen, P.G. Bruce, Hybrid electrolytes with 3D bicontinuous ordered ceramic and polymer microchannels for all-solid-state batteries, *Energy Environ. Sci.* 11 (2018) 185–201. https://doi.org/10.1039/C7EE02723K

[75] Q. Chen, R. Xu, Z. He, K. Zhao, L. Pan, Printing 3D gel polymer electrolyte in lithium-ion microbattery using stereolithography, *J. Electrochem. Soc.* 164 (2017) A1852. https://doi.org/10.1149/2.0651709JES

[76] Z. Liu, Q. Yu, Y. Zhao, R. He, M. Xu, S. Feng, S. Li, L. Zhou, L. Mai, Silicon oxides: A promising family of anode materials for lithium-ion batteries, *Chem. Soc. Rev.* 48 (2019) 285–309. https://doi.org/10.1039/C8CS00441B

[77] Y. Gu, A. Wu, H. Sohn, C. Nicoletti, Z. Iqbal, J.F. Federici, Fabrication of rechargeable lithium ion batteries using water-based inkjet printed cathodes, *J. Manuf. Process.* 20 (2015) 198–205. https://doi.org/10.1016/J.JMAPRO.2015.08.003.

[78] S. Wang, Y. Yu, S. Luo, X. Cheng, G. Feng, Y. Zhang, Z. Wu, G. Compagnini, J. Pooran, A. Hu, All-solid-state supercapacitors from natural lignin-based composite film by laser direct writing, *Appl. Phys. Lett.* 115 (2019) 083904. https://doi.org/10.1063/1.5118340

[79] J. Ding, K. Shen, Z. Du, B. Li, S. Yang, 3D-Printed hierarchical porous frameworks for sodium storage, *ACS Appl. Mater. Interfaces.* 9 (2017) 41871–41877. https://doi.org/10.1021/ACSAMI.7B12892/ASSET/IMAGES/LARGE/AM-2017-12892B_0005.JPEG

[80] S. Wang, Z.S. Wu, S. Zheng, F. Zhou, C. Sun, H.M. Cheng, X. Bao, Scalable fabrication of photochemically reduced graphene-based monolithic micro-supercapacitors with superior energy and power densities, *ACS Nano.* 11 (2017) 4283–4291. https://doi.org/10.1021/ACSNANO.7B01390/ASSET/IMAGES/NN-2017-01390A_M005.GIF

[81] R.M. Hensleigh, H. Cui, J.S. Oakdale, J.C. Ye, P.G. Campbell, E.B. Duoss, C.M. Spadaccini, X. Zheng, M.A. Worsley, Additive manufacturing of complex micro-architected graphene aerogels, *Mater. Horizons.* 5 (2018) 1035–1041. https://doi.org/10.1039/C8MH00668G

[82] J. Bauer, A. Schroer, R. Schwaiger, O. Kraft, Approaching theoretical strength in glassy carbon nanolattices, *Nat. Mater.* 15 (2016) 438–443. https://doi.org/10.1038/NMAT4561

[83] J. Sha, Y. Li, R. Villegas Salvatierra, T. Wang, P. Dong, Y. Ji, S.K. Lee, C. Zhang, J. Zhang, R.H. Smith, P.M. Ajayan, J. Lou, N. Zhao, J.M. Tour, Three-dimensional printed graphene foams, *ACS Nano.* 11 (2017) 6860–6867. https://doi.org/10.1021/ACSNANO.7B01987/ASSET/IMAGES/LARGE/NN-2017-019872_0007.JPEG

[84] L.X. Duy, Z. Peng, Y. Li, J. Zhang, Y. Ji, J.M. Tour, Laser-induced graphene fibers, *Carbon N. Y.* 126 (2018) 472–479. https://doi.org/10.1016/J.CARBON.2017.10.036

[85] D. Xuan Luong, A.K. Subramanian, G.A. Lopez Silva, J. Yoon, S. Cofer, K. Yang, P. Samora Owuor, T. Wang, Z. Wang, J. Lou, P.M. Ajayan, J.M. Tour, D.X. Luong, G.A. Lopez Silva, J. Yoon, T. Wang, Z. Wang, J.M. Tour, A.K. Subramanian, P.S. Owuor, J. Lou, P.M. Ajayan, S. Cofer, K. Yang, Laminated object manufacturing of 3D-printed laser-induced graphene foams, *Adv. Mater.* 30 (2018) 1707416. https://doi.org/10.1002/ADMA.201707416

[86] J. Yang, Q. Cao, X. Tang, J. Du, T. Yu, X. Xu, D. Cai, C. Guan, W. Huang, 3D-Printed highly stretchable conducting polymer electrodes for flexible supercapacitors, *J. Mater. Chem. A.* 9 (2021) 19649–19658. https://doi.org/10.1039/D1TA02617H

[87] V.G. Rocha, E. García-Tuñón, C. Botas, F. Markoulidis, E. Feilden, E. D'Elia, N. Ni, M. Shaffer, E. Saiz, Multimaterial 3D printing of graphene-based electrodes for electrochemical energy storage using thermoresponsive inks, *ACS Appl. Mater. Interfaces.* 9 (2017) 37136–37145. https://doi.org/10.1021/ACSAMI.7B10285

[88] J. Li, F. Ye, S. Vaziri, M. Muhammed, M.C. Lemme, M. Östling, Efficient inkjet printing of graphene, *Adv. Mater.* 25 (2013) 3985–3992. https://doi.org/10.1002/ADMA.201300361

[89] L.T. Le, M.H. Ervin, H. Qiu, B.E. Fuchs, W.Y. Lee, Graphene supercapacitor electrodes fabricated by inkjet printing and thermal reduction of graphene oxide, *Electrochem. Commun.* 13 (2011) 355–358. https://doi.org/10.1016/J.ELECOM.2011.01.023

[90] C. Zhao, C. Wang, R. Gorkin, S. Beirne, K. Shu, G.G. Wallace, Three dimensional (3D) printed electrodes for interdigitated supercapacitors, *Electrochem. Commun.* 41 (2014) 20–23. https://doi.org/10.1016/J.ELECOM.2014.01.013

[91] C.W. Foster, M.P. Down, Y. Zhang, X. Ji, S.J. Rowley-Neale, G.C. Smith, P.J. Kelly, C.E. Banks, 3D printed graphene based energy storage devices, *Sci. Rep.* 7 (2017). https://doi.org/10.1038/SREP42233

[92] S. Hyeon Park, M. Kaur, D. Yun, W. Soo Kim, *Hierarchically Designed Electron Paths in 3D Printed Energy Storage Devices*, Langmuir. 34 (2018) 28. https://doi.org/10.1021/acs.langmuir.8b02404

12 Two-dimensional Molybdenum Disulphide-based Materials

Synthesis, Modification and Applications in Supercapacitor Technology

Dipanwita Majumdar and Munmun Mondal

12.1 INTRODUCTION

12.1.1 Importance of Supercapacitor Technology

The sky-high issues of the energy crisis to meet the all-round daily demands and needs of modern civilization as well as to address the environment pollution concerns have motivated the scientific community worldwide to exercise renewable and sustainable energy generation schemes (Kyriakopoulos & Arabatzis, 2016). Such green energy productions stimulate the necessity for well-organized and diligent energy conversion and storage devices, which can momentarily supply energy at the immediate point of requirement to full effectiveness (Akram et al., 2020). In this aspect, electrochemical energy storage technology, comprising rechargeable batteries, supercapacitors and their hybrids, has been one of the most flourishing fields that have emerged in recent times. Supercapacitors, in contrast to bulky and short-lived batteries, can be miniatured, and offer fast charging time, unhurried discharging time, greater power density, longer durability and easy portability (Yan et al., 2014). Unfortunately, the practical applications of these supercapacitor devices have not yet fully accomplished their full success in industrial sectors owing to poor energy storage capability, limited operating voltage, and high self-discharge rates (Liu et al., 2021). Consequently, researchers worldwide are very much focused on developing schemes by adopting well-controlled synthesis of nanomaterials with well-defined physico-chemical properties which can successfully redress the current challenges in this technological field (Huang et al., 2019). Even efforts are being made in designing hybrids of batteries

DOI: 10.1201/9781003318859-12

and supercapacitors in the forms of outstanding supercapatteries which club together the good qualities of the individual systems for overall better outputs (Majumdar et al., 2020). Such developments in electrochemical energy storage outline as well as recommend the promotion of clean energy carrier systems to supplement or even replace the need for non-eco-friendly non-renewable fossil fuels in the coming days (Wang et. al 2017; Chen, 2017).

12.2 BASICS OF SUPERCAPACITOR TECHNOLOGY

The basic components of a typical supercapacitor device, as shown in Figure 12.1(a) (trivial model), comprise two large surface area-based electrodes separated by a semi-porous membrane (separators), and ionically connected by electrolyte, the entire set-up remaining sealed inside an insulating jacket (Majumdar et al., 2020; González et al., 2016). Technically, the energy accumulation and conversion efficacies of these devices are steered by energy capacity and power deliverance competences defined by the terms – energy density or specific energy and power densities or specific power respectively. Their magnitudes are dependent on parameters such as gravi-metric/area/volumetric capacitance, internal resistances, voltage window, etc., which are calculated from the data obtained from cyclic voltammetry (CV), galvanostatic charging/ discharging (GCD) studies and Electrochemical Impedance Spectroscopy (EIS) analyses (Najib & Erdem, 2019; Yaseen et al., 2021). Systematic and rigorous investigations have confirmed that the overall electrochemical performances are pre-dominantly guided by the nature, type and chemistry of electrode and electrolyte

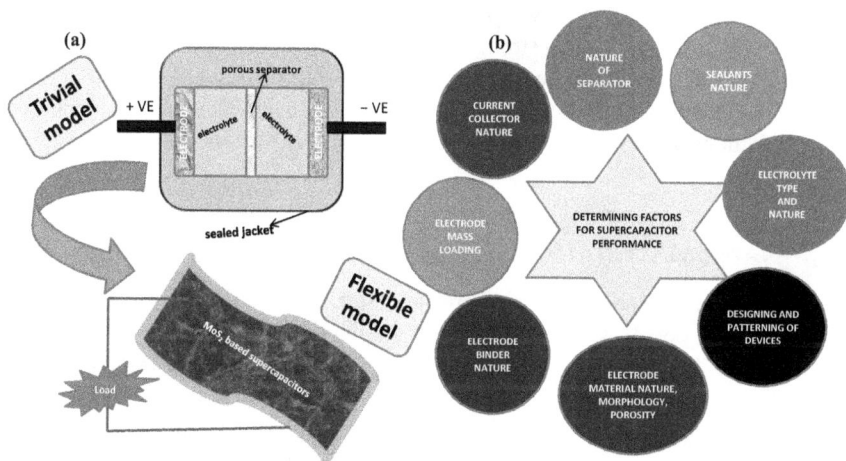

FIGURE 12.1 (a) Diagrammatic representation of the evolution of typical supercapacitor set-up (trivial model) to currently focussed designs (flexible model) for integrating with wearable electronics (b) Schematic representation showing the different components of the supercapacitor devices, different varieties of electrodes and electrolytes employed and the main parameters that dictate the overall performance of the supercapacitors.

materials, although contributions from other structural components, as schematically indicated in Figure 12.1(b), are also vital and cannot be taken frivolously.

The electrode materials employed in supercapacitors can be categorized into two kinds: pseudocapacitive and electrical double-layer capacitive materials (Majumdar & Ghosh, 2021; Majumdar & Bhattacharya, 2017). Pseudocapacitors, comprising different metallic compounds embracing oxides, sulphides, nitrides, carbides, etc., conducting polymers, and other nanosystems, show greater efficacy towards energy density owing to faradaic reversible redox reaction process compared to the electrical double layer capacitors composed of nano-carbon materials like graphene, glassy carbons, carbon nanotubes, g-c_3N_4, etc., where charge accumulation is governed by charge separation at the electrode/electrolyte interfaces (Majumdar, 2019; Majumdar et al., 2019; Majumdar et al., 2017). However, the former experiences sluggish kinetics and irreversibility of electrochemical reactions with exceptionally low electrochemical constancy. Not to mention that, although the characteristics of electrolytes play a significant role in charge accumulation kinetics and storage, the modification of performances, largely, can be more comprehensively carried out through proper fabrication of electrode materials. Additionally, in recent years, with the advent of wearable electronics, designing of next-generation supercapacitors with excellent flexibility and mechanical strength are urged that maintain unaltered electrochemical responses even on exposure to different mechanical deformation conditions and under extreme weather conditions (Sundriyal & Bhattacharya, 2020). Accordingly, scientists are continuously focussing on designing and developing new smart and diligent electrode materials with large surface area, optimized morphology and high mechanical tenacity, by implementing facile, cost-effective and well-controlled synthetic schemes of nanomaterials, which can deliver high capacitance, better charge transport, large cyclic efficacy, wide operating potential, and electrochemical stability as well as excellent mechanical strength and durability, to be effectively deployed for designing supercapacitors to be employed in powering miniatured and feather-light wearable electronics, as indicated in Figure 12.1(a) (Majumdar, 2021; Majumdar, 2022; Majumdar, 2019).

12.3 IMPORTANCE OF MOLYBDENUM DISULPHIDE NANOMATERIALS

Transition metal sulphides such as copper sulphides, nickel sulphides, tin sulphides, zinc sulphides, cobalt sulphides, ferrous sulphides, molybdenum and tungsten sulphides, etc., have been comprehensively explored as an electrochemically active material for supercapacitor performance (Geim & Grigorieva, 2013). Molybdenum disulphide (MoS_2) is one of the most popular earth-abundant representative transition metal sulphides, with unique physicochemical properties, layered structure and electronic properties, which has been widely investigated in various other sectors of modern-day technology (Li & Zhu, 2015). It is the uniqueness of MoS_2 in terms of easy fabrication, different oxidation states with polymorphism, distinctive surface and edge states, exclusive layered morphology, large specific surface area, tremendous physical and chemical stability, and distinctive high electrochemical properties have placed them as highly demanded nanomaterials for energy storage and conversion systems (Li & Zhu, 2015; Theerthagiri et al., 2017).

12.4 STRUCTURE OF MOLYBDENUM DISULPHIDE

Morphological studies have revealed that mono-layered MoS_2 comprised of S–Mo–S sandwiched atomic arrangements that remain stacked to adjacent layers vide van der Waals interactions (Toh et al., 2017; Kim & Lee, 2018). Depending on the different coordination modes of Mo and S atoms, MoS_2 phases have been characterized as 2H, 3R and 1T respectively, where the digit is indicative of the number of layers in the crystallographic unit cell, and the latter designates the type of symmetry exhibited; **T** stands for tetragonal (D_{3d} group), **H** represents hexagonal (D_{3h} group), and **R** denotes rhombohedral (C^5_{3v} group), as indicated in Figure 12.2 (Toh et al., 2017).

The 1T form possesses metallic behaviour, while the other forms exhibit semiconducting behaviour (Zarach et al., 2022). The 2H and 3R types are naturally obtainable while 1T type is found in the metastable state, but spontaneously gets converted to the 2H phase (Shi et al., 2018). Thus, the 2H phase, being the commonest existing state of MoS_2, has drawn extensive consideration for its potential applications for energy conversion and storage. Nevertheless, the inferior intrinsic activity and inertness of the basal plane of 2H–MoS_2 have substantially restricted the application in

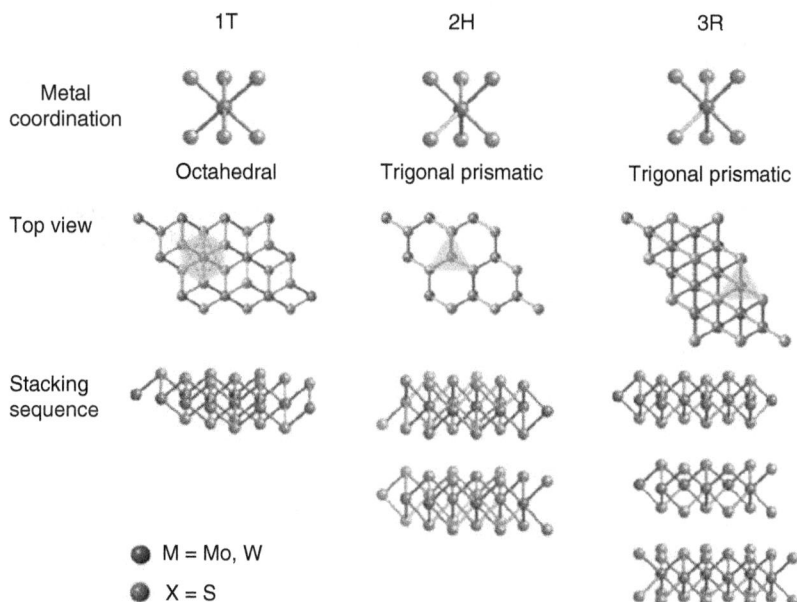

FIGURE 12.2 Model representation of structural unit cells, metal coordination and stacking sequences of different phases of transition metal dichalcogenides in tetragonal symmetry (1T), hexagonal symmetry (2H) and rhombohedral symmetry (3R) environment. (Toh et al., 2017).

energy conversion and storage fields (Ramadoss et al., 2014). Accordingly, numerous strategies have been adopted for enhancing the intrinsic activity of 2H–MoS_2. For example, the introduction of defects into the MoS_2 structure can bring about the generation of unsaturated coordination of Mo or S atoms, which can, in turn, activate the chemical responsiveness of the basal plane (Liang et al., 2021). The rational design of MoS_2-based heterostructures can also play a significant role in the enhancement of the physicochemical properties, and the abundant interfaces existing in the heterostructures can optimize the electronic configuration of the hybrid through the charge transfer across the interfaces (Han et al., 2022). Accordingly, various bottom-up as well as top-down preparation approaches have been adopted to plan and fabricate the frameworks of MoS_2 and MoS_2-based nanomaterials, such as chemical vapour deposition (CVD) method, hydrothermal method, ion intercalation techniques, etc to obtain different MoS_2 morphologies such as mono/few layers, nanospheres, nanoflowers, porous nanostructure, etc. The main objective behind the designing and fabrication lies in the successful synthesis of MoS_2 nanomaterials, which can possess high electrical conductivity, large surface area, abundant electroactive sites, possess ion channels to facilitate ion diffusion and accelerate the electrolytes penetration, thereby achieving improved performance for energy conversion and storage (Zhang et al., 2016).

12.5 SYNTHESIS OF MOLYBDENUM DISULPHIDE FOR SUPERCAPACITOR APPLICATIONS

There are several methods of preparing molybdenum disulphide nanostructures for electrochemical energy storage applications as outlined in Figure 12.3. In general, the synthetic methodology has been classified into two major categories, namely, top-down and bottom-up approaches, which have been discussed in the following section (Sun et al., 2017; Han et al., 2019).

a) Top-down or Liquid-phase Exfoliation
In this approach, a multilayered component can be converted into a single-layered one simply by overcoming the inter-sheet cohesive forces vide micromechanical techniques. One of the popular methods involves the production of few-layered

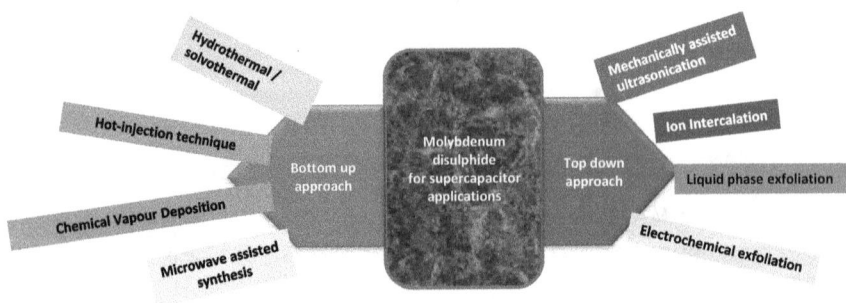

FIGURE 12.3 Some popular methods of preparing molybdenum disulphide-based nanomaterials for electrochemical energy storage applications.

exfoliated MoS$_2$ flakes that can be easily peeled off from the bulk MoS$_2$ using sticky tapes and placed over a variety of substrates. The existence of van der Waals forces between the substrate and the MoS$_2$ adheres the latter on the former's surface itself and the procedure is repeated to obtain random shapes and sizes of MoS$_2$ for constructing high-quality devices with superior energy performances (Magda et al., 2015).

Exfoliation of MoS$_2$ in solutions can be carried out by means of two strategies. The first one involves mechanically assisted ultrasonication, bubbling, grinding, and stirring, and the second one involves intercalation of ions or even using suitable surfactants (Ciesielski & Samorì, 2014). In the first case, the choice of a good solvent is essential as depending on the solvent surface tension and material's surface energy, the extent of exfoliation will occur. Thus, MoS$_2$ exfoliation is only feasible in N-methyl-2-pyrrolidone when the solvent surface tension matches with the surface energy to overcome van der Waals forces existing between the layers (Lou & Ajayan, 2015). Liquid-phase exfoliation is beneficial as its yield is appreciable but the efficiency is inferior in comparison to the tape-assisted exfoliation. It is a relatively straightforward method where merely some bulk materials in organic solvents are ultrasonicated for exfoliation purposes (Nicolosi et al., 2013; Jawaid et al., 2015). Thus, to improve the exfoliation efficiency, the second strategy involving the intercalation of ions in between the layers of the materials is executed, which results in layer reduction. The technique is highly solvent-dependent. MoS$_2$ nanosheets can also be prepared by the surfactant-assisted exfoliation method using sodium chlorate in sodium solution. The method opened up the scope of carrying out the delaminating process of MoS$_2$ layers under ambient conditions. Besides, it is scalable and also promotes the preparation of good-quality films, hybrids, and composites as well (Smith et al., 2015). Similarly, the electrochemical exfoliation technique has also been employed for the synthesis of nanohybrids of MoS$_2$. Particularly it is advantageous in one-step fabrication involving electrochemical exfoliation with simultaneous deposition of the other component with a high degree of scalability (Ali et al., 2020). Further, the electrochemical approach is nowadays widely accepted compared to other traditional synthetic techniques, as it is rapid and saves time, creates zero explosion threat, releases no hazardous chemicals, and thus avoids environmental waste formation (Rong et al., 2022).

b) Bottom-up Approach
For the preparation of MoS$_2$ nanostructures, the bottom-up approach mostly involves hydrothermal/ solvothermal, microwave-assisted synthesis and hot injection, as well as chemical vapor deposition techniques. In the hydrothermal approach, high temperature and pressure are created in a closed vessel (autoclave) by heating an aqueous reaction mixture (Huo et al., 2020). Hydrothermal methods, by far, are among the commonest and most attractive approaches to synthesizing varied MoS$_2$ nanostructures usually prepared by combinations of molybdenum precursors such as MoO$_3$, NH$_4$Mo$_7$O$_{24}$, Na$_2$MoO$_4$, etc. and sulphur sources (sulphide salts, thiourea, thioacetamide, sulphur powder, etc.) under characteristic reaction conditions (Peng et al., 2001). The selection of appropriate precursors, nature of surfactants, solvent, duration of reaction, and acidity of the medium can precisely determine and assist in tuning the morphology, structure, size and crystalline phase of MoS$_2$. Solvothermal

methods are analogous to hydrothermal with the exception that organic solvents, instead of water, are employed here. Various morphologies, such as dozens-layered MoS_2 nanosheets, few-layered MoS_2 nanosheets, ultra-small MoS_2 nanosheets along with nanorods, hollow spheres, flower-like morphologies built up through assembling of nanosheets units, etc. has already been accomplished by this technique, based on the Mo/S mole ratio, reaction time, temperature, pH of the reaction medium, which have shown notable electrochemical responses to be applied for energy storage electrode fabrications.

The microwave-assisted technique for synthesizing MoS_2 is facile, secure, less time-consuming and energy efficient. Nanotubes and fullerene-like MoS_2 nanoparticles can be created using this processing pathway. For instance, during calcination of amorphous MoS_2 powders for two hours at 600°C in an argon atmosphere, after 200 s of microwave irradiation, Structural analysis confirmed randomly oriented fullerene-like structure formation whereas a longer time (600 s) of irradiation, the morphology of MoS_2 analogous to nanotube and fullerene were produced (Panigrahi & Pathak, 2011). In the hot-injection method, injection of a cold precursor containing stock solution is carried out into the hot high-boiling organic mixture (Savjani et al., 2016). Here, the size and morphology of the materials are determined by organic ligands and also pH, and concentration of reactants. Studies reveal that controlling the reaction time results in the variation of MoS_2 morphologies, from quantum dots to nanosheets. The formation mechanism follows a bottom-up process, starting from quantum dots to small pieces and finally to large-area nanosheets (Liu et al., 2017; Zheng et al., 2017). Another, important method under this section involves the chemical vapor deposition (CVD) process. Here, the substrate is exposed to some volatile precursors so that the vapours spontaneously decompose on the substrate surface to produce the desire nanostructures. It is highly suitable for designing 2D layered MoS_2 and MoS_2-based heterostructures for vivid applications (Zheng et al., 2017). By using this method, pure and high-quality, size-controllable MoS_2 can be prepared for energy storage applications. Common starting precursors involve molybdenum-based compound powder, molybdenum sulphide powder, deposited molybdenum-based film and ammonium molybdates depending on the desired morphology required for applications (Lee et al., 2012). The typical reactions involved in the fabrication of molybdenum disulphide nanostructures during the CVD process are as per the equations (1-3) stated below:

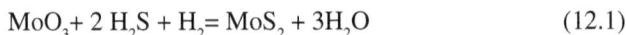

$$MoO_3 + 2\,H_2S + H_2 = MoS_2 + 3H_2O \qquad (12.1)$$

$$8\,MoO_3 + 24\,H_2S = 8\,MoS_2 + 24\,H_2O + S_8 \qquad (12.2)$$

$$16\,MoO_3 + 7S_8 = 16\,MoS_2 + 24\,SO_2 \qquad (12.3)$$

This method is used to prepare wafer-scale, high-quality and crystalline monolayer MoS_2 films which are used in various electronic and optoelectronic devices (Kim et al., 2019; Xu et al., 2019). However, issues related to the optimization of atomic properties, active edges, surface area, defects concentration and distribution, and stoichiometry controls, are major challenges of the CVD technique. Moreover, small production, low yield, random growth and limited substrates are hurdles that still need to be addressed. As far as thermal methods are concerned, the method of thermolysis is adopted to obtain MoS_2 by simple thermal decomposition of $(NH_4)_2MoS_4$ although this is less popular due to its inability for in-situ control of shapes and dimensions of the decomposed products. Besides, thermal annealing is often executed under an inert environment to achieve the desired morphology and structure for MoS_2 nanomaterials often in combination with other preparation methods (Liu et al., 2012). The pros and cons of different conventional techniques used for the nanostructure synthesis of MoS_2 have been listed in Table 12.1. Each of the strategies exploited for the fabrication of MoS_2 nanostructures has been used in the enhancement of physicochemical performance, even though, precise control over unique morphology and microstructure, conductivity, active sites, and surface area, is yet to be accomplished. Many modification methods, such as introducing defects, doping hetero-atoms, functionalization for interlayer spacing expansion, phase engineering, activation of the basal plane, and heterostructures design, have been adopted as alternative approaches for upgrading the electrochemical performances. Often combinations of

TABLE 12.1
Advantages and Disadvantages of the Synthetic Approaches Adopted for Obtaining MoS₂ Nanostructures

Techniques	Benefits	Limitations
Chemical Vapor Deposition	High-quality products Controlled size/layers Uniform film with low porosity	Low product yield High temperature requirement Limited substrates availability Expensive instrumentations High technical skills requirement for operation Emits toxic gases as by-products
Hydrothermal	Simple to operate Single-step procedure Environmentally benign Good dispersibility of products in solution	Low product yield Residues are treated as wastes Uncontrolled size High temperature and pressure
Mechanical exfoliation	High product yield Simple to operate	Long process time Organic solvent usage required
Electrochemical synthesis	High product yield, Simple to operate	Expensive instrumentation Irregular shaped product formation Uncontrolled size production
Annealing	High product yield Simple to operate	High energy consumption Uncontrolled morphology/size
Ion/Solvent/Surfactant-assisted exfoliation	Simple to operate	Long process time Residues are treated as wastes Low yield

methods have been considered for producing desired MoS_2-based nanocomposites with varieties of organic and inorganic materials. Therefore, rational design and construction of MoS_2-based nanostructures are definitely promising ways to achieve high performances in energy storage fields.

12.6 MODIFICATION OF MOS$_2$ FOR UPGRADING THE ELECTROCHEMICAL PERFORMANCES

To improve the electrochemical performance of MoS_2-based nanomaterials, chemical modifications/functionalization have always been performed and encouraged to upgrade the physicochemical parameters such as composition, particle dimensions, morphology, conductivity, and electronic structure of MoS_2 (Benavente et al., 2002). Such modifications include exfoliation/expansion of layered-MoS_2 through the intercalation process to form single-layered MoS_2 nanosheets, heteroatom doping, engineering of MoS_2 lattice defects, designing of hetero-structured MoS_2, and others, which have been employed successfully for developing high-quality electrode materials for energy production, conversion, and storage applications, as schematically illustrated in Figure 12.4.

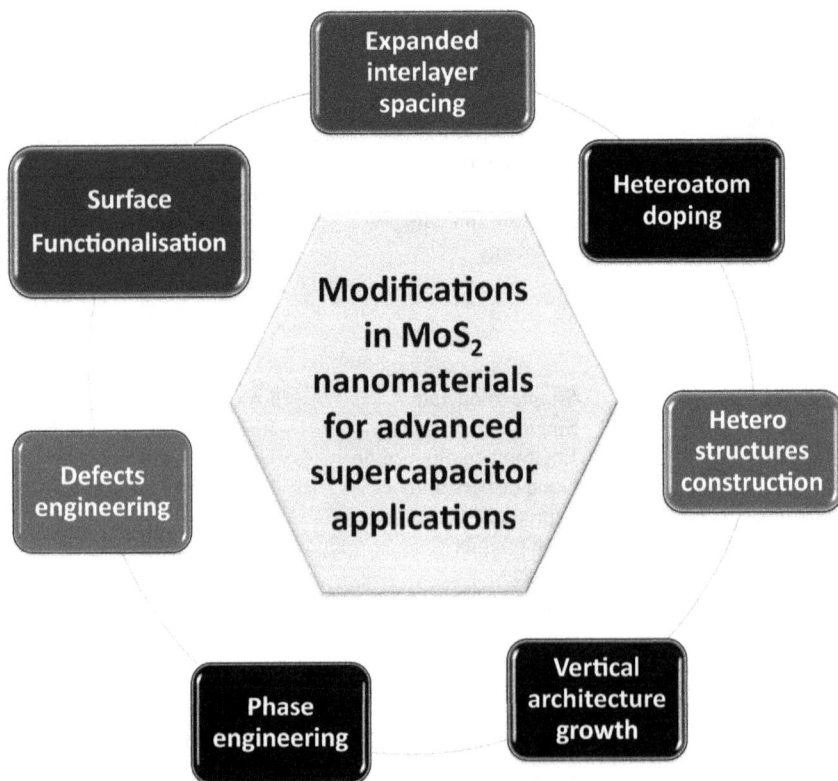

FIGURE 12.4 Some of the common strategies adopted to boost the electrochemical performance of MoS_2-based nanomaterials.

To illustrate, the stacked three atomic layers (S-Mo-S) of 2H phase MoS_2 with inter-layer spacing of 6.15 Å, held by van der Waals interactions, can be productively expanded to about 9 Å by introducing NH_4^+ ions (Liu et al., 2015). DFT studies have further revealed that the ion intercalation activation energy for electrochemical kin-etics could be minimized due to the expanded interlayer spacing of 2H MoS_2 (Lin et al., 2016), (Rasamani et al., 2017). Thus, expansion of MoS_2 interlayer spacing improves the ion insertion/desertion processes through lowering the diffusion energy barrier. The intercalaters often play an important role in boosting the interlayer con-ductivity significantly as well (Liang et al., 2015).

Again, simple routed techniques such as mechanical exfoliation and sonication-assisted exfoliation have been proposed for scalable exfoliation of MoS_2 to achieve mono- or few-layered MoS_2 materials. Such exfoliated single- or few-layered MoS_2 nanosheets have distinctive electronic, mechanical, electrochemical and sur-face features that are not typically owned by their corresponding bulk counterparts (Manzeli et al., 2017). For instance, N– methyl–2–pyrrolidone has been intercalated into the inner layers of bulk MoS_2 to form exfoliated MoS_2 befitted with superior elec-trochemical energy storage responses. Typically, the intercalation ions between the interlayer of MoS2 can react with the solvent to form hydrogen, leading to expanded interlayer spacing (Krishnamoorthy et al., 2016; Nguyen et al., 2015). Additionally, through the introduction of edge defects, S/Mo vacancies, lattice dislocations, and grain boundary modifications, the intrinsic inertness of the MoS_2 basal plane can be activated. Such atomic arrangement adaptations can modulate the electron con-figuration to improve the electrochemical activity (Yao et al., 2019). Defect engin-eering drives exposure of specific crystal planes that otherwise remain veiled due to aggregation of MoS_2 nanosheets, thereby upgrading their propensity towards elec-trochemical reactions. Furthermore, phase engineering for MoS_2 from the 2H phase into the 1T phase has been confirmed by various strategies such as intercalation of alkali /transition metals, heteroatom doping, laser beam irradiation, etc., which have triggered the formation of phase with higher conducting abilities, enlarged interlayer spacings as well as larger density of active sites compared to the precursor phase. For the purpose of the exfoliation of bulk to few layered MoS_2, various organo-lithium compounds, like methyl lithium, n–butyl lithium etc., have been exploited for inter-calating Li^+ bulk MoS_2 layers which also simultaneously induced transition from 2H to 1T phase (Wei et al., 2019). Thus, single-layered 2H MoS_2 nanocrystals by inter-calating K^+ ions into the MoS_2 layers produce a K_xMoS_2 system that gets transformed from semiconducting to a more metallic state with increasing potassium intercalation (Andersen et al., 2012).

Alternatively, the occurrence of poor in-plane basal plane conductivity in MoS_2, as well as the preferential existence of active sites at the edges of MoS_2 nanosheets, has driven the approach of designing well-oriented, edge-rich MoS_2 nanostructures containing a high density of unsaturated atoms and enlarged interlayer spacings accomplished via vertical growth of MoS_2 nanosheets, which can be directly correlated to their physicochemical performances (Xie et al., 2016). Such designing has been very effective in prohibiting restacking and aggregation of MoS_2 nanosheets, thereby revealing abundant active edge sites and high structural stability as well as contributing to the formation of smooth and facile charge transfer pathways, all together ensuing outstanding reversible capacity, rate capability, and long cycle life in electrochemical energy storage devices. Another promising, facile yet cost-effective approach involves the introduction of heteroatoms as dopants into the MoS_2 crystal structure to optimize the electronic configuration as well as the overall physico-chemical performances (Isacfranklin et al., 2022). The incorporation of dopants creates defects in the basic plane that enhance the basic plane activity significantly. Apart from single-component doping, dual heteroatom doping strategies have also been performed to exhibit distinct electronic structure change through a synergistic effect (Wang and Liu et al., 2017). Besides the above approaches, numerous kinds of MoS_2-based nanocomposites through coupling of MoS_2 nanomaterials with suitable conducting substrates have been framed for synergistically enhancing the electrical conductivity, and mechanical stability through forbidding of MoS_2 nanosheets aggregations, generating high density of electroactive sites have been targeted. Thus, 3D metal foam, graphene, carbon nanotube, carbon materials, conducting polymers, and other highly conductive substrates have been hybridized with molybdenum disulphide to support MoS_2-based nanocomposites as promising catalytic systems based on their enriched interfaces, high conductivity, large surface area and the optimized electronic configurations (Pumera et al., 2014).

12.7 RECENT ADVANCEMENTS OF MOLYBDENUM DISULPHIDE-BASED NANOMATERIALS FOR SUPERCAPACITOR APPLICATIONS

12.7.1 Undoped or Pristine MoS_2-based Nanomaterials as Supercapacitor Electrodes

In this section, we discuss the recent developments of pristine MoS_2-based supercapacitors. Studies have confirmed that molybdenum disulphide in pristine form can attain an extraordinary gravimetric as well as areal and volumetric capacitance, the latter being a more important parameter while designing flexible and wearable electronic gadgets (Samy & Moutaouakil, 2021). Various morphologies have been tried to optimize the capacitance of this material via different synthetic approaches. For instance, a few years ago, spherically clustered MoS_2 nanostructures through the assembling of MoS_2 nanosheets were prepared via a facile hydrothermal approach that showed a maximum specific capacitance of 122 Fg^{-1} with better energy and power densities (Ilanchezhiyan et al., 2015). In another approach, uniform MoS_2 nanosheets are assembled to form three-dimensional nanospheres via a simple hydro-thermal procedure using SiO_2 nanosphere templates. The resultant MoS_2 nanospheres demonstrated high energy storage capability as observed from the recorded specific capacitance as high as 683 Fg^{-1} at a current density of 1 Ag^{-1} and 85.1% capacitance retention efficiency even after 10000 charging/discharging cycles. Such morphology ensured a larger surface area, affording plentiful surface active sites for charge transfer and also shortened the diffusion path length of electrolyte ions facilitating ultrafast charging/discharging kinetics (Gao et al., 2018). Similarly, the flower-shaped morphology of MoS_2 was produced via a hydrothermal approach by Zhou et al. that recorded a specific capacitance of 122 Fg^{-1} at a current density of 1 Ag^{-1} in 1 M M KCl electrolyte (Zhou et al., 2014). A few years ago, Geng et al. obtained a very interesting composite of metallic-MoS_2-H_2O comprising phase-pure metallic MoS_2 with water via hydrothermal technique where the layer of water molecules at the sides prevented restacking of the MoS_2 layers besides augmenting the number of accessible active sites for ion diffusion process, as depicted in Figure 12.5(a) (Geng et al., 2017). The resultant nanocomposite electrode was tested in different aqueous electrolytes as shown in Figure 12.5 (b) and accordingly showed its highest capacitive response in Li_2SO_4 electrolyte, owing to easy percolation of small-sized lithium ions inside the water-coupled metallic MoS_2 structure relative to the other large sized alkali metal ions (Geng et al., 2017).

Similarly, Wang et al. adopted a sacrificial template synthesis approach to develop hollow morphology of molybdenum disulphide for advanced energy storage characteristics. They employed $MnCO_3$ nanospheres as sacrificial templates for

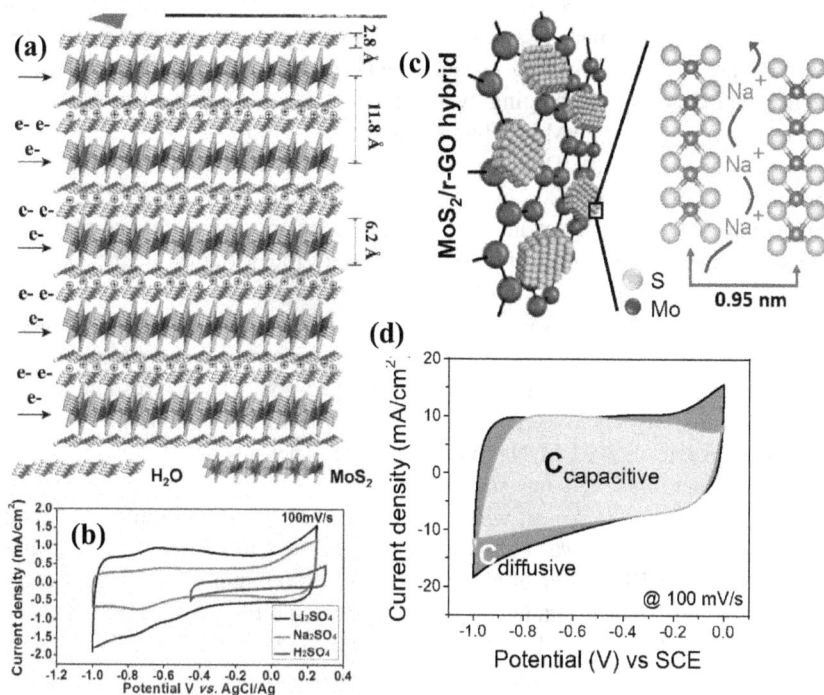

FIGURE 12.5 (a) Cyclic voltammograms for 2D water-coupled metallic MoS_2 electrodes in three kinds of electrolytes recorded at the voltage scan rate of 100 mVs^{-1}. (b) Schematic illustration of the multilayered 2D water-coupled metallic MoS_2 based symmetric supercapacitor (Geng et al., 2017). (c) Schematic representation of intercalation of ammonium ion into the interlayer spacing of MoS_2/r-GO hybrid (d) Cyclic voltammogram highlighting the segregation of capacitive and diffusion-controlled charge storage contributions recorded for MoS_2/r-GO hybrid at the voltage scan rate of 100 mVs^{-1} (Sarkar et al., 2019).

synthesizing the hierarchical hollow nanospheres of MoS_2 which delivered a capacitance was 142 Fg^{-1} @ current density of 0.59 Ag^{-1} with upgraded cyclic performance of 92.9% even after completing 1000 charging/discharging cycles (Wang & Ma et al., 2015). Again it is noteworthy that the capacitance and the cyclic stability are significantly influenced by the number of layers, thickness, and crystalline phase of MoS_2 nanostructures (Yun et al., 2020). As already established, high hydrophilicity and conducting 1T phase of MoS_2 relative to semiconducting 2H or 3R phases would be ideal for designing high-performance supercapacitors, have considerably driven worldwide researchers to explore and understand the plausible mechanism of charge storage in 1T MoS_2 phases (Acerce et al., 2015). Subsequent examinations in different aqueous electrolytes unveiled the efficient intercalation of protons and different alkali metal ions to capacitance values ranging from ~400 to ~700 Fcm^{-3}. The layered MoS_2 film could also be suitably applied to an operating voltage as high as 3.5 V with organic electrolytes, thereby exhibiting outstanding volumetric energy and power densities as well as stable Coulombic efficiencies of more than 95% beyond 5000 charging/discharging cycles (Acerce et al., 2015). A monolayer mixture of 2H and 1T-MoS_2 (with 40% 1T phase) via solvothermal technique was reported to attain gravimetric capacitance of 366.9 Fg^{-1} at 0.5 Ag^{-1} in 6 M M KOH electrolyte as well as nearly 92% capacitance retaining capacity on completing continuous 1000 charging/discharging cycles much superior to that of the few-layered mixtures as well as pristine MoS_2 phases (Jiang et al., 2015). Likewise, the cyclic stability of a slim (~ 250 nm) 2D layer of MoS_2 film electrode recorded cyclic efficiency of nearly 97% even after 5000 cycles compared to that of a one-micron-thick 3D MoS_2 electrode (Choudhary et al., 2015). A few years ago, Falola and co-researchers experimented with electrodeposited MoS_2 to study the variation of the cyclic stability in neutral electrolytes with different loadings/thicknesses. They demonstrated a simple electrochemical pathway for obtaining amorphous MoS_2 sheets from ammonium tetrathiomolybdate precursor via cyclic voltammetry, which was subsequently annealed under an inert atmosphere of argon gas to obtain the corresponding crystalline phase that displayed exceptional specific capacitance of 360 Fg^{-1} at 10 mVs^{-1} voltage scan rate in 1 M M Na_2SO_4 electrolyte. They observed that films with a thickness greater than 250 nm showed a maximum capacitance retaining ability of ~ 90–100% for 1000 cycles, while for the film thickness ~ 50 nm, even though the observed capacitance was comparatively high, the capacitance retaining ability dropped down to 87% (Falola et al., 2016). Thus, an optimum loading of the active material was important to attain the utmost cyclic stability of the electrode during prolonged potential cycling. Lately, Manuraj et al. developed brush-like structures of MoS_2 on Ni foam via a hydrothermal approach that displayed gravimetric capacitance as high as 766 Fg^{-1} at 1 mVs^{-1} voltage scan rate in 1 M KOH electrolyte. Such exploration also confirmed the methodology of obtaining high mass loading of MoS_2 nanostructures for designing superior energy storage electrodes (Manuraj et al., 2020).

Accordingly, the examinations were executed in different electrolytes including ionic liquids. The experiment recorded appreciable volumetric capacitance of

118 Fcm^{-3} and gravimetric capacitances of 42 Fg^{-1}, respectively, due to exfoliation of MoS$_2$ interlayer through ionic liquid intercalation. The authors also proposed that such larger interlayer spacing accelerated the ion movements and triggered faster-charging responses (Liang et al., 2021). A micro-supercapacitor which was designed through spraying of MoS$_2$ nanosheets on a Si/SiO$_2$ chip followed by laser patterning was found to display exceptional electrochemical performance compared to graphene-based micro-supercapacitors. The observed outstanding areal capacitance of 8 mFcm^{-2} and volumetric capacitance 178 Fcm^{-3} values have surely opened up new windows in the development of portable and flexible micro-electronic devices (Cao et al., 2013). A recent work based on MoS$_2$ quantum sheets (QSs) with two-dimensional size con-finement both in lateral and vertical orientations, was obtained via a salt-assisted ball milling process. The designed electrode material demonstrated a high capacitance of 162 Fg^{-1} originating from combinations of quantum capacitances and electrochem-ical capacitances that conferred high specific energy as well as long cycle life as the ultimate electrochemical output (Nardekar et al., 2020). The MoS$_2$ layered structure analogous to graphite exhibits two key crystal characteristics: a basal plane and an edge plane. The edge plane of MoS$_2$ being more electroactive than the basal plane, edge-oriented nanowall films, similar to those of carbon nanotubes, were fabricated by chemical vapour deposition technique that demonstrated high energy-storing capability.

Thus, intensive electrochemical investigations on different morphologies, crystal structure, layer thickness, and exfoliated phases in pristine MoS$_2$ have revealed their exceptional high potentiality as a smart supercapacitor material; nonetheless, at prac-tical levels, a different scenario is usually observed. Often, very unsatisfactory energy performances are recorded primarily due to the massive volume alterations during ion insertion/de-insertion processes as well as easy restacking and aggregation of the nanosheets during fast and several thousands of charging/ discharging processes that ultimately result in tremendously poor capacitive responses as well as cyclic instability. Moreover, very low conductivity between the adjacent van der Waals bonded layers in the stable phase of MoS$_2$ drastically suppresses their overall electrochemical effect-iveness. Hence, adjusting the microcosmic morphology through heterojunction for-mation in the course of combination with other suitable nanomaterials would be an effective way to combat these important and non-avoidable issues.

12.8 NANOCOMPOSITES OF MOS$_2$ AS SUPERCAPACITOR ELECTRODES

Molybdenum disulphide has been judiciously blended with a wide variety of materials including nano-carbons such as functionalized graphenes, carbon nanotubes, carbon spheres, g-C$_3$N$_4$, etc., as well as conjugated polymers and varied metallic compounds to form nanocomposites with superior nano-heterojunction properties that success-fully contribute to boost up the charge storing capacities as well as promote fabri-cation of flexible energy devices that may be used as powering sources in various wearable electronics (Wang, & Chen et al., 2017).

12.8.1 MoS₂/Nano-carbon-based Nanocomposites as Supercapacitor Electrodes

MoS₂/nano-carbon composite electrodes have been primarily developed to supplement the poor electrical conductivity of intrinsic MoS_2 as well as provide skeletal support for the formation of diverse nanoheterostructures, which may subsequently promote superior electroactive site exposure as well as broad channels for smooth electrolytic ion mobility. Accordingly, various carbon nanomaterials such as carbon nanofibers, functionalized graphene, carbon nanotubes, and carbon spheres, have been used for the desired nanocomposite electrode fabrications. Thus, Huang et al. designed carbon aerogel/MoS_2 nanocomposite electrodes that recorded gravimetric capacitance of 260 Fg^{-1} at a current density of 1 Ag^{-1} with appreciable cyclic efficacy of 92.4% over 1,500 GCD cycles. The introduction of carbon aerogel plates, in this case, offered a highly conductive path, as well as served as a large surface area support for MoS_2 that promoted better electrode/electrolyte interfacial interactions besides facilitating smooth and ultrafast flowing of electrolyte ions during charge-discharge processes (Huang & Wang et al., 2015). Additionally, porous and tubular C/MoS_2 nanocomposites were framed by Hu and co-researchers using porous anodic aluminium oxide as nano-templating material that displayed gravimetric capacitance of 210 Fg^{-1} at a current density of 1 A g^{-1} along with appreciable cycling performance. Such improvement in electrochemical responses was attributed to the formation of porous morphology-based nanocomposite that promoted rapid diffusion of ions during charging and discharging processes, minimized the internal resistances and facilitated faster electronic transport (Hu et al., 2013). In another work, combination of the multi-walled carbon nanotubes (MWCNT) with MoS_2 to form the MoS_2/MWCNT nanocomposites ensured uniform loading of MoS_2 nanosheets on the surface of three-dimensionally interconnected CNTs that eventually provided excellent charge transfer network as well as broad electrolyte diffusion channels besides confronting the unwanted structural collapse and aggregation of redox-active materials during fast and out-numbered charging–discharging sequences. As a result, the as-prepared electrode material displayed much improved gravimetric capacitance of 402 Fg^{-1} at a current density of 1 Ag^{-1}, much higher than that of neat MWCNTs and plain MoS_2, respectively, thereby implying the existence of synergistic influence among the components in the nanocomposite. This also supported the cause of an outstanding cycling stability of 81.9% even after 10,000 continuous charging/discharging cycles at a current density of 1 A g^{-1} (Chen et al., 2018). To unveil the influence of the carbon material structure on the charge storage process in MoS_2-based supercapacitors, 2H MoS_2 were hydrothermally synthesized and combined with commercially purchased CNTs and graphene nanoflakes to fabricate desired nanocomposites The recorded capacitances were observed to vary with the structure of the carbon material. Among them, the composite MoS_2/CNT/GNF (1:1:1) composite showed optimum specific capacitance as well as good capacitance retention behaviour. In-depth analysis revealed that the preferential addition of CNTs to MoS_2 improved the capacitance, ascribed to their small diameters and large surface area relative to graphene nanoflakes (Ali & Metwalli et al., 2018). Lately, a MoS_2 flower-shaped nanostructure was combined with 3D graphene for fabricating a prototype supercapacitor through stacking of MoS_2 nanoflowers over 3D graphene. The resultant supercapacitor displayed high energy density of 24.59 WhKg^{-1}, and a power density of 8.8 WKg^{-1}, along with a wide operating window ~2.7 V (−1.5 to + 1.2 V) (Singh et al., 2017). A MoS_2-reduced graphene oxide nanocomposite as fabricated by Sarkar and co-researchers via a hydrothermal approach using thiourea, MoS_2, and rGO reactants was found to record superior electrochemical responses (Sarkar et al., 2019). Here, thiourea assisted in the intercalation of the ammonium ions along with r-GO in between the MoS_2 nanosheets, as depicted in Figure 12.5(c), for easy intercalation/de-intercalation of sodium ions from electrolyte, which again, in turn, satisfactorily boosted the capacitance, rate capacity, and cycling stability of the resultant nanocomposite. In addition, the authors also carried out rigorous electrochemical analysis/interpretation to ensure that the open and expanded multilayered structure MoS_2/r-GO largely promoted capacitive-charge-storage processes geared up the insertion/de-insertion process of sodium ions, as reflected in Figure 12.5(d) (Sarkar et al., 2019; Zhan & Liu et al., 2018). To improve the adherence forces among the components, Yuan et al., reported 2D layered MoS_2 structures with 3D nitrogen-doped graphene aerogel having large surface area and high conductivity. The resultant hybrid aerogel electrode demonstrated an appreciable rise in specific capacitance (532 Fg^{-1} at a current density of 1 Ag^{-1}) relative to pristine MoS_2 nanosheets and nitrogen-doped graphene aerogels alone, in addition to superior cycling stability (93.6% capacitance retention efficacy even after completing 10000 cycles at a high current density of 10 Ag^{-1}) (Yuan et al., 2019). In another approach of innovative novel synthesis strategies for controlling the nanometre-scale dimensions, regulating the injection temperature in hot-injection thermolysis, successfully modulated the active edge-density in the molybdenum disulphide nanosheets. Using this approach, Savjani et al. produced pure and high-quality, size-controlled, single-layered 1T-MoS_2 nanosheets capped by oleylamine, which on blending with graphene produced superior supercapacitive material demonstrating appreciable areal capacitance as well as tremendously good cycling stability owing to improved highly controlled lateral dimensions and thicknesses of the modified MoS_2 nanosheets (Savjani et al., 2016). In recent work, Weng et al. proposed a hybrid supercapacitor using a pseudocapacitive core of MoS_2 nanosheets and an electrostatic double-layer capacitive shell composed of microporous carbons. The so-obtained uniform mesoporous carbon-coated MoS_2 nanosheets displayed 1T-phase and delivered a capacitance of 189 Fg^{-1} @ a current density of 1 Ag^{-1}. An improved cyclic performance (98% capacitance retaining capacity after 6000 cycles) was observed which is attributed to the combination of improved structural stability and amplified electrochemical charge storing capability (Weng & Wang, et al., 2015).

12.8.2 MoS_2/CONDUCTING POLYMER-BASED NANOCOMPOSITES AS SUPERCAPACITOR ELECTRODES

Conducting polymer nanomaterials such as polyanilines, polypyrroles, etc., have been widely used in electronics owing to their superior charge transport properties as well as elevated pseudocapacitive responses which have made them highly popular in supercapacitor technology over the decades. Nonetheless, these nanosystems frequently suffer from structural instability as observed by their extremely poor cyclic performances. Thus, hybridizing them with structurally stable, highly potential electroactive nanomaterials like MoS_2 would definitely boost the overall capacitive responses as well as avoid issues related to polymer instability issues (Ren, et al., 2015; Ma et al., 2013; Yang et al., 2016; Tang & Wang et al., 2015). Conversely, in combinations with conjugated polymers, MoS_2 layers' restacking, and improvement in structural as well as electronic properties would be collectively augmented (Savjani et al., 2016; Huang & Wang, et al., 2013). Thus, to illustrate, polyaniline-MoS_2 nanocomposite prepared by combinations of hydrothermal and in-situ polymerization techniques was fund to demonstrate appreciable gravimetric capacitance (575 Fg^{-1} at a current density of 1 Ag^{-1}) as well as excellent long-term cyclic stability, attributed to the positive synergistic effect of the components responsible for the overall improvement in the electrochemical performance. The graphene-like MoS_2 subunits provided remarkable 2D conductive skeletal supports that provided the platform for larger contact surfaces as well as enhanced the electrolytic accessibility of PANI redox-active sites, thereby increasing the capacitance as well as rate capacity of the composite system (Huang & Wang et al., 2013). In another exploration, regulated growth of PANI nanowire arrays on the tubular MoS_2 surface was executed to provide scope for optimizing the capacitive performance of the derived electrodes. Thus, the optimum loaded (MoS_2/PANI-60) hybrid electrode, expectedly, displayed a maximum gravimetric capacitance output of 552 Fg^{-1} at a current density of 0.5 Ag^{-1} as well as high rate capability (Ren et al., 2015). Analogous to PANI, polypyrroles (PPy) have also been widely popularized in the sector of energy conversion and storage and accordingly have also been exploited for designing versatile nanocomposites for advanced supercapacitive operations. Thus, polypyrrole-MoS_2 (PPy/MoS_2) nanocomposites framed via combinations of hydrothermal and intercalative in-situ polymerization techniques displayed an utmost gravimetric capacitance of 553.7 Fg^{-1}, which was retained up to 90% of its initial value even after 500 cycles at a current density of 1 Ag^{-1}, thus showing improved structural stability relative to the pristine components (Ma et al., 2013). Likewise, Yang et al. reported the rational synthesis of carbon shell-coated PANI on 1T MoS_2 monolayers for energy storage applications. The best composition electrode with a ~3 nm carbon shell recorded a notable gravimetric capacitance of 678 Fg^{-1} measured at a voltage scan rate of 1 mVs^{-1}, advanced cyclic stability responses as well as superior rate performance owing to the synergistic influence of PANI nanostructure and thin, uniformly carbon coated 1T MoS_2 substrates (Yang et al., 2016). Similarly, Tang and co-researchers afforded a scalable solution-based synthesis approach for developing ultrathin films of PPy on 2D MoS_2 monolayers with advanced supercapacitive performances. The resultant heterostructure demonstrated exceptionally high gravimetric capacitance of 695 Fg^{-1} at a current density of 0.5 Ag^{-1} with high rate capacity even at elevated discharge current densities, thereby demonstrating their admirable rate performances (Tang & Wang et al., 2015). Zhao et al. demonstrated the fabrication of metallic 1T-MoS_2/PANI heterostructures with unique alternating arrangement through an electrostatic attraction-induced self-assembly technique, where the negatively charged metallic 1T-MoS_2 monolayers formed by chemical exfoliation were electrostatically held by positively charged PANI resulted from acidification. Such interactions greatly improved the electron and ion transports within the nanocomposites besides conferring high-quality structural stability as observed from the cyclic tests carried out in the aqueous electrolytes. Thus, high specific capacitance and capacitance retention up to 91% at 10 A/g after 2000 cycles was accordingly recorded (Zhao & Ang et al., 2017).

To overcome the issues related to the low capacity of MoS_2 owing to the overlaying of the layers and the poor electric conductivity, covalent functionalization of MoS_2 has been highly encouraged through sandwich-like 4-aminophenyl functionalized MoS_2/polyaniline (MoS_2–NH_2/PANI) nanocomposites formation via in-situ growth of PANI on MoS_2–NH_2 templates. The optimized nanocomposite electrode demonstrated a good capacitive response, thus recording an utmost specific capacitance of 326.4 Fg^{-1} at a current density of 0.5 Ag^{-1} compared to neat components as well as the long-term cyclic stability, as obtained from the electrochemical results depicted in Figure 12.6(a-c) respectively. The results thus confirm the existence of strong adhesive forces among the components and show the different electrochemical analyses carried out (Zeng & Li et al., 2019).

A few years ago, Ansari and co-researchers adopted an easy and facile approach to scheme a nanocomposite using mechanically exfoliated MoS_2 nanosheets with PANI via in-situ chemical oxidative polymerization technique. The resultant nanocomposite electrode displayed a specific capacitance of 510.12 Fg^{-1} at a current of 1 Ag^{-1}, almost twice that of the value for the analogous nanocomposite prepared using bulk metal sulphide, assuring the presence of synergistic influence between the components that led to the augmented electrical conductivity and improved supercapacitive response (Ansari et al., 2017).

12.8.3 MoS₂/METAL COMPOUND-BASED NANOCOMPOSITES AS SUPERCAPACITOR ELECTRODES

The strategy behind the formation of nano-heterojunction with various metal/metal oxides or metal sulphides and molybdenum disulphide is to introduce structural stability of MoS_2 as well as improve the electrochemical performance of supercapacitor electrodes. For instance, the Ti/TiO_2/MoS_2 coaxial fibres composed of a fabricated TiO_2 buffer layer and MoS_2 nanosheets developed on Ti nanowires using hydrothermal technique exhibited flexible structural properties, delivering gravimetric capacitance of 230.2 Fg^{-1} with an appreciable energy density of 2.70 Wh kg^{-1}. In addition, these composite fibres were successfully processed to produce stretchable, spring-shaped, self-powering devices which could successfully withstand large deformations without compromising its electrochemical performances, as highlighted in Figure 12.7 (Li & Li et al., 2017).

The urge to accomplish high capacitance, energy and power density is responsible for engineering nano-architecture-based electrodes through the intimate integration of electroactive materials with the current collectors. Such rational designing ensured proper employment of the superior conductivity of the inner metal core as well as large electrochemically active area of the outer surfaces. Accordingly, free-standing

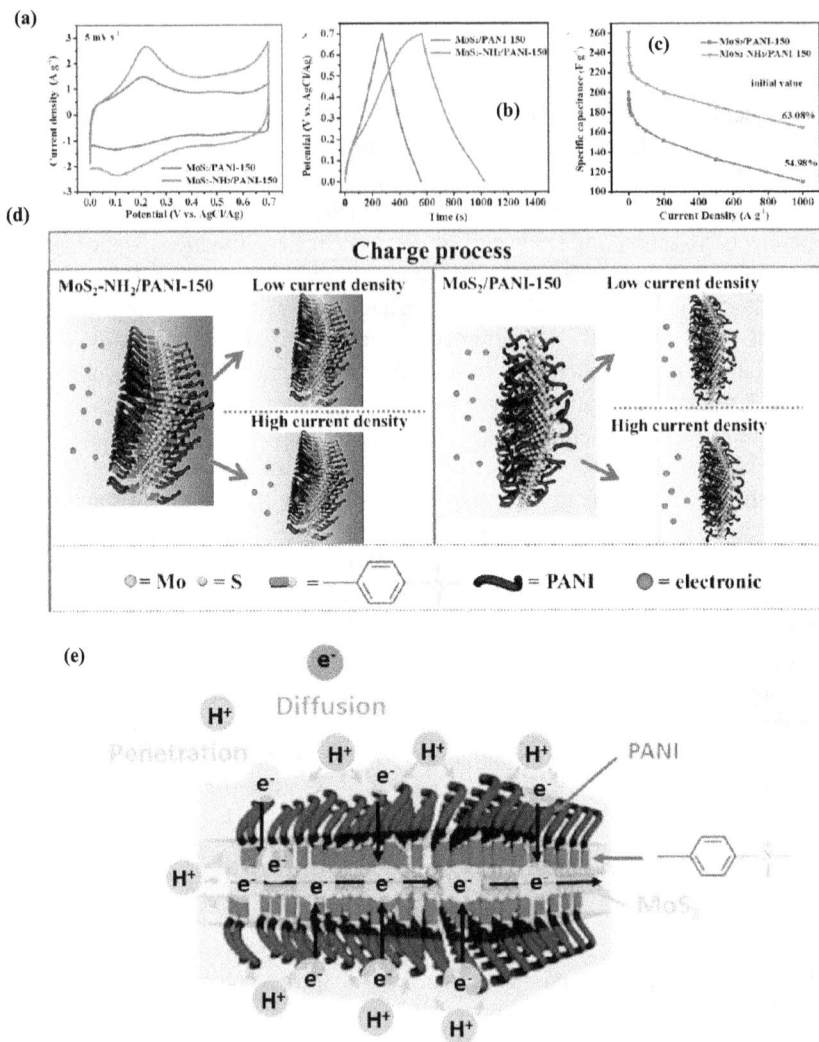

FIGURE 12.6 (a) Comparative cyclic voltammograms of MoS$_2$/PANI-150 (non-functionalized) and MoS$_2$-NH$_2$/PANI-150 (functionalized) recorded at voltage scan rate of 5 mVs^{-1} (b) Comparative galvanostatic charging/discharging profiles of (non-functionalized) MoS$_2$/PANI-150 and (functionalized) MoS$_2$-NH$_2$/PANI -150 recorded at current density of 0.5 Ag^{-1}; (c) Rate capacity of both samples with varying current densities (d) Illustration of the charging process in both the samples (e) Electrochemical mechanism of the sandwich-like MoS$_2$-NH$_2$/PANI nanosheets electrode (Zeng & Li et al., 2019).

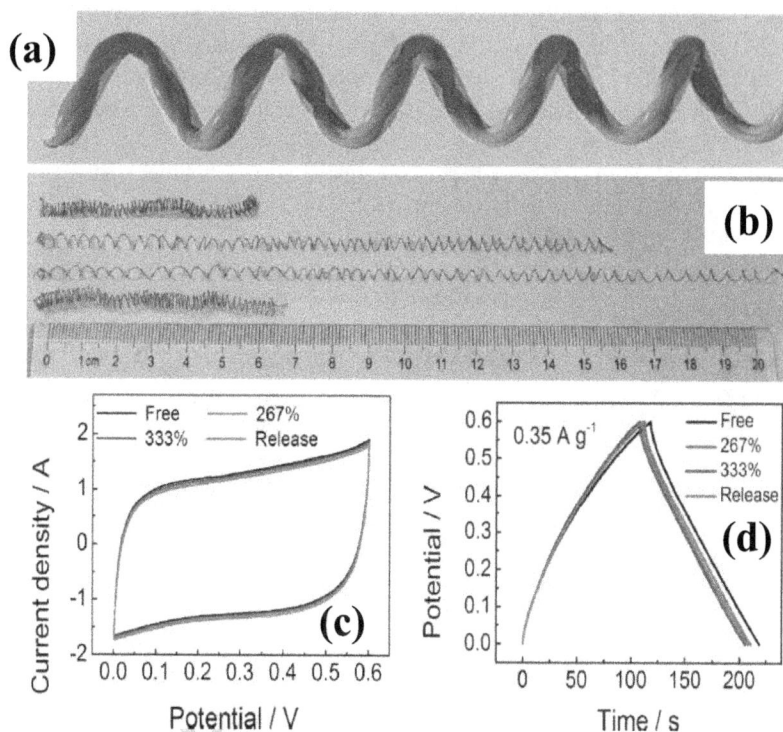

FIGURE 12.7 (a) Photographic illustration of the collectively twisted spring–like supercapacitor fabricated using Ti/TiO$_2$/MoS$_2$ fiber electrodes. (b) Digital photographs of the spring–like supercapacitor at different deformation states. (c) and (d) showing corresponding CV and GCD profiles of the spring-like supercapacitor at different stages of deformation (Li & Li et al., 2017).

binder-free electrodes were framed by Kim and co-workers through the construction of MoS$_2$ sheets on molybdenum foil using a hydrothermal technique that delivered a specific capacitance of 192.7 Fg^{-1} at a current density was 1 mAcm^{-2} (Krishnamoorthy & Veerasubramani et al., 2016). Similarly, the benefits of core-shell heterostructures in energy storage have been employed for fabricating hierarchical carbon-coated Ni3S2@MoS2 double-core-shell nanorods along with a facial hydrothermal technique using carbon-coated nickel nanorods as the starting template. It is well accepted that core-shell nanostructures with tailored composition and functionalities significantly improve both the conductivity as well as charge storage capability in electrodes. They have been judiciously employed herein to ensure that the conductive core promotes material charge conduction and skeletal support while the porous shell upgrades the electrode/electrolyte interfacial activities, increases the electroactive sites exposure and serves as a buffer for the massive volume alterations during fast charging/discharging processes. Thus, the C@Ni3S2@MoS2 hetero-structured electrodes rightfully exhibited a capacitance as high as 1544 Fg^{-1} at a current density of 2 Ag^{-1} in addition to appreciable cyclic stability (capacitance retaining efficacy of 92.8% @ a current density of 20 Ag^{-1} even after passing 2000 charging/discharging cycles (Li & Yang et al., 2016). Another good instance of the combination of different metal sulphides with different textures employed to befit the beneficial effects of core conductivity and flexible protective features of shells has been illustrated by Zhang et al. Initially, the authors confirmed a higher capacitive response (~1.6 times) for amorphous MoS_2 than that of the crystalline counterpart. Further, considering the advantages of independent tunability of core diameter and shell thickness, they had engineered electrodes using crystalline Ni_3S_4 core@ MoS_2 amorphous shell heterostructures via a facile one-pot chemical technique that delivered a specific capacitance as high as 1440.9 Fg^{-1} @ a 2 Ag^{-1} current density along with appreciable capacitance retention efficacy of 90.7% on completing 3000 cycles recorded at a high current density of 10 Ag^{-1} (Zhang & Sun et al., 2015). Furthermore, high surface-area-based Ni3S2@ MoS_2 core-shell nanorod arrays grown on Ni current collectors were designed by Wang's group by adopting a green and scalable one-step approach where the heterostructures delivered a capacitance of 848 Fg^{-1}, almost double that of pure Ni_3S_2 (425 Fg^{-1}) as well as 97% capacitance retaining efficacy even after completing 2000 cycles (Wang & Chao, et al., 2014). Analogously, Wang et al. recently reported the fabrication of a heterojunction using two kinds of metal sulphides, namely, MoS_2 and SnS_2, via a hydrothermal approach. The obtained 2D SnS_2/MoS_2 nanostructure delivered an appreciable capacitance of 466.6 Fg^{-1} @ a 1 Ag^{-1} current density. It was thus established that such composite formulation effectively minimized the agglomeration and restacking issues associated with MoS_2 as well as ensured rapid electronic transport and buffering volume adjustments leading to high rate capacity and long cyclic efficiency correspondingly (Wang & Ma et al., 2015).

In the recent past, Liang's group fabricated MoS_2-Co_3O_4 nanocomposites using laser ablation in liquids and an aging-induced phase transformation technique to develop ultrafine Co_3O_4 nanoparticles coated MoS_2 nanosheets. The resultant MoS_2-Co_3O_4 heterostructures displayed high energy storage characteristics as well as electrochemical stability much superior to that of neat MoS_2 or bare Co_3O_4 respectively (Liang & Tian et al., 2015). Lately, MoS_2 nanosheet/MnO_2 nanowire heterostructures were designed to explore the influence of heterogeneous interface on the energy storage response. Results confirmed that there was a remarkable improvement in charge storage response when the metal current collector (CC) was in contact with MnO_2 instead of the MoS_2 side, clearly indicating the unidirectional as well as unimpeded charge transport across the CC/MnO_2/MoS_2 heterogeneous interface in comparison to CC/MoS_2/MnO_2 channel that indicated a blocking effect on the electron transport. Such insights surely confirm the decisive role of appropriate heterojunction fabrication for the optimization of kinetics in electrochemical energy storage devices (Liao et al., 2018). Analogously, MnO_2 nanowires were grown over synthesized MoS_2 nanosheets by single-step hydrothermal technique. The optimized composition of the resultant nanocomposite displayed synergistic interaction between the MoS_2 nanosheets and MnO_2 nanowires that resulted in a decent upgradation of overall electrochemical responses (Sahoo et al., 2022).

12.8.4 MoS₂-based Ternary Nanocomposites as Supercapacitor Electrodes

In the last few years, scientists have frequently designed ternary nanocomposites through the blending of three components in appropriate proportions, which would promote synergism among them to accomplish appropriate dimensions, crystalline structure, conductivity, mass loading and electrolyte selectivity to provide advanced capacitance, charge transfer kinetics through utmost utilization of electroactive materials, energy and power densities, in addition to improved mechanical stability and environmental adaptability (Majumdar, 2021). Ternary nanocomposites of molybdenum disulphide /reduced graphene oxide@ polyaniline (MoS_2/RGO@PANI) were constructed using combinations of facile and cost-effective hydrothermal and chemical polymerization techniques as schematically represented in Figure 12.8(a). Such preparation strategy ensured uniform coverage of PANI, as observed from the morphological studies shown in Figure 12.8(b), on the outer surface of molybdenum disulphide/reduced graphene oxide sheets, resulting in intimate interactions between the components materials that synergistically resulted in outstanding capacitive response (1224 Fg^{-1} @ a current density of 1 Ag^{-1}), outstanding rate capacity (capacitance falls to only 721 Fg^{-1} @ a high current density of 20 Ag^{-1}) and exceptional cyclic efficacy (82.5% capacitance retention at the end of 3000 loops of charging/discharging cycles), also clearly reflected from the graphs in Figure 12.8(c-e) correspondingly (Li & Zhang et al., 2016).

In another strategy, molybdenum disulphide/ polyaniline/ reduced graphene oxide aerogel was designed via a multi-step synthetic approach such that MoS_2/PANI remained well-dispersed on reduced graphene oxide. The optimized ternary MoS_2/PANI/rGO showed notable supercapacitive response in terms of specific capacitance and its retaining capability relative to binary analogue MoS_2/PANI (Sha, et al., 2016). Very recently, a ternary nanocomposite comprising of inter-layered structure of graphene and MoS_2, encapsulating PANI nanorods was designed for developing smart heterostructures for accomplishing outstanding supercapacitive responses, particularly in terms of cyclic stability. The resultant ternary nanocomposite recorded improved cyclic stability as observed from its recovering ability towards capacitance retention (~ 98.11%) relative to pristine PANI (~40%) and binary composites (~ 60–96%). Such upgraded responses relative to binary composites were attributed to the high charge transport phenomenon prevailing at the electrodes/electrolyte interface of the ternary nanocomposite electrode (Palsaniya, et al., 2018). In another approach, bifunctional transition metals (Mn, Co, Ni) doped $SnO2@MoS2$ nanocomposites for enhanced energy storage response were framed via facile hydrothermal technique. Among them, the Mn-doped $SnO2@MoS2$ nanocomposite electrode displayed the highest gravimetric capacitance of 242 Fg^{-1} at a 0.50 Ag^{-1} current density. In addition, it also displayed high capacitance retention efficacy (~ 83.95 %) even after completing 5,000 uninterrupted charging/ discharging cycles. Analysis inferred that the existence of mixed ionic states of Mn greatly assisted in enhancing the specific capacitance. Besides, the presence of a greater number of Mn ions within the lattice sites of SnO_2 considerably enhanced the electroactive sites, upgraded the electrical conductivity,

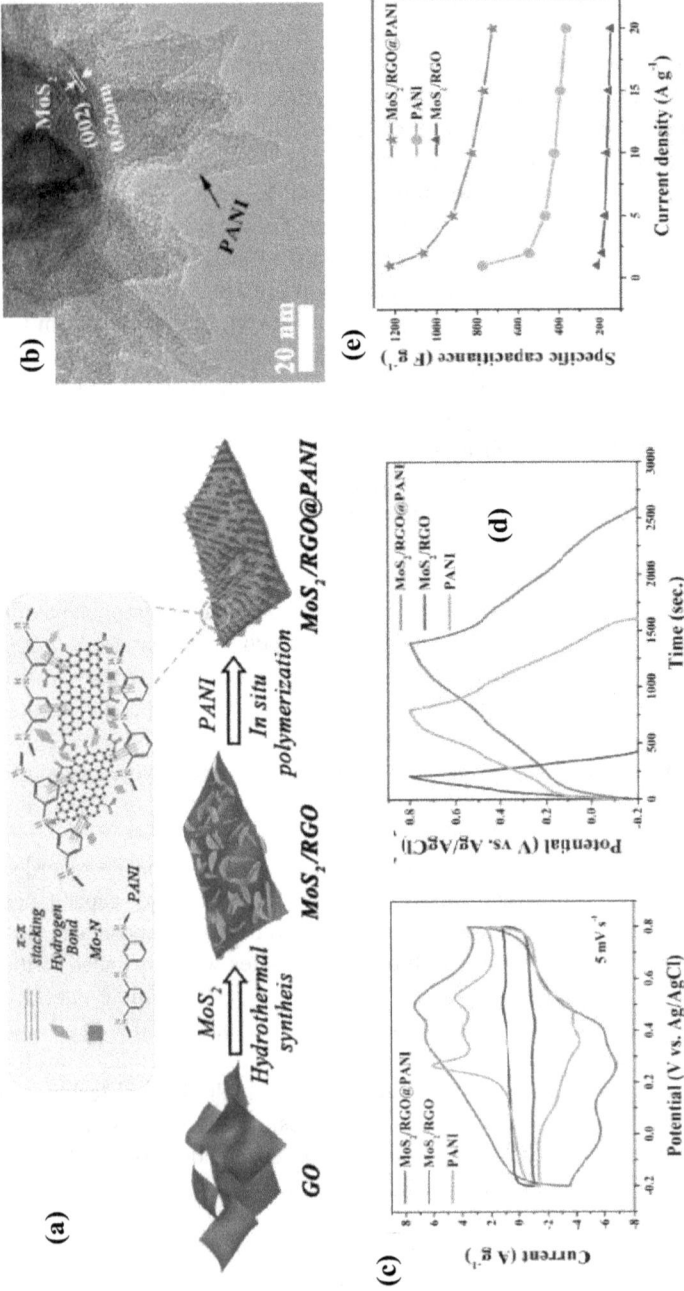

FIGURE 12.8 (a) Schematic representation of the preparation of the ternary nanocomposite MoS₂/RGO@PANI (b) TEM image indicating the morphology of the MoS2/RGO@PANI composite. (c) CV profiles of PANI, MoS₂/RGO and MoS₂/RGO@PANI composite recorded in 1M 1 M H₂SO₄ electrolyte at a potential scan rate of 5 mVs⁻¹. (d) Charging/discharging profiles of PANI, MoS₂/RGO, and MoS₂/RGO@PANI composite at a current density of 1 Ag⁻¹. (e)Cyclic stability of MoS₂/RGO@PANI composite obtained over 3000 cycles at 10 Ag⁻¹ (Li & Zhang et al., 2016).

and minimized the ion diffusion corridor, all positively contributing to the electro-chemical performance as a whole (Asaithambi et al., 2021). Lately, nanoflower-like MoS_2 was developed on the surface of MWCNTs/PANI nano-stem via a combination of facile in-situ polymerization and hydrothermal techniques. The resultant hierarch-ical structure promoted good interfacial contact at the electrode/electrolyte bound-aries for promoting adequate capacitive charge storage. It is found that the specific capacitance of the obtained MWCNTs/PANI/MoS_2 hybrid is 542.56 Fg^{-1} at a current density of 0.5 Ag^{-1} along with improved rate capability and cyclic stability (Zhang & Liao et al., 2018). Very recently, MoS_2/PANI/functionalized carbon cloth has been successfully designed via a drop-casting technique that registered remarkably long cyclic life as well as high specific capacitances superior to those of MoS_2/CC and pure PANI/CC electrodes respectively (Wang & Lv et al., 2021).

12.9 CONCLUSION

For the last few years, tremendous research has come up regarding 2D MoS_2 and MoS_2-based materials possessing fascinating physical and chemical properties for energy conversion and storage applications. The chapter outlines the various syn-thesis processes and modification strategies adopted for regulating and fine-tuning of the electronic structure, defects introduction, controlling phase structure, and constructing heterostructures, which in turn, have opened up new channels for designing smart MoS_2-based nanomaterials with advanced electrochemical responses. Varied MoS_2-based nanomaterials because of their easy fabrication, tunable elec-trical conductivity, hydrophilicity, high active surface area, and expandable interlayer separation, have marked their capabilities as versatile and promising candidates for supercapacitor applications. MoS_2 nanosheets offer enormous opportunities for facile exfoliation, and fabrication of functional hybrid nanostructures with a wide range of materials including organic and biomaterials, carbonaceous materials, metal-organic frameworks, metallic compounds such as oxides, chalcogenides, etc for developing unique and distinctive physicochemical features that definitely trigger the manufac-turing of smart electroactive nanomaterials for energy storage applications at com-mercial level. Nonetheless, these nanomaterials are yet to meet the desired marketable outputs due to the following existing challenges.

Facile and cost-effective fabrication strategies for developing pure and high-quality MoS_2-based nanomaterials are in high demand for shortening the gap between laboratory research and commercial applications. More in-depth understandings and elementary knowledge of the variables that activate the phase transformation for enhancing the electrical transport properties of low conducting phases of MoS_2 are also largely urged. Furthermore, suitable morphological engin-eering, interfacial modifications and structural control of these nanomaterials are considered necessary to be carried out for better electrochemical competence. Again, MoS_2-based supercapacitor technology is facing a number of challenges in terms of productivity and large-scale material fabrication with the desired proper-ties, embracing issues of evading unwanted impurities, upgraded conductivity, and mechanical and electrochemical stability. Significant progress has been made in the endorsement of the defect engineering of MoS_2-based materials and thus, integration of the required modifications in these systems has to be profitably executed for developing stable, smart and diligent electrode materials for high-performing energy storage devices.

Nonetheless, allied research is still in the nascent stage and there are a lot of theoretical simulations and experimental explorations yet to be conducted ahead. Further, its future prospect is enormous as precursors of MoS_2 are economic, readily obtainable and efficient structural and morphological adaptations that are readily feasible and experimentally attainable using simple and economic solution chemistry relative to other 2D nanomaterials, which have placed this material at a higher priority in the commercial field of energy applications. In addition, it is highly inspiring that the rapid advancement of R&D of related and supporting technologies can sincerely assist in overcoming the existing deficiencies. Hopefully, with the rising quantity of fundamental research on MoS_2, the current drawbacks and unstable features will be adequately addressed related to its implications in the vast field of electrochemical energy storage and will possibly open up new directions of industrial applications for similar metal sulphides/selenides ($MoSe_2$, $MoTe_2$, WS_2, WSe_2, etc.) systems in the near future.

ACKNOWLEDGEMENTS

DM acknowledges Chandernagore College for giving permission for honorary research work. DM also acknowledges SPD for constant motivation and inspiration for continuing research work.

REFERENCES

Acerce, M., Voiry, D., & Chhowalla, M. (2015). Metallic 1T phase MoS2 nanosheets as supercapacitor electrode materials. *Nature Nanotechnology*, *10*(4), 313–318. https://doi.org/10.1038/NNANO.2015.40

Akram, U., Nadarajah, M., Shah, R., & Milano, F. (2020). A review on rapid responsive energy storage technologies for frequency regulation in modern power systems. *Renewable and Sustainable Energy Reviews*, *120*, 109626. https://doi.org/10.1016/j.rser.2019.109626

Ali, B. A., Metwalli, O. I., Khalil, A. S., & Allam, N. K. (2018). Unveiling the effect of the structure of carbon material on the charge storage mechanism in MoS2-based supercapacitors. *ACS Omega*, *3*(11), 16301. https://doi.org/10.1021/acsomega.8b02261

Andersen, A., Kathmann, S. M., Lilga, M. A., Albrecht, K. O., Hallen, R. T., & Mei, D. (2012). First-principles characterization of potassium intercalation in hexagonal 2H-MoS2. *Journal of Physical Chemistry C*, *116*(2), 1826. https://doi.org/10.1021/jp206555b

Ansari, S. A., Fouad, H., Ansari, S. G., Sk, M. P., & Cho, M. H. (2017). Mechanically exfoliated MoS2 sheet coupled with conductive polyaniline as a superior supercapacitor electrode material. *Journal of Colloid and Interface Science*, *504*, 276. https://doi.org/10.1016/j.jcis.2017.05.064

Asaithambi, S., Sakthivel, P., Karuppaiah, M., Balamurugan, K., Yuvakkumar, R., Thambidurai, M., & Ravi, G. (2021). Synthesis and characterization of various transition metals doped

SnO2@ MoS2 composites for supercapacitor and photocatalytic applications. *Journal of Alloys and Compounds*, *853*, 157060. https://doi.org/10.1016/j.jallcom.2020.157060

Benavente, E., Santa Ana, M. A., Mendizábal, F., & González, G. (2002). Intercalation chemistry of molybdenum disulfide. *Coordination Chemistry Reviews*, *224*(1–2), 87. https://doi.org/10.1016/S0010-8545(01)00392-7

Cao, L., Yang, S., Gao, W., Liu, Z., Gong, Y., Ma, L., Shi, G.,Lei, S., Zhang, Y., Zhang, S., Vajtai, R., & Ajayan, P. M. (2013). Direct laser-patterned micro-supercapacitors from paintable MoS2 films. *Small*, *9*(17), 2905. https://doi.org/10.1002/smll.201203164

Chao Yang, Zhongxin Chen, Imran Shakir, Yuxi Xu & Hongbin Lu. (2014). Rational synthesis of carbon shell coated polyaniline/MoS 2 monolayer composites for high-performance supercapacitors. *Nano Research*, *9*, 951. https://doi.org/10.1007/s12274-016-0983-3, https://www.sciencedirect.com/science/article/pii/S2095495620306768, https://doi.org/10.1016/j.jechem.2020.09.041

Chen, G. Z. (2017). Supercapacitor and supercapattery as emerging electrochemical energy stores. *International Materials Reviews*, *62*(4), 173. .https://doi.org/10.1080/09506608.2016.1240914

Chen, X., Ding, J., Jiang, J., Zhuang, G., Zhang, Z., & Yang, P. (2018). Preparation of a MoS 2/carbon nanotube composite as an electrode material for high-performance supercapacitors. *RSC Advances*, *8*(52), 29488. https://doi.org/10.1039/c8ra05158e

Choudhary, N., Patel, M., Ho, Y. H., Dahotre, N. B., Lee, W., Hwang, J. Y., & Choi, W. (2015). Directly deposited MoS 2 thin film electrodes for high performance supercapacitors. *Journal of Materials Chemistry A*, *3*(47), 24049..https://doi.org/10.1039/C5TA08095A

Ciesielski, A., & Samorì, P. (2014). Graphene via sonication assisted liquid-phase exfoliation. *Chemical Society Reviews*, *43*(1), 381. https://doi.org/10.1039/c3cs60217f

Das, K., & Majumdar, D. (2022). Prospects of MXenes/graphene nanocomposites for advanced supercapacitor applications. *Journal of Electroanalytical Chemistry*, *905*, 115973. https://doi.org/10.1016/j.jelechem.2021.115973

Falola, B. D., Wiltowski, T., & Suni, I. I. (2016). Electrodeposition of MoS2 for charge storage in electrochemical supercapacitors. *Journal of The Electrochemical Society*, *163*(9), D568–D574. https://doi.org/10.1149/2.0011610jes

Gao, Y. P., Huang, K. J., Wu, X., Hou, Z. Q., & Liu, Y. Y. (2018). MoS2 nanosheets assembling three-dimensional nanospheres for enhanced-performance supercapacitor. *Journal of Alloys and Compounds*, *741*, 174–181. https://doi.org/10.1016/j.jallcom.2018.01.110

Geim, A. K., & Grigorieva, I. V. (2013). Van der Waals heterostructures. *Nature*, *499*(7459), 419–425. https://doi.org/10.1038/nature12385

Geng, X., Zhang, Y., Han, Y., Li, J., Yang, L., Benamara, M., Chen, L., & Zhu, H. (2017). Two-dimensional water-coupled metallic MoS2 with nanochannels for ultrafast supercapacitors. *Nano Letters*, *17*(3), 1825–1832. https://doi.org/10.1021/acs.nanolett.6b05134

González, A., Goikolea, E., Barrena, J. A., & Mysyk, R. (2016). Review on supercapacitors: Technologies and materials. *Renewable and Sustainable Energy Reviews*, *58*, 1189–1206. https://doi.org/10.1016/j.rser.2015.12.249

Han, L., Luo, J., Zhang, R., Gong, W., Chen, L., Liu, F., Ling, Y., Dong, Y., Yong, Z., Zhang, Y., Wei, L., Zhang, X., Zhang, Q., & Li, Q. (2022). Arrayed heterostructures of MoS2 nanosheets anchored TiN nanowires as efficient pseudocapacitive anodes for fiber-shaped ammonium-ion asymmetric supercapacitors. *ACS Nano*, *16*(9), 14951–14962. https://doi.org/10.1021/acsnano.2c05905

Han, Y., Chatti, M., Ge, Y., Wang, C., Chao, Y., Simonov, A. N., & Wallace, G. G. (2019). Binder-Free electrodes derived from interlayer-expanded MoS2 nanosheets on carbon cloth with a 3D porous structure for Lithium storage. *ChemElectroChem*, *6*(8), 2338–2343. https://doi.org/10.1002/celc.201900270

Hu, B., Qin, X., Asiri, A. M., Alamry, K. A., Al-Youbi, A. O., & Sun, X. (2013). Synthesis of porous tubular C/MoS2 nanocomposites and their application as a novel electrode material for supercapacitors with excellent cycling stability. *Electrochimica Acta*, *100*, 24–28. https://doi.org/10.1016/j.electacta.2013.03.133

Huang, K. J., Wang, L., Liu, Y. J., Wang, H. B., Liu, Y. M., & Wang, L. L. (2013). Synthesis of polyaniline/2-dimensional graphene analog MoS2 composites for high-performance supercapacitor. *Electrochimica Acta*, *109*, 587–594. https://doi.org/10.1016/j.electa cta.2013.07.168

Huang, K. J., Wang, L., Zhang, J. Z., & Xing, K. (2015). MoS2-based nanocomposites for electrochemical energy storage. Journal of *Electroanalytical Chemistry*, *752*, 33–40. https://doi.org/10.1016/j.jelechem.2015.06.005

Huang, S., Zhu, X., Sarkar, S., & Zhao, Y. (2019). Challenges and opportunities for supercapacitors. *APL Materials*, *7*(10), 100901. https://doi.org/10.1063/1.5116146

Huo, J., Ge, R., Liu, Y., Guo, J., Lu, L., Chen, W., Liu, C., Gao, H..& Liu, H. (2020). Recent advances of two–dimensional molybdenum disulfide based materials: Synthesis, modification and applications in energy conversion and storage. *Sustainable Materials and Technologies*, *24*, e00161. https://doi.org/10.1016/j.susmat.2020.e00161

Ilanchezhiyan, P., Kumar, G. M., & Kang, T. W. (2015). Electrochemical studies of spherically clustered MoS2 nanostructures for electrode applications. *Journal of Alloys and Compounds*, *634*, 104–108. http://dx.doi.org/10.1016/j.jallcom.2015.02.082

Isacfranklin, M., Princy, L. E. M., Rathinam, Y., Kungumadevi, L., Ravi, G., Al-Sehemi, A. G., &Velauthapillai, D. (2022). Rare earth-doped MoS2 for supercapacitor application. *Energy & Fuels*, *36*(12), 6476–6482. https://doi.org/10.1021/acs.energyfu els.2c00536

Jawaid, A., Nepal, D., Park, K., Jespersen, M., Qualley, A., Mirau, P., Drummy,L. F., & Vaia, R. A. (2016). Mechanism for liquid phase exfoliation of MoS2. *Chemistry of Materials*, *28*(1), 337–348. https://doi.org/10.1021/acs.chemmater.5b04224

Jiang, L., Zhang, S., Kulinich, S. A., Song, X., Zhu, J., Wang, X., & Zeng, H. (2015). Optimizing hybridization of 1T and 2H phases in MoS2 monolayers to improve capacitances of supercapacitors. *Materials Research Letters*, *3*(4), 177–183. https://doi.org/10.1080/21663831.2015.1057654

Kar, T., Godavarthi, S., Pasha, S. K., Deshmukh, K., Martínez-Gómez, L., & Kesarla, M. K. (2022). Layered materials and their heterojunctions for supercapacitor applications: A review. *Critical Reviews in Solid State and Materials Sciences*, *47*(3), 357–388. https://doi.org/10.1080/10408436.2021.1886048

Kim, H. U., Kim, M., Jin, Y., Hyeon, Y., Kim, K. S., An, B. S., Yang,C. W., Kanade, V., Moon, J. Y., Yeom, G. Y., Whang, D., Lee, J. H., & Kim, T. (2019). Low-temperature wafer-scale growth of MoS2-graphene heterostructures. *Applied Surface Science*, *470*, 129–134. https://doi.org/10.1016/j.apsusc.2018.11.126

Kim, J. & Lee, Z. (2018). Phase transformation of Two-Dimensional transition metal dichalcogenides. *Applied Microscopy*, *48*, 43–48. http://dx.doi.org/10.9729/AM.2018.48.2.43

Krishnamoorthy, K., Pazhamalai, P., Veerasubramani, G. K., & Kim, S. J. (2016). Mechanically delaminated few layered MoS2 nanosheets based high performance wire type solid-state symmetric supercapacitors. *Journal of Power Sources*, *321*, 112–119. https://doi.org/10.1016/j.jpowsour.2016.04.116

Krishnamoorthy, K., Veerasubramani, G. K., Pazhamalai, P., & Kim, S. J. (2016). Designing two dimensional nanoarchitectured MoS2 sheets grown on Mo foil as a binder free electrode for supercapacitors. *Electrochimica Acta*, *190*, 305–312. https://doi.org/10.1016/j.electacta.2015.12.148

Kyriakopoulos, G. L., &Arabatzis, G. (2016). Electrical energy storage systems in electricity generation: Energy policies, innovative technologies, and regulatory regimes. *Renewable and Sustainable Energy Reviews*, *56*, 1044–1067. https://doi.org/10.1016/j.rser.2015.12.046

Lee, Y. H., Zhang, X. Q., Zhang, W., Chang, M. T., Lin, C. T., Chang, K. D., Yu,Y. C., Wang,.J. T. W., Chang, C. S., Li, L. J., & Lin, T. W. (2012). Synthesis of large-area MoS2 atomic layers with chemical vapor deposition. *Advanced Materials*, *24*(17), 2320–2325. https://doi.org/10.1002/adma.201104798

Li, L., Yang, H., Yang, J., Zhang, L., Miao, J., Zhang, Y., Sun, C., Huang, W., Dong, X., & Liu, B. (2016). Hierarchical carbon@ Ni 3 S 2@ MoS 2 double core–shell nanorods for high-performance supercapacitors. *Journal of Materials Chemistry A*, *4*(4), 1319–1325. https://doi.org/10.1039/C5TA08714G

Li, X., & Zhu, H. (2015). Two-dimensional MoS2: Properties, preparation, and applications. *Journal of Materiomics*, *1*(1), 33–44. https://doi.org/10.1016/j.jmat.2015.03.003

Li, X., Li, X., Cheng, J., Yuan, D., Ni, W., Guan, Q., Gao, L., & Wang, B. (2016). Fiber-shaped solid-state supercapacitors based on molybdenum disulfide nanosheets for a self-powered photodetecting system. *Nano Energy*, *21*, 228–237. https://doi.org/10.1016/j.nanoen.2016.01.011

Li, X., Zhang, C., Xin, S., Yang, Z., Li, Y., Zhang, D., & Yao, P. (2016). Facile synthesis of MoS2/reduced graphene oxide@ polyaniline for high-performance supercapacitors. *ACS Applied Materials & Interfaces*, *8*(33), 21373–21380. https://doi.org/10.1021/acsami.6b06762

Liang, D., Tian, Z., Liu, J., Ye, Y., Wu, S., Cai, Y., & Liang, C. (2015). MoS2 nanosheets decorated with ultrafine Co3O4 nanoparticles for high-performance electrochemical capacitors. *Electrochimica Acta*, *182*, 376–382. https://doi.org/10.1016/j.electacta.2015.09.085

Liang, Q., Zhang, Q., Zhao, X., Liu, M., & Wee, A. T. (2021). Defect engineering of two-dimensional transition-metal dichalcogenides: applications, challenges, and opportunities. *ACS Nano*, *15*(2), 2165–2181. https://doi.org/10.1021/acsnano.0c09666

Liang, Y., Yoo, H. D., Li, Y., Shuai, J., Calderon, H. A., Robles Hernandez, F. C., Grabow, L. C., & Yao, Y. (2015). Interlayer-expanded molybdenum disulfide nanocomposites for electrochemical magnesium storage. *Nano Letters*, *15*(3), 2194–2202. https://doi.org/10.1021/acs.nanolett.5b00388

Liang, Z., Zhao, C., Zhao, W., Zhang, Y., Srimuk, P., Presser, V., & Feng, G. (2021). Molecular understanding of charge storage in MoS2 supercapacitors with ionic liquids. *Energy & Environmental Materials*, *4*(4), 631–637. https://doi.org/10.1002/eem2.12147

Liao, X., Zhao, Y., Wang, J., Yang, W., Xu, L., Tian, X.,Shuang, Y., Owusu, K. A., & Mai, L. (2018). MoS 2/MnO2 heterostructured nanodevices for electrochemical energy storage. *Nano Research*, *11*, 2083–2092. https://doi.org/10.1007/s12274-017-1826-6

Lin, X., Li, W., Dong, Y., Wang, C., Chen, Q., & Zhang, H. (2016). Two-dimensional metallic MoS2: A DFT study. *Computational Materials Science*, *124*, 49–53. https://doi.org/10.1016/j.commatsci.2016.07.020

Liu, K. K., Zhang, W., Lee, Y. H., Lin, Y. C., Chang, M. T., Su, C. Y., Chang, C. S., Li, H., Shi, Y., Zhang, H., Lai, C. S., & Li, L. J. (2012). Growth of large-area and highly crystalline MoS2 thin layers on insulating substrates. *Nano Letters*, *12*(3), 1538–1544. https://doi.org/10.1021/nl2043612

Liu, K., Yu, C., Guo, W., Ni, L., Yu, J., Xie, Y., Wang, Z., Ren, Y., & Qiu, J. (2021). Recent research advances of self-discharge in supercapacitors: Mechanisms and suppressing strategies. *Journal of Energy Chemistry*, *58*, 94–109. https://doi.org/10.1016/j.jec hem.2020.09.041

Liu, Q., Li, X., He, Q., Khalil, A., Liu, D., Xiang, T., Wu, X., & Song, L. (2015). Gram-scale aqueous synthesis of stable few-layered 1T-MoS2: Applications for visible-light-driven photocatalytic hydrogen evolution. *Small*, *11*(41), 5556–5564. https://doi.org/10.1002/ smll.201501822

Liu, Y., Zhong, Q., Chen, K., Zhou, J., Yang, X., & Chen, W. (2017). Morphologies controllable synthesis of MoS 2 by hot-injection method: From quantum dots to nanosheets. *Journal of Materials Science: Materials in Electronics*, *28*, 13633–13637. https://doi. org/10.1007/s10854-017-7204-z

Ma, G., Peng, H., Mu, J., Huang, H., Zhou, X., & Lei, Z. (2013). In situ intercalative polymerization of pyrrole in graphene analogue of MoS2 as advanced electrode material in supercapacitor. *Journal of Power Sources*, *229*, 72–78. https://doi.org/10.1016/j.jpows our.2012.11.088

Magda, G. Z., Pető, J., Dobrik, G., Hwang, C., Biró, L. P., &Tapasztó, L. (2015). Exfoliation of large-area transition metal chalcogenide single layers. *Scientific Reports*, *5*(1), 14714. https://doi.org/10.1038/srep14714

Majumdar, D. (2019). Polyaniline nanocomposites: Innovative materials for supercapacitor applications–PANI nanocomposites for supercapacitor applications. In *Polymer Nanocomposites for Advanced Engineering and Military Applications*, IGI Global, 220–253. https://doi.org/10.4018/978-1-7998-8591-7.ch026

Majumdar, D. (2019). Ultrasound-assisted synthesis, Exfoliation and functionalisation of graphene derivatives. *Graphene Functionalization Strategies: From Synthesis to Applications*, 63–103. https://doi.org/10.1007/978-981-32-9057-0_3

Majumdar, D. (2021). Review on current progress of MnO2-based ternary nanocomposites for supercapacitor applications. *ChemElectroChem*, *8*(2), 291–336. https://doi.org/10.1002/ celc.202001371

Majumdar, D. (2022). Application of microbes in synthesis of electrode materials for supercapacitors. *Application of Microbes in Environmental and Microbial Biotechnology*, 39–92. https://doi.org/10.1007/978-981-16-2225-0_2

Majumdar, D., & Bhattacharya, S. K. (2017). Sonochemically synthesized hydroxy-functionalized graphene–MnO2 nanocomposite for supercapacitor applications. *Journal of Applied Electrochemistry*, *47*, 789–801. https://doi.org/10.1007/s10800-017-1080-3

Majumdar, D., & Ghosh, S. (2021). Recent advancements of copper oxide based nanomaterials for supercapacitor applications. *Journal of Energy Storage*, *34*, 101995. https://doi.org/ 10.1016/j.est.2020.101995

Majumdar, D., Baugh, N., & Bhattacharya, S. K. (2017). Ultrasound assisted formation of reduced graphene oxide-copper (II) oxide nanocomposite for energy storage applications. *Colloids and Surfaces A: Physicochemical and Engineering Aspects*, *512*, 158–170. https://doi.org/10.1016/j.colsurfa.2016.10.010

Majumdar, D., Mandal, M., & Bhattacharya, S. K. (2019). V2O5 and its carbon-based nanocomposites for supercapacitor applications. *ChemElectroChem*, *6*(6), 1623–1648. https://doi.org/10.1002/celc.201801761

Majumdar, D., Mandal, M., & Bhattacharya, S. K. (2020). Journey from supercapacitors to supercapatteries: Recent advancements in electrochemical energy storage systems. *Emergent Materials*, *3*, 347–367. https://doi.org/10.1007/s42247-020-00090-5

Manuraj, M., Nair, K. K., Unni, K. N., & Rakhi, R. B. (2020). High performance supercapacitors based on MoS2 nanostructures with near commercial mass loading. *Journal of Alloys and Compounds*, *819*, 152963. https://doi.org/10.1016/j.jallcom.2019.152963

Manzeli, S., Ovchinnikov, D., Pasquier, D., Yazyev, O. V., &Kis, A. (2017). 2D transition metal dichalcogenides. *Nature Reviews Materials*, *2*(8), 1–15. http://doi.org/10.1038/natrevm ats.2017.33

Najib, S., & Erdem, E. (2019). Current progress achieved in novel materials for supercapacitor electrodes: Mini review. *Nanoscale Advances*, *1*(8), 2817–2827. https://doi.org/10.1039/ C9NA00345B

Nardekar, S. S., Krishnamoorthy, K., Pazhamalai, P., Sahoo, S., Mariappan, V. K., & Kim, S. J. (2020). Exceptional interfacial electrochemistry of few-layered 2D MoS2 quantum sheets for high performance flexible solid-state supercapacitors. *Journal of Materials Chemistry A*, *8*(26), 13121–13131. https://doi.org/10.1039/D0TA01156H

Nguyen, E. P., Carey, B. J., Daeneke, T., Ou, J. Z., Latham, K., Zhuiykov, S., &Kalantar-zadeh, K. (2015). Investigation of two-solvent grinding-assisted liquid phase exfoliation of layered MoS2. *Chemistry of Materials*, *27*(1), 53–59. https://doi.org/10.1021/ cm502915f

Nicolosi, V., Chhowalla, M., Kanatzidis, M. G., Strano, M. S., & Coleman, J. N. (2013). Liquid exfoliation of layered materials. *Science*, *340*(6139), 1226419. https://doi.org/10.1126/ science.1226419

Palsaniya, S., Nemade, H. B., & Dasmahapatra, A. K. (2018). Synthesis of polyaniline/ graphene/MoS2 nanocomposite for high performance supercapacitor electrode. *Polymer*, *150*, 150–158. https://doi.org/ 10.1016/j.polymer.2018.07.018

Panigrahi, P. K., & Pathak, A. (2011). A novel route for the synthesis of nanotubes and fullerene-like nanostructures of molybdenum disulfide. *Materials Research Bulletin*, *46*(12), 2240–2246. https:// doi. org/ 10. 1016/j. materresbu ll. 2011. 09. 003

Peng, Y., Meng, Z., Zhong, C., Lu, J., Yu, W., Yang, Z., & Qian, Y. (2001). Hydrothermal synthesis of MoS2 and its pressure-related crystallization. *Journal of Solid State Chemistry*, *159*(1), 170–173. https://doi.org/10.1006/jssc.2001.9146

Pumera, M., Sofer, Z., & Ambrosi, A. (2014). Layered transition metal dichalcogenides for electrochemical energy generation and storage. *Journal of Materials Chemistry A*, *2*(24), 8981–8987. https://doi.org/10.1039/C4TA00652F

Ramadoss, A., Kim, T., Kim, G. S., & Kim, S. J. (2014). Enhanced activity of a hydrothermally synthesized mesoporous MoS 2 nanostructure for high performance supercapacitor applications. *New Journal of Chemistry*, *38*(6), 2379–2385. https://doi.org/10.1039/ C3NJ01558K

Rasamani, K. D., Alimohammadi, F., & Sun, Y. (2017). Interlayer-expanded MoS2. *Materials Today*, *20*(2), 83–91. https://doi.org/10.1016/j.mattod.2016.10.004

Ren, L., Zhang, G., Yan, Z., Kang, L., Xu, H., Shi, F., Lei, Z., & Liu, Z. H. (2015). Three-dimensional tubular MoS2/PANI hybrid electrode for highrate performance supercapacitor. *ACS Applied Materials & Interfaces*, *7*(51), 28294–28302. https://doi. org/10.1021/acsami.5b08474

Sadan, M., Houben, L., & Naveh, D. (2018). Growth mechanisms and electronic properties of vertically aligned MoS2. *Scientific Reports*, *8*(1), 16480. http://doi.org/10.1038/s41 598-018-34222-z

Sahoo, D., Shakya, J., Choudhury, S., Roy, S. S., Devi, L., Singh, B., Ghosh, S., & Kaviraj, B. (2022). High-performance MnO2 Nanowire/MoS2 nanosheet composite for a sym-metrical solid-state supercapacitor. *ACS Omega*, *7*(20), 16895–16905. https://doi.org/ 10.1021/acsomega.1c06852

Samy, O., & El Moutaouakil, A. (2021). A review on MoS2 energy applications: Recent developments and challenges. *Energies*, *14*(15), 4586. https://doi.org/10.3390/en14154586

Sarkar, D., Das, D., Das, S., Kumar, A., Patil, S., Nanda, K. K., Sarma, D. D., & Shukla, A. (2019). Expanding interlayer spacing in MoS2 for realizing an advanced supercapacitor. *ACS Energy Letters*, *4*(7), 1602–1609. https://doi.org/10.1021/acsenergylett.9b00983

Savjani, N., Lewis, E. A., Bissett, M. A., Brent, J. R., Dryfe, R. A., Haigh, S. J., & O'Brien, P. (2016). Synthesis of lateral size-controlled monolayer 1 H-MoS2@ Oleylamine as supercapacitor electrodes. *Chemistry of Materials*, *28*(2), 657–664. https://doi.org/10.1021/acs.chemmater.5b04476

Sha, C., Lu, B., Mao, H., Cheng, J., Pan, X., Lu, J., & Ye, Z. (2016). 3D ternary nanocomposites of molybdenum disulfide/polyaniline/reduced graphene oxide aerogel for high performance supercapacitors. *Carbon*, *99*, 26–34. https://doi.org/10.1016/j.carbon.2015.11.066

Shen, J., He, Y., Wu, J., Gao, C., Keyshar, K., Zhang, X., Yang, Y., Ye, M., Vajtai, R., Lou, J., & Ajayan, P. M. (2015). Liquid phase exfoliation of two-dimensional materials by directly probing and matching surface tension components. *Nano Letters*, *15*(8), 5449–5454. https://doi.org/10.1021/acs.nanolett.5b01842

Shi, S., Sun, Z., & Hu, Y. H. (2018). Synthesis, stabilization and applications of 2-dimensional 1T metallic MoS 2. *Journal of Materials Chemistry A*, *6*(47), 23932–23977. https://doi.org/10.1039/C8TA08152B

Singh, K., Kumar, S., Agarwal, K., Soni, K., Ramana Gedela, V., & Ghosh, K. (2017). Three-dimensional graphene with MoS 2 nanohybrid as potential energy storage/transfer device. *Scientific Reports*, *7*(1), 1–12. https://doi.org/10.1038/s41598-017-09266-2

Smith, R. J., King, P. J., Lotya, M., Wirtz, C., Khan, U., De, S., O'Neill, A., Duesberg, G. S., Grunlan, J. C., Moriarty, G., Chen, J., Wang, J., Minett, A. I., Nicolosi, V., & Coleman, J. N. (2011). Large-scale exfoliation of inorganic layered compounds in aqueous surfactant solutions. *Advanced Materials*, *23*(34), 3944–3948. https://doi.org/10.1002/adma.201102584

Stern, C., Grinvald, S., Kirshner, M., Sinai, O., Oksman, M., Alon, H., Meiron, O. E., Bar-Sadan, M, Houben, L, Naveh, D. (2018). Growth mechanisms and electronic properties of vertically aligned MoS₂. *Scientific Reports*, *8(1)*,16480. https://doi.org/10.1038/s41598-018-34222-z

Sun, J., Li, X., Guo, W., Zhao, M., Fan, X., Dong, Y., Xu, C., Deng, J., & Fu, Y. (2017). Synthesis methods of two-dimensional MoS2: A brief review. *Crystals*, *7*(7), 198. https://doi.org/10.3390/cryst7070198

Sundriyal, P., & Bhattacharya, S. (2020). Textile-based supercapacitors for flexible and wearable electronic applications. *Scientific Reports*, *10*(1), 1–15. https://doi.org/10.1038/s41598-020-70182-z

Tang, H., Wang, J., Yin, H., Zhao, H., Wang, D., & Tang, Z. (2015). Growth of polypyrrole ultrathin films on MoS2 monolayers as high-performance supercapacitor electrodes. *Advanced Materials*, *27*(6), 1117–1123. https://doi.org/10.1002/adma.201404622

Theerthagiri, J., Senthil, R. A., Senthilkumar, B., Polu, A. R., Madhavan, J., & Ashokkumar, M. (2017). Recent advances in MoS2 nanostructured materials for energy and environmental applications–a review. *Journal of Solid State Chemistry*, *252*, 43–71. https://doi.org/10.1016/j.jssc.2017.04.041

Toh, R. J., Sofer, Z., Luxa, J., Sedmidubský, D., & Pumera, M. (2017). 3R phase of MoS 2 and WS 2 outperforms the corresponding 2H phase for hydrogen evolution. *Chemical Communications*, *53*(21), 3054–3057. https://doi.org/10.1039/C6CC09952A

Wang, F., Wu, X., Yuan, X., Liu, Z., Zhang, Y., Fu, L., Zhu, Y., Zhou, Q., Wu, Y., & Huang, W. (2017). Latest advances in supercapacitors: From new electrode materials to novel device designs. *Chemical Society Reviews*, *46*(22), 6816–6854. https://doi.org/10.1039/C7CS00205J

Wang, J., Chao, D., Liu, J., Li, L., Lai, L., Lin, J., & Shen, Z. (2014). Ni3S2@ MoS2 core/shell nanorod arrays on Ni foam for high-performance electrochemical energy storage. *Nano Energy*, *7*, 151–160. https://doi.org/10.1016/j.nanoen.2014.04.019

Wang, L., Ma, Y., Yang, M., & Qi, Y. (2015). Hierarchical hollow MoS2 nanospheres with enhanced electrochemical properties used as an electrode in supercapacitor. *Electrochimica Acta*, *186*, 391–396. https://doi.org/10.1016/j.electacta.2015.10.130

Wang, L., Ma, Y., Yang, M., & Qi, Y. (2015). One-pot synthesis of 3D flower-like heterostructuredSnS 2/MoS 2 for enhanced supercapacitor behavior. *Rsc Advances*, *5*(108), 89069–89075. https://doi.org/10.1039/C5RA16300E

Wang, T., Chen, S., Pang, H., Xue, H., & Yu, Y. (2017). MoS2-based nanocomposites for electrochemical energy storage. *Advanced Science*, *4*(2), 1600289. https://doi.org/10.1002/advs.201600289

Wang, Y., Liu, S., Hao, X., Zhou, J., Song, D., Wang, D., Hou, L., & Gao, F. (2017). Fluorine-and nitrogen-codoped MoS2 with a catalytically active basal plane. *ACS Applied Materials & Interfaces*, *9*(33), 27715–27719. https://doi.org/10.1021/acsami.7b06795

Wang, Y., Lv, X., Zou, S., Lin, X., & Ni, Y. (2021). MoS 2/polyaniline/functionalized carbon cloth electrode materials for excellent supercapacitor performance. *RSC Advances*, *11*(18), 10941–10950. https://doi.org/10.1039/d0ra09126j

Wei, S., Cui, X., Xu, Y., Shang, B., Zhang, Q., Gu, L., Fan, X., Zheng, L., Hou, C., Huang, H.,Wen, S., & Zheng, W. (2018). Iridium-triggered phase transition of MoS2 nanosheets boosts overall water splitting in alkaline media. *ACS Energy Letters*, *4*(1), 368–374. https://doi.org/10.1021/acsenergylett.8b01840

Weng, Q., Wang, X., Wang, X., Zhang, C., Jiang, X., Bando, Y., & Golberg, D. (2015). Supercapacitive energy storage performance of molybdenum disulfide nanosheets wrapped with microporous carbons. *Journal of Materials Chemistry A*, *3*(6), 3097–3102. https://doi.org/10.1039/C4TA06303A

Xie, X., Makaryan, T., Zhao, M., Van Aken, K. L., Gogotsi, Y., & Wang, G. (2016). MoS2 nanosheets vertically aligned on carbon paper: A freestanding electrode for highly reversible sodium-ion batteries. *Advanced Energy Materials*, *6*(5), 1502161. https://doi.org/10.1002/aenm.201502161

Xu, X., Das, G., He, X., Hedhili, M. N., Fabrizio, E. D., Zhang, X., & Alshareef, H. N. (2019). High-performance monolayer MoS2 films at the wafer scale by two-step growth. *Advanced Functional Materials*, *29*(32), 1901070. https://doi.org/10.1002/adfm.201901070

Yan, J., Wang, Q., Wei, T., & Fan, Z. (2014). Recent advances in design and fabrication of electrochemical supercapacitors with high energy densities. *Advanced Energy Materials*, *4*(4), 1300816. http://doi: 10.1002/aenm.201300816

Yang, C., Chen, Z., Shakir, I., Xu, Y., & Lu, H. (2016). Rational synthesis of carbon shell coated polyaniline/MoS$_2$ monolayer composites for high-performance supercapacitors. *Nano Research*, *9*, 951. https://doi.org/10.1007/s12274-016-0983-3

Yao, K., Xu, Z., Huang, J., Ma, M., Fu, L., Shen, X., Li, J., & Fu, M. (2019). Bundled defect-rich MoS2 for a high-rate and long-life sodium-ion battery: Achieving 3D diffusion of sodium ion by vacancies to improve kinetics. *Small*, *15*(12), 1805405. https://doi.org/10.1002/smll.201805405

Yaseen, M., Khattak, M. A. K., Humayun, M., Usman, M., Shah, S. S., Bibi, S., Hasnain, B. S. U., Ahmad, S. M., Khan, A., Shah, N. Tahir, A., & Ullah, H. (2021). A review of supercapacitors: Materials design, modification, and applications. *Energies*, *14*(22), 7779. https://doi.org/10.3390/en14227779

Yuan, Y., Lv, H., Xu, Q., Liu, H., & Wang, Y. (2019). A few-layered MoS2 nanosheets/nitrogen-doped graphene 3D aerogel as a high performance and long-term stability supercapacitor electrode. *Nanoscale*, *11*(10), 4318–4327. https://doi.org/10.1039/C8NR05620J

Yun, Q., Li, L., Hu, Z., Lu, Q., Chen, B., & Zhang, H. (2020). Layered transition metal Dichalcogenide-Based Nanomaterials for Electrochemical Energy Storage. *Advanced Materials*, *32*, 1903826. https://doi.org/10.1002/adma.201903826

Zarach, Z., Szkoda, M., Trzciński, K., Łapiński, M., Trykowski, G., & Nowak, A. P. (2022). The phenomenon of increasing capacitance induced by 1T/2H-MoS2 surface modification with Pt particles–Influence on composition and energy storage mechanism. *Electrochimica Acta*, *435*, 141389. https://doi.org/10.1016/j.electacta.2022.141389

Zeng, R., Li, Z., Li, L., Li, Y., Huang, J., Xiao, Y., Yuan, K., & Chen, Y. (2019). Covalent connection of polyaniline with MoS$_2$ nanosheets toward ultrahigh rate capability supercapacitors. *ACS Sustainable Chemistry & Engineering*, *7*(13), 11540–11549. https://doi.org/ 10.1021/acssuschemeng.9b01442

Zhan, C., Liu, W., Hu, M., Liang, Q., Yu, X., Shen, Y.,Lv, R.,Kang, F., & Huang, Z. H. (2018). High-performance sodium-ion hybrid capacitors based on an interlayer-expanded MoS2/rGO composite: Surpassing the performance of lithium-ion capacitors in a uniform system. *NPG Asia Materials*, *10*(8), 775–787. https://doi.org/10.1038/s41 427-018-0073-y

Zhang, R., Liao, Y., Ye, S., Zhu, Z., & Qian, J. (2018). Novel ternary nanocomposites of MWCNTs/PANI/MoS2: Preparation, characterization and enhanced electrochemical capacitance. *Royal Society Open Science*, *5*(1), 171365. https://doi.org/10.1098/rsos.171365

Zhang, X., Lai, Z., Tan, C., & Zhang, H. (2016). Solution-processed two-dimensional MoS2 nanosheets: Preparation, hybridization, and applications. *Angewandte Chemie International Edition*, *55*(31), 8816–8838. https://doi.org/10.1002/anie.201509933

Zhang, Y., Sun, W., Rui, X., Li, B., Tan, H. T., Guo, G., Madhavi, S., Zong, Y., & Yan, Q. (2015). One-pot synthesis of tunable crystalline Ni3S4@ amorphous MoS2 core/shell nanospheres for high-performance supercapacitors. *Small*, *11*(30), 3694-3702. https://doi.org/10.1002/smll.201403772

Zhao, C., Ang, J. M., Liu, Z., & Lu, X. (2017). Alternately stacked metallic 1T-MoS2/polyaniline heterostructure for high-performance supercapacitors. *Chemical Engineering Journal*, *330*, 462–469. https://doi.org/10.1016/j.cej.2017.07.129

Zheng, J., Yan, X., Lu, Z., Qiu, H., Xu, G., Zhou, X., Wang, P., Pan, X., Liu, K., & Jiao, L. (2017). High-mobility multilayered MoS2 flakes with low contact resistance grown by chemical vapor deposition. *Advanced Materials*, *29*(13), 1604540. https://doi.org/ 10.1002/adma.201604540

Zhou, X., Xu, B., Lin, Z., Shu, D., & Ma, L. (2014). Hydrothermal synthesis of flower-like MoS2 nanospheres for electrochemical supercapacitors. *Journal of Nanoscience and Nanotechnology*, *14*(9), 7250–7254. https://doi.org/10.1166/jnn.2014.8929

13 Recent Advances in Mesoporous Metal-based Materials for Energy Production, Storage, and Conversion

Athira A R and Xavier T S

13.1 INTRODUCTION

The advent of mesoporous (mp) materials constitutes a significant scientific advancement that has captured considerable interest owing to their exceptional properties, including an exceedingly high surface area, modifiable pore size, and a relatively narrow pore size distribution. Considerable effort has been directed towards synthesizing mesoporous materials, which offer extensive cavities for accommodating guest species. This is unlike conventional microporous materials utilized in industry, whose pore size is constrained between 50 to 2 nm. Consequently, these materials have broadened the scope of their applicability, including adsorption, [1–3], separation of large molecules,[4,5] catalysis, [6–8] drug and DNA delivery, [9–11], sensors, [12,13]functional devices, [14,15] and fuel cells [16,4]. The larger-sized mesopores, in particular, significantly facilitate guest species transport, thus enhancing access to reaction sites.

Recent research has rapidly extended to non-siliceous mesoporous materials with diverse framework compositions, such as metals, metal oxides, sulfides, inorganic-organic hybrid materials, carbons, and polymers. Despite successful preparation reports of non-siliceous mesoporous materials, preserving structural integrity remains challenging since the mesopores tend to collapse during pore wall crystallization and template removal. Mesoporous metals have garnered significant attention due to their remarkable physicochemical properties, such as high thermal and electrical conductivity. These properties offer potential applications that other mesoporous materials cannot achieve. Both experimental and theoretical studies demonstrate that the morphology and structure of mesoporous metals significantly influence their physical and chemical properties, rendering them promising candidates in biosensors, surface-enhanced Raman scattering, surface plasmon resonance, hydrogen storage, and catalysis.

DOI: 10.1201/9781003318859-13

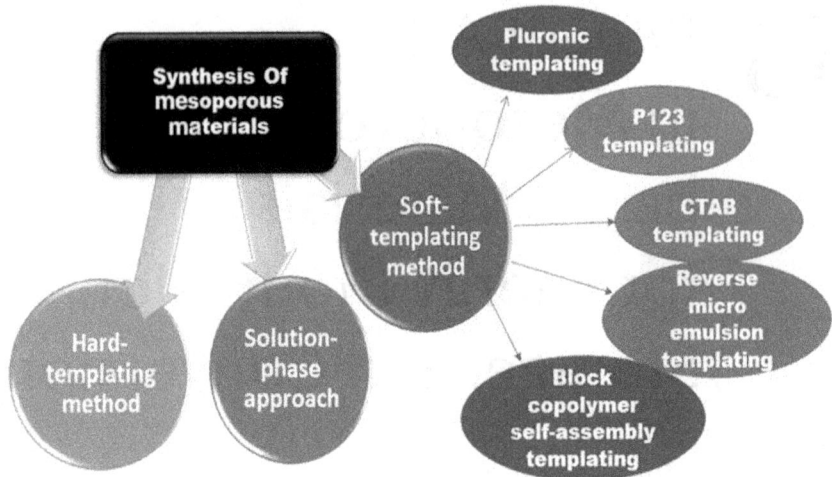

FIGURE 13.1 Synthesis of mesoporous metal oxides.

The precise modulation of properties in mesoporous metals necessitates a rational design approach. By manipulating the shape and morphology of a material, a diverse range of novel porous nanostructures with unique properties can be synthesized, including dendritic metals that exhibit randomly oriented nanopores in contrast to the oriented mesoporous materials. The synthesis of mesoporous structures typically involves templating methods (i.e., hard or soft), while wet-chemical methods are utilized to fabricate dendritic nanostructures. Recent advancements in the field have successfully prepared various nanoporous metals (i.e., microporous and mesoporous metals) with varying sizes and shapes under different conditions. The general preparation methods are consolidated in Figure 13.1.

13.2   PREPARATION OF MESOPOROUS MATERIALS

13.2.1   Hard-templating Method

The hard-templating method is a widely used technique for synthesizing mesoporous metal-based materials. This method uses a sacrificial template, such as a block copolymer or a surfactant, to create a porous structure in the metal oxide material. The template is then removed, leaving behind a well-defined mesoporous structure. Mesoporous materials have attracted considerable interest due to their unique properties, including high surface area, large pore volume, and tunable pore size. These properties make them ideal for various applications, including catalysis, adsorption, energy storage, and drug delivery [17] The hard-templating method uses a sacrificial template to create a porous structure in the metal oxide material. The template can be a surfactant or a block copolymer, depending on the desired properties of the final material.

The basic steps involved in the hard-templating method are as follows [18];

1 The template is added to a solution containing the metal precursor.
2 The metal precursor reacts with the template to form a composite material.
3 The composite material is then heated to a high temperature, which causes the template to decompose and leave behind a porous structure in the metal oxide material.
4 The metal oxide material is washed and dried to remove any residual template.

A sacrificial template creates a well-defined porous structure in the metal oxide material. The pore size, pore volume, and surface area of the material can be tuned by adjusting the template's properties. Two main types of templates can be used in the hard-templating method: Surfactants and block copolymers. Surfactants are molecules that have a hydrophilic head and a hydrophobic tail. The surfactant molecules assemble into a micelle structure when added to a metal precursor solution. The metal precursor then reacts with the micelle to form a composite material. The micelle structure is then removed by heating, leaving a porous structure in the metal oxide material. Block copolymers are composed of two or more different types of polymer chains. The block copolymer self-assembles into a well-defined structure when added to a solution containing a metal precursor. The metal precursor then forms a composite material with the block copolymer. The block copolymer structure is then removed by heating, leaving a porous structure in the metal oxide material [19].

13.3 ADVANTAGES AND LIMITATIONS OF HARD-TEMPLATING METHOD

The hard-templating method has several advantages over other methods of synthesizing mesoporous materials. First, using a sacrificial template creates a well-defined porous structure in the metal oxide material. Second, the pore size, pore volume, and surface area of the material can be tuned by adjusting the template's properties. Third, the method is relatively simple and can be performed using standard laboratory equipment. However, the hard-templating method also has some limitations. First, the method can be time-consuming, as heating must remove the template. Second, the method can be expensive, as some templates, such as block copolymers, can be costly. Finally, the method can be limited by the solubility of the metal precursor in the chosen solvent.

The hard-templating process was initially employed for synthesizing mesoporous carbon. In recent years a lot of mesopores metals, metal oxides, and metal sulfides have been synthesized successfully through the hard-template method. In most reports, mesoporous silica with a robust framework and high thermal stability is a hard template for synthesizing metal replicas. Peng Mei et al. recently reported mesoporous cobalt phosphide (*meso*-CoP) prepared by a hard-templating method using SBA-15 found application in oxygen evolution reaction (OER) [20]. A facile and efficient approach towards synthesizing 3D-ordered mesoporous ternary nitrides $NiCo_2N$ has been reported by Ali Saad using a silica KIT-6 template for OER application [21]. Y. Shi et al. synthesized highly ordered mesoporous crystalline $MoO2$

materials with a continuous Ia3d mesostructure using phosphomolybdic acid as a precursor and mesoporous silica KIT-6. These materials exhibited reversible lithium storage capacity [21].

13.4 SOFT-TEMPLATING METHOD

Soft-templating is a method that is commonly used to create mesoporous metal-based materials. This method uses a surfactant template, typically made from a block copolymer, to create a porous structure within a metal-based material. The surfactant template acts as a soft scaffold, dictating the size and shape of the pores created within the material. The surfactant template can be manipulated to control the size and shape of the pores, which in turn influences the properties of the resulting material. The mixture is then heated or processed to form the final mesoporous material. The advantage of the soft-templating method is that it creates highly ordered mesoporous materials with a high degree of control over their pore size and shape. This, in turn, allows for fine-tuning the material's properties, such as its surface area, pore volume, and chemical reactivity. Additionally, the resulting material is typically more flexible and robust because the method uses a surfactant rather than a hard template. Mesoporous metal-based materials made using the soft-templating method have a variety of potential applications, including catalysis, gas storage, and drug delivery. For example, mesoporous metal oxides created using this method are effective catalysts for various chemical reactions. Additionally, mesoporous metals have been used as electrode materials in batteries and supercapacitors due to their high surface area and electrical conductivity.

13.4.1 PLURONIC TEMPLATING

Pluronic is a type of triblock copolymer that is commonly used as a surfactant template in soft-templating methods. The pluronic is mixed with the metal precursor and a solvent. The mixture is then heated to form a gel, which is further processed to create the final mesoporous material. The advantage of using pluronic is that it can form highly ordered mesoporous structures with a high degree of control over the pore size and shape [22].

13.4.2 P123 TEMPLATING

P123 is another type of triblock copolymer that is commonly used as a surfactant template in soft-templating methods. The P123 is mixed with the metal precursor and a solvent. The mixture is then heated and aged to form a gel, which is further processed to create the final mesoporous material. The advantage of using P123 is that it can form mesoporous structures with high structural stability [23].

13.4.3 CTAB TEMPLATING

Cetyltrimethylammonium bromide CTAB is a cationic surfactant that is commonly used as a surfactant template in soft-templating methods. The CTAB is mixed with the metal precursor and a solvent. The mixture is then heated to form a gel, which is further processed to create the final mesoporous material. The advantage of using CTAB is that it can form highly ordered mesoporous structures with a high degree of control over the pore size and shape. The resulting materials have a high degree of chemical stability [24].

13.4.4   Reverse Microemulsion Templating

In this method, a reverse microemulsion is used as the surfactant template. The microemulsion comprises a water-in-oil solution, with the metal precursor dissolved in the water phase. The surfactant template stabilizes the water droplets, which are cross-linked to form a porous structure. The advantage of this method is that it can create highly uniform and well-ordered mesoporous structures [25].

13.4.5   Block Copolymer Self-assembly Templating

This method uses a block copolymer to self-assemble into a micelle structure, which serves as the surfactant template. The metal precursor is then added to the micelle solution, which is further processed to create the final mesoporous material. The advantage of this method is that it allows for a high degree of control over the pore size and shape, and the resulting materials have a high degree of uniformity [26].

13.5   SOLUTION-PHASE APPROACH

The solution-phase approach is a commonly used method for preparing mesoporous metal-based materials. The general process involves the following steps [27–29]:

1. Preparation of the metal precursor solution
2. Addition of a surfactant and a co-surfactant to the metal precursor solution
3. Formation of micelles and self-assembly of the micelles to form a mesophase
4. Addition of a reducing agent to the mesophase to reduce the metal ions to metal nanoparticles
5. Removal of the surfactant template to obtain mesoporous metal-based materials
6. Preparation of the metal precursor solution

The first step in the solution-phase approach is preparing the metal precursor solution. Metal salts, such as chlorides, nitrates, and sulfates, are commonly used as metal precursors. The choice of metal precursor depends on the desired metal composition of the final product. For example, platinum chloride ($PtCl_2$) can be used as the metal precursor to prepare mesoporous platinum-based materials. Once the metal precursor solution is prepared, a surfactant and a co-surfactant are added to the solution. Surfactants are molecules with a hydrophilic head and a hydrophobic tail, which can self-assemble into micelles in an aqueous solution. Co-surfactants are small molecules that help to increase the stability of the micelles. The choice of surfactant and co-surfactant depends on the desired pore size of the final product. Commonly used surfactants include cetyltrimethylammonium bromide (CTAB), pluronic P123, and Brij-58. Co-surfactants such as ethanol and 1-propanol are commonly used in combination with surfactants.

After adding the surfactant and co-surfactant, the solution is heated at proper temperature and pressure (hydrothermal reactor) [30–31] and stirred to promote micelle formation. Micelles are formed when the surfactant molecules self-assemble into spherical or cylindrical structures with the hydrophilic head facing outward and the hydrophobic tail facing inward. The micelles then self-assemble to form a mesophase, a liquid crystal phase with a well-defined pore structure. The next step is adding a reducing agent to the mesophase to reduce the metal ions to metal nanoparticles. Commonly used reducing agents include sodium borohydride ($NaBH_4$), hydrazine (N_2H_4), and hydrogen gas (H_2). The choice of reducing agent depends on the metal nanoparticles' desired metal composition and size. Finally, the surfactant template is removed by calcination or solvent extraction to obtain mesoporous metal-based materials. Calcination involves heating the sample at a high temperature to burn off the surfactant template, leaving behind the mesoporous metal-based material. Solvent extraction involves using a solvent to dissolve the surfactant template and leaving the mesoporous metal-based material behind.

Mesoporous nanomaterials are considered to be better candidates for energy applications extending from energy production to storage. The diverse roles of mp materials in energy applications are shown in Figure 13.2.

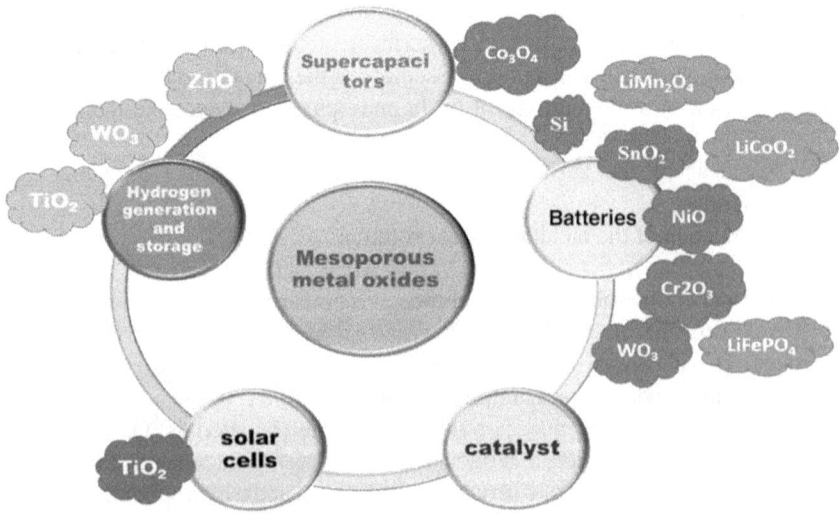

FIGURE 13.2 Mesoporous metal oxides for vivid energy applications-electrodes for supercapacitors and batteries.

13.6 MESOPOROUS METAL-BASED MATERIALS FOR SUPERCAPACITORS

For applications needing high power densities, lengthy cycle times, and quick charge/discharge rates, supercapacitors (SCs) have arisen as an alternative energy storage solution. Due to their distinctive qualities, such as high surface area, tunable pore size, and high-conductivity, mesoporous metal-based materials have been investigated as prospective options for SC electrodes. Mesoporous metal-based SC materials have been made utilizing various techniques, including hydrothermal, sol-gel, and template-assisted approaches. Nickel, cobalt, iron, and manganese are used most frequently to produce mesoporous metal-based materials. Mesoporous materials based on nickel, cobalt, iron, copper, and manganese have recently drawn much attention because of their high specific capacitance and superior electrochemical performance. Due to their high theoretical capacitance, strong conductivity, and low cost, the materials mentioned above, alone, in combination, or as oxides, sulfides, or doped with other materials, are reported as viable candidates for SC electrodes [32,33] Sol-gel and template-assisted procedures have both been used to manufacture mesoporous NiO. Mesoporous NiO electrodes have been shown to have outstanding cycle stability and great electrochemical performance.

According to recent research by Song et al., co-doped boron, nitrogen, and sulfur atoms in nickel-based sulfide composites increase the amount of electrochemical energy storage sites and speed up electrochemical reactions, enhancing charge storage [34]. According to Chenghao Huang and colleagues, the amorphous Ni-Co hydroxide (NiCo-OH) has specific capacities of 888 and 662 C g^{-1} when measured at 1 and 50 A g^{-1}, respectively [35]. Fangxiang and colleagues created a composite electrode in their study using porous mSiO2@Ni3S2/NiS2 functionalized with carbon and boron nitride (BN). Compared to Hg/HgCl$_2$, this electrode showed a high specific potential (V) of 1.8 V and a high reversible capacity of about 449.7 F g^{-1} at 1 g A g^{-1}. The electrode also demonstrated excellent rate capability, achieving a maximum energy density of 202.5 Wh Kg $^{-1}$ at a power density of 959.2 W kg $^{-1}$, as well as a capacity of 81 F g $^{-1}$ at 20 g A g $^{-1}$ [36]. NiOx@NMC nanocomposite was made through hydrothermal technique employing a costless smoked cigarette filter as a nitrogenous carbon source. The resultant NiOx@NMC-NC exhibits outstanding mesoporous (12 nm) characteristics and a high surface area of about 1843 m^2 g^{-1}. According to Ubaidullah et al., it displayed exceptional electrochemical performance and stability up to 4000 cycles as an SC, with a specific capacitance of 740 F g $^{-1}$ at 1 A g^{-1} in 6 M KOH [37].

Co$_3$O$_4$, a transition-metal oxide, has outstanding electrochemical characteristics with a theoretical specific capacitance and good reversibility, making it a highly desirable electrode material for SCs [38]. However, the reversibility of Co$_3$O$_4$ is compromised by capacitive deterioration at high current densities [39,40], resulting in an actual specific capacitance significantly lower than its theoretical value. Co$_3$O$_4$ can only be applied in certain circumstances in SCs as a result. However, it has been demonstrated that controlling the micromorphology of Co$_3$O$_4$ improves its electrochemical performance. Co$_3$O$_4$ nanofibers [41], layered Co$_3$O$_4$ [42], Co$_3$O$_4$ nanoparticles[43], Co$_3$O$_4$ nanorod arrays [44], Co$_3$O$_4$ core-shell[45], Co$_3$O$_4$ porous nanowires[46], and hollow coral-shaped Co_3O_4 have all been made in recent years using a variety of methods [47]. Li and colleagues created a 3D independent electrode heterostructure made of sea urchin-like $Co(OH)_2$ microspheres placed atop nickel foam using a one-step hydrothermal technique. The capacitance was determined to be 1916 F g^{-1} at 10 mA cm^{-2} after its electrochemical performance was examined. After 5000 charge and discharge cycles at 80 mA cm^{-2} current density, 79.3% of the initial capacitance was preserved. However, some $Co(OH)2$ microspheres, transitioning from clear sea urchin-like structures to unnoticeable rod-like and stacked plate-like CoOOH due to compositional and structural changes during charging and discharging, can be blamed for the drop in capacitance [48].

Compared to single-metal oxides, ternary transition-metal oxides—including two distinct cations have been reported to have better conductivity and capacitive activity [49]. This is because they increase electronic conductivity and offer more active sites for redox processes. In particular, they demonstrate much enhanced electrochemical characteristics compared to binary transition-metal oxides like Co_3O_4. Due to the connection of two transition metals, ternary transition-metal oxides like $MnCo_2O_4$, $NiCo_2O_4$, $ZnCo_2O_4$, etc., produce a synergistic effect that improves their electrochemical properties [50]. Due to their high electronic conductivity, two orders of magnitude higher than equivalent oxides, transition-metal sulfides can replace cobalt-containing ternary metal oxides in SC applications. Researchers have been investigating their derivatives to improve the electrochemical performance of these metal oxides—sulfides' high conductivity results from the transition metals' close resemblance in valence state to the metals. The flexibility of the structure, which helps prevent structural disintegration of the electrodes due to interlayer elongation and promotes electron transport, is further enabled by sulfur's reduced electronegativity compared to oxygen. Combining two or more sulfides can further increase the electrical characteristics of transition-metal sulfides, resulting in richer redox reactions due to the more varied states, narrower optical band gaps, and higher chemical stability of bimetallic sulfides compared to single-metal sulfides. With better capacitance, multivalent redox processes, and higher conductivity than single-metal oxides, transition-metal sulfides like Co-Mo-S and $NiCo_2S_4$ have great potential for SC applications.

The material capacity and power density have increased due to developments in SC manufacturing technology. Kore et al. [51] used a robust solvent-deficient method to synthesize an electrode material based on Fe_2O_3 utilizing the precursors $Fe(NO_3)_3$ $9H_2O$ and NH_4HCO_3, which increased capacitive performance. In contrast to untreated precursors, washing the Fe_2O_3 NPs produced results with increased energy density, surface area, and power density. After 2000 cycles, both untreated and treated Fe2O3 showed stability, with the treated variant retaining 84% of its specific capacitance. The poor electrical conductivity of iron oxide materials has long been a problem, but recent research has shown promise in overcoming this issue. Polypyrrole, iron oxide, and vanadium oxide hydrate were mixed by Maitra et al. [52] to create a supercapacitor electrode material with excellent performance. The electrode showed 95% cyclic stability, 38.2 Wh kg^{-1} energy density, and 1202 F g^{-1} specific capacitance. Additionally, mp-Fe_2O_3 NPs produced using a reliable, solvent-free method displayed a specific capacitance of 469 F g^{-1} and kept 84% of this value after 2,000 cycles. Notably, it has been demonstrated that dumping one metal oxide onto another with differing electrochemical and morphological characteristics enhances the electrochemical performance of SCs [53]. Using inexpensive, ecologically acceptable iron oxide as a negative electrode material for synthesis is one area of SC development that has advanced. Due to the low capacitance of carbon-based materials, negative electrode materials were historically primarily based on carbon materials in alkaline electrolytes. Iron oxide has several advantages over carbon-based materials when used as an electrode in SCs. Despite this, iron oxide's preparation technique has made it difficult to use it in SCs. Numerous methods have been tried to address these issues; however, the majority have failed to deliver the needed particular capacity.

Mohamed et al. [54] have successfully created a mesoporous-Fe_3O_4 film by a straightforward in situ hydrothermal procedure for use as a negative electrode in SCs to overcome these difficulties. This procedure has increased the Fe_3O_4 film's stability and capability, leading to an increased specific surface area of 247 m^2 g^{-1}. Over a larger range of current densities of 1 g A g^{-1} and 50 g A g^{-1}, the film's capacity remained constant at 221 C g^{-1} and 154 C g^{-1}, respectively. Compared to electric-double-layer capacitors, using iron-based materials in SCs shows promise because of their superior electrochemical capabilities. Significant advancements in SC's energy density while preserving power density have been accomplished recently. Several methods have been used, including asymmetric SCs containing two different electrodes acting as power and energy sources. The significant capacitance mismatch between the anode and cathode, resulting from the naturally low electric double-layer capacitance in carbon nanomaterials, poses difficulties for such asymmetric SCs. As a result, these capacitors' ability to store energy is diminished. Creating anode materials with equivalent capacitance is crucial to overcoming this constraint and enhancing the rate and capacity of asymmetric SCs.

Manganese-based mesoporous materials have also been explored as potential candidates for SC electrodes. Mesoporous manganese oxide (MnO_2) and manganese phosphate ($Mn_3(PO_4)_2$) have been synthesized using sol-gel and template-assisted methods. Mesoporous MnO_2 and $Mn_3(PO_4)_2$ electrodes have demonstrated high specific capacitance and good cycling stability. Due to its wide availability, environmental friendliness, and high theoretical capacitance (1400 F g^{-1}), MnO_2, a transition-metal oxide, has drawn much attention. MnO_2 has a higher energy density than other transition-metal oxides due to its larger operating potential window. Due to their higher surface area and reduced inactive volume, three-dimensional porous nanostructures of MnO_2 have recently grown in popularity. These structures may reduce ion diffusion resistance and improve transport kinetics, increasing energy density and power capacity.

Several composites based on MnO_2 have been developed, such as composites with precious metals (gold, silver) [55–57], transition metals (Cu, Ni, Mn) [58–61], graphene[62,63], carbon nanotubes (CNT) [64,65], porous carbon [66,67], and others. MnO_2-based materials have shown promising results in enhanced electrical conductivity and increased specific surface area, as demonstrated by experimental and computational studies, and have been extensively reviewed from different perspectives. The 3D conductive network allowed for efficient electron/ion transport while in close contact with Mn_3O4/MnO_2, improving electrode reaction kinetics. To address issues of poor reversibility and stability observed in single MnO_2 and Co_3O_4 electrodes, Adaikalam et al. [68] utilized N and S double-doped GO sheets. This approach enhanced the material's mechanical strength and electrical conductivity, enabling better ion penetration into the electrode. Additionally, the electrochemical performance of MnO_2-based materials was further improved by hybridizing them with oxygenates. Gao et al. [69] designed CoFe2O4@MnO2 nanoarrays on a nickel foam framework, which facilitated MnO_2 growth and the enhancement of electronic/ionic conductivity. Finally, Jia et al. [70] developed a ZnCo2O4@MnO2 core-shell structure, in which the generation of MnO_2 films increased the specific surface area and efficiently promoted rapid ion transport.

Due to their excellent electrochemical performance and availability, recent research has focused on developing mesoporous metal-based materials for SC applications, particularly Ni, Co, Fe, and Mn-based materials. These materials have a high surface area, enhanced ion diffusion, and high conductivity, improving capacitance and charge/discharge rates. Mesoporous nickel-based materials, such as NiO/NiCoOx composites and $NiCo_2O_4$ nanosheets, have shown high specific capacitance and cycling stability. Cobalt-based materials, such as $Co3O_4$ and $Co(OH)_2$, have also shown potential, with mesoporous Co_3O_4 nanowires and Co(OH)2 nanosheets exhibiting high specific capacitance and good rate capability. Iron-based materials, such as Fe_3O_4 and Fe_2O_3, and manganese-based materials, such as MnO_2 and Mn_3O_4, have also been studied, demonstrating high specific capacitance and good cycling stability. These recent advancements in mesoporous metal-based materials for SCs show promise for developing efficient and sustainable energy systems.

13.7 MESOPOROUS METAL-BASED MATERIALS FOR BATTERIES

Numerous battery applications, such as lithium-ion, sodium-ion, and zinc-ion batteries, have used mesoporous metal-based materials. Mesoporous metal-based materials have been employed as anodes and cathodes in lithium-ion batteries. Mesoporous materials based on nickel have been utilized as cathodes, whereas mesoporous materials based on titanium have been employed as anodes. Mesoporous cobalt-based materials have been employed as anodes in sodium-ion batteries. Both lithium-ion and sodium-ion batteries have utilized mesoporous manganese-based materials as cathodes. Mesoporous nickel-based materials have been employed as anodes in zinc-ion batteries.

13.7.1 Mesoporous Anode Materials

Metallic Li is, without a doubt, the anode material with the largest specific capacity and negative potential [71]. However, safer solutions are needed because of the inherent safety issues. Unfortunately, most anode materials offer a potential around 1 V higher than metallic Li, which results in a considerable drop in cell voltage. Elemental anodes have a high volumetric capacity that is comparable to Li's.

However, they experience significant volume changes during lithiation because their lattice structures are inadequate to hold the added Li ions.

On the other hand, metal oxides have more stable lattice structures throughout cycling, but their larger lattices have significantly lower specific and volumetric capacities. Modelling by Park et al. demonstrates how mesoporous materials' spacious topology minimizes volume growth during lithiation [72]. Mesoporosity can make a space set aside for volume expansion, but this may not be enough and could lead to partial or total structural failure.

Anodes and cathodes are examined equally concerning a Li/Li^+ reference electrode in a two-electrode cell. As a result, an electrode performs just like a cathode material but with a much lower positive potential. Therefore, labelling charge/discharge directions resembles the same standard as cathode materials. However, anode materials have the opposite charge/discharge direction compared to cathode materials. In a complete cell, lithiation of the cathode and de-lithiation of the anode occurs during charging, but these two processes alternate during discharging [73]. The anode capacity is commonly given as the specific capacity, though labeling is unnecessary. Due to the high coulombic efficiency of most anode materials, charging and discharging specific capacities are comparable. However, some irreversible lithiation processes, especially in nanostructured materials, cause lithium to be stored in the anode and inaccessible during de-lithiation. Therefore, claiming a significant amount of intercalated lithium is meaningless if the same amount can't be removed. Even though most mesoporous materials have high initial lithiation, this is a drawback. There have been misunderstandings in the literature as a result of some writers who wrongly quoted high specific capacities for irreversible lithium storage. Instead of just considering the irreversible intercalation during the first cycle, the quantity of lithium that an anode material can reversibly store should be used to calculate the storage capacity [74].

13.7.2 TIN AND ITS ALLOYS

Using tin and its alloys as anodes in lithium-ion batteries has shown great possibilities. Nevertheless, the substantial volume change that occurs when tin is used as an anode material presents many significant difficulties. Due to insufficient free space, lithium ions must be intercalated into the metallic lattice structure of Sn, which causes a drastic volume change of more than 200%t for the entire system. Researchers have explored using mesoporous Sn, which has enough room to handle the volume changes through the mesopores, to solve this problem. One usual method for producing mesoporous Sn is using a silica template. Due to Sn's low melting point and reactivity during the template removal process, this technique is ineffective for Sn. Instead, researchers have found that after removing the template, mesoporous SnO_2 can form and then be transformed into mesoporous Sn-C [75]. Furthermore, mesoporous Sn-C has been investigated as an anode material due to its promising potential [76]. Ordered mesoporous Sn-C has been produced using liquid crystal templates, and it has been discovered to significantly improve the rate capability and cyclability of Sn anodes [77]. Researchers have looked into utilizing Sn alloys and mesoporous Sn to reduce the volume changes during lithiation. Using an SBA-15 silica template, Chen et al. created an ordered mesoporous Sn-Cu anode that displayed no capacity fading during the first 20 cycles [78]. After 200 cycles of long-term cycling, capacity fading was around 50%. Park et al. discovered that alloying Sn anodes with Co further improved their cyclability; the anode showed 20% capacity fading during the first 10 cycles but maintained a specific capacity of roughly 600 mAh g^{-1} over 50 cycles [79].

13.7.3 SILICON

Previous studies have shown that silicon is a good candidate for anodes with its large specific capacity (4200 mAh g^{-1}) and lower positive potential than Li/Li$^+$ [80]. The issue with silicon anodes is that their volume increases significantly (by over 300%) during Li intercalation, which causes mechanical breakage and poor cyclability [81]. The silica template approach, which uses silica as both a silicon supply and a mesoporous template, offers one possible answer. As an illustration, molten Mg can convert SiO$_2$ to silicon [82,83]. The resulting Si anode has remarkable battery performance compared to a typical nano-Si electrode. Adding a carbon conducting agent has demonstrated excellent cyclability with full capacity retention over 100 cycles (roughly 1500 mAh g^{-1} from the second to the 100th cycle) [83]. However, a different investigation using a comparable Si/C anode revealed a capacity reduction of almost 50% between the second and the 100th cycle [84].

13.7.4 SNO$_2$

According to earlier studies, Wang and colleagues synthesized an ordered SnO$_2$ anode using an SBA-15 silica template, which had a greater specific capacity than other nanostructured SnO$_2$ forms [85]. Despite the possibility of improving battery performance with a 3D interconnected SnO$_2$ design, the existing cyclability is insufficient for most practical purposes [86]. According to Shon et al., incomplete removal of the silica templates can result in residual silica species remaining in the SnO$_2$ structure, which can enhance battery performance by increasing specific capacity and cyclability [87].

13.7.5 NIO

Due to its typical pseudocapacitive behavior occurring at a less positive voltage, nickel oxide (NiO) has been regarded as a possible pseudocapacitive anode material. Researchers have created many methods to synthesize mesoporous NiO with ordered holes at various scales to improve its performance. For instance, Liu et al. created ordered mesoporous NiO using the silica template approach; this material had a much greater specific capacity than commercial bulk NiO (188 mA h g^{-1}) at 0.1C [88]. They attributed the enhanced performance to the mesostructured material's decreased activation energy.

Zheng et al. described a different strategy for creating a 3D interconnected macro-mesoporous NiO based on tiny nanodots on a parent Ni foam . This technique produced ordered pores at two different scales (300 nm and 3 nm). A NiO material with 3 nm pores has a 5 times higher specific capacity than one with 30 nm pores. Furthermore, after 10,000 cycles, this mesoporous NiO anode still had 93% of its initial capacity, above 800 F g^{-1}. A 3 nm mesoporous NiO grown on a copper foil produced a 10% lower specific capacity at 1 g A g^{-1}. Still, a 90 % lower specific capacity at 10 g A g^{-1}, demonstrating the significance of the parent Ni porous foam for high-rate capabilities [89].

13.7.6 CR$_2$O$_3$

In a study by Dupont and colleagues, an intriguing behavior that wasn't seen in the bulk counterpart of mesoporous Cr$_2$O$_3$ was seen during the charge and discharge of that material [90]. The researchers hypothesized that metallic chromium nanoparticles with a diameter of up to 10 nm developed and were enmeshed in a Li$_2$O matrix during the discharging process. In contrast, the organic components helped form a solid electrolyte interphase in the bulk of Cr$_2$O$_3$ (SEI). In the process of charging, the metallic chromium nanoparticles were subsequently oxidized. The freedom of small chunks of the lattice network in the delicate walls of the mesopores, which opens up a brand-new window for the electrochemical reaction, is the special quality of mesoporous Cr$_2$O$_3$.

On the other hand, the bulk material's stiff network traps most of the lattice units, and the only process that could account for this is the counterions' solid-state diffusion. The mesoporous version of the Cr$_2$O$_3$ anode performed electrochemically substantially better than the bulk material. In a different investigation, Liu et al. created two ordered mesoporous Cr$_2$O$_3$ materials using the silica templates SBA-15 and KIT-6. These materials produced specific capacities double that of bulk Cr$_2$O$_3$ at 521 and 540 mAh g^{-1}, respectively [91].

13.7.7 Co$_3$O$_4$

Lithium cobalt oxide cathode compositions have frequently been prepared using Co$_3$O$_4$, a promising anode material [92]. Researchers have looked into mesoporous, nanowire, nanoparticle, and microparticle forms of Co$_3$O$_4$. Bruce and his co-workers created mesoporous Co3O4 for the first time using a typical mesoporous silica template. They showed that mesoporous Co$_3$O$_4$ anodes perform much better than other types of Co$_3$O$_4$ in terms of specific capacity and cyclability [93]. However, Lin et al. found that mesoporous Co$_3$O$_4$'s specific capacity and rate capability can decrease as pore length increases. This might be because the center region of the pores is difficult for diffusing Li ions to access, especially at high speeds when solid-state diffusion is not the primary mode of transport [94]. Several parameters must be considered to maximize battery performance, including the size and distribution of mesopores.

13.7.8　WO$_3$

According to recent research, tungsten is a pricey option for use in lithium batteries, but it's still less expensive than several other materials under consideration. Yoon et al. created a high-conductivity, mesoporous WO$_3$ to solve this problem, demonstrating a specific capacity of 748 mA h g^{-1} at a rate of 0.1C [95]. The Li/W ratio of 6.5 in this anode material allows for a volumetric capacity of up to 1500 mA h cm^{-3}, comparable to that of metallic Li. In contrast, porous materials frequently have poor volumetric capacity (i.e., 2000 mA h cm^{-3}). A mesoporous Mo-doped WO$_2$ that Chen et al. reported on was able to deliver a specific capacity of 635 mA h g^{-1} throughout 70 cycles without seeing any discernible capacity reduction is another encouraging development. Due to the edge lattice's main function, doping's effect on mesoporous materials can be complex [96].

13.8　CATHODE MATERIALS

The performance of lithium-ion batteries depends on the cathode materials. Among the possible cathode options are manganese oxides and β-MnO$_2$, which has drawn much interest. However, only a small quantity of Li can be intercalated into β-MnO$_2$ because of its narrowest channels, which restricts its potential [97-99]. Xia and his coworkers created a mesoporous β-MnO$_2$ with crystalline walls that can successfully intercalate Li ions to overcome this restriction [100]. Jiao and Bruce further enhanced the β-MnO$_2$ mesoporous material's battery performance and attained a remarkable specific capacity of 284 mA h g^{-1}, equivalent to Li$_{0.92}$MnO$_2$ [100]. Recent studies by Li et al. showed that β-MnO$_2$'s specific capacity and rate capability are superior to both electrochemically active MnO$_2$ nanorods and commercial manganese oxide [101]. According to the study, β-MnO$_2$ is a promising material for lithium-ion batteries.

13.8.1　LiMn$_2$O$_4$

It is suggested that after producing a mesoporous manganese oxide and removing the silica template, thermal lithiation should be carried out to synthesize LiMn$_2$O$_4$. Unfortunately, the temperatures at which manganese oxide is generally lithiated are too high for the mesopores to stay stable, frequently causing their collapse. However, Xia and his coworkers successfully created mesoporous LiMn$_2$O$_4$ using a low-temperature lithiation procedure using the traditional silica hard-template method. The finished cathode material had good battery performance [102].

13.8.2　LiCoO$_2$

According to Bruce and colleagues, the synthesis of mesoporous lithium cobalt oxide (LiCoO$_2$) can be accomplished using the silica template approach [103]. The intermediate cobalt oxide won't thermally degrade with this procedure since the lithiation temperature is low enough to prevent it. Using SBA-15 and KIT-6 silica templates, Bruce et al. also reported the synthesis of mesoporous Co$_3$O$_4$, which was later converted to mesoporous LiCoO$_2$ by lithiation at 400°C after the silica template was removed.

13.8.3 TiO$_2$

Due to its potential applications, one-dimensional porous titanium oxide has been widely produced using various techniques, such as anodizing a Ti -electrode. Mesoporous TiO$_2$ preparation, however, has proven to be challenging. The initial claim was made in 1995; it was later shown that mesoporous TiO$_2$ could not be produced using conventional template-based procedures because the lattice structure collapses during thermal crystallization [104,105]. Triblock copolymer templates successfully created mesoporous TiO2 [106–109].

The KIT-6 silica template method for producing ordered mesoporous TiO2 results in a three-dimensional hierarchical structure and provides a great opportunity for Li diffusion [110]. The energy density of this 3D mesoporous material was up to 200% higher than that of other easily accessible nanostructured titanium oxides [111]. The mesoporous TiO2 made by KIT-6 was also found to have a substantially higher proton conductivity than other forms of TiO$_2$. According to a detailed examination of the TiO$_2$ crystallization before removing the silica template, complete crystallization is impossible with the KIT-6 silica template due to the strong interaction between the crystal and the wall of the silica cages in SBA-15 [112]. The surfactant-based template method [112,113] can be used to introduce a second metal into the lattice framework of TiO$_2$ to increase structural stability. This approach works effectively for electroactive materials where dopants are required to enhance electrochemical behavior but are ineffective for pure TiO$_2$ compounds. Utilizing alumina hard templates, one-directional ordered mesoporous TiO$_2$ can be created [114].

13.8.4 LiFePO$_4$

According to recent research [115], the sluggish transfer of ions and electrons is hampered by the widespread use of LiFePO$_4$ as a cost-effective cathode material with reliable battery performance. Various techniques are commonly employed to address this issue, such as applying conductive carbon coatings or increasing surface area. A study using the soft template method with the surfactant pluronic P123 resulted in mesoporous FePO$_4$ with exceptional performance [116–117]. The mesoporous structure exhibited lower charge transfer resistance, leading to high battery performance, achieving up to 90% of the theoretical capacity [117]. The creation of 2D-ordered mesoporous transition-metal phosphates, such as FePO$_4$, Mn$_3$(PO$_4$)$_2$, and Co$_3$(PO$_4$)$_2$, in a nonpolar solvent, was also accomplished by researchers [118]. The mesoporous FePO$_4$ nanoflakes made using this technique in this work showed a minor capacity loss of around 15% during the early cycles. Still, they maintained an average capacity of about 140 mA h g^{-1} throughout 100 cycles.

Mesoporous metal-based materials hold great promise for battery applications. The ordered structure of the mesopore walls allows for a larger accessible surface area, providing better access to the internal walls. Additionally, the mesopore walls offer the opportunity to maintain a more crystalline lattice structure, as opposed to the surface of nanoparticles, often composed of defects. Furthermore, the vertical alignment of the mesopore walls facilitates perpendicular diffusion towards the surface unit cells and beyond, making it a more efficient option for battery applications. Despite the difficulty of preparing electroactive materials with such small particle sizes, the benefits of using mesoporous materials for battery applications far outweigh the challenges.

13.9   MESOPOROUS METAL-BASED MATERIALS FOR HYDROGEN GENERATION AND STORAGE

Developing sustainable and efficient energy sources is one of our most pressing global challenges. Hydrogen-based systems have great potential among the various renewable energy technologies due to their high energy density and environmental benefits. However, the practical application of hydrogen energy is limited by the difficulty of storing and transporting hydrogen gas. Researchers have been exploring using mesoporous metal-based materials for hydrogen generation and storage to address this challenge. These materials have unique structural properties, such as high surface area, tunable pore size, and controllable surface chemistry, which make them promising candidates for catalyzing hydrogen production from renewable sources and storing hydrogen safely and efficiently. In this context, this paper aims to provide an overview of recent advancements in the development and characterization of mesoporous metal-based materials for hydrogen generation and storage, focusing on their potential applications in the context of a sustainable energy future. Researchers are exploring new metal-based materials for photocatalytic applications, such as mesoporous metal oxides (e.g., ZnO, TiO_2, WO_3), metal sulfides (e.g., CdS, ZnS), and metal nitrides (e.g., TiN, Zn_3N_2). These materials have unique properties that can enhance their photocatalytic activity.

Metal NPs have been frequently used as supports for immobilizing in the dehydrogenation of FA using mesoporous silica materials, especially those with ordered mesoporous structures. Due to its well-defined big mesopores of around 6 nm, high specific surface area, and high pore volume [119–122], SBA-15 has emerged as the most popular alternative among them. Before adding metal species, various organic functional species, such as amine groups and carbon nitride, have been grafted into the mesopores of SBA-15 to produce small metal NPs that are constrained inside the matrix. These functional groups introduce some basic sites that improve the catalytic activity of FA dehydrogenation and act as chelating agents to stabilize metal precursors and prevent the overgrowth and aggregation of produced metal NPs.

For instance, 3-aminopropyltriethoxysilane was used to surface functionalize an amine- SBA-15-supported ultrasmall Pd NPs catalyst [120]. After the reduction in an H_2 environment at 250°C, HAADF-STEM measurements showed that the average Pd NP size in Pd@SBA-15-amine was only 1.9 nm, and the Pd NPs were entirely contained within the mesoporous channels. However, the Pd NPs in the SBA-15 support without grafting amines were considerably bigger. Part of the Pd NPs was found on the SBA-15 surface, indicating that the amine groups had drastically decreased the Pd diameters. With an initial TOF of 293 h^{-1} at room temperature, the resultant Pd@SBA-15-amine catalyst demonstrated remarkable activity for FA dehydrogenation. The Pd species could be stabilized using a variety of organoamine groups. The authors of this study investigated the impact of various amine types on the size and functionality of Pd nanoparticles (NPs) supported by SBA-15 catalysts for FA dehydrogenation. They discovered that Pd NPs functionalized with primary and secondary amines had nearly identical sizes, coming in at 1.5 nm. However, the addition of tertiary amine caused the size of Pd NPs to grow to 2.9 nm. This was explained by the fact that tertiary amines interact with metals less strongly than primary and secondary amines.

According to scientists, the primary amine-functionalized Pd@SBA-15 catalyst had a much higher turnover frequency (TOF) value than the secondary and tertiary amine-functionalized catalysts. Variations in the hydrophilicity/hydrophobicity of the SBA-15 support and electronic and steric interactions between the amine groups and Pd NPs caused this. The usage of various varieties of mesoporous silica, including periodic mesoporous, KIE-6 [123,124], KIE-11 [125], MCM-41 [126], KCC-1 [127], and organosilica [128] as supports for attaching metal NPs for FA dehydrogenation, was also studied by the scientists. These results imply that the performance of metal NPs in catalytic processes can be significantly influenced by the choice of support material.

13.9.1 Metal Nanocatalysts Supported by Other Porous Materials

Recent reports show various other porous materials have been utilized to immobilize metal nanoparticles (NPs) for catalytic purposes. These materials include porous organic polymers (POPs) [129], organo-silica nanotubes [130], porous chitosan-graphene-oxide aerogels [131], and covalent triazine frameworks [132–134]. The porous structure of these supports enhances the formation of small metal NPs, resulting in highly active nanocatalysts for hydrogen (H_2) production from formic acid (FA) dehydrogenation under mild conditions.

Active metal species have recently been anchored to hybrid porous supports to improve catalytic performance further. For instance, Jiang and colleagues described a wet-chemical technique for immobilizing an ultrasmall AuPd-MnOx nanocomposite on a ZIF-8-reduced graphene-oxide (rGO) bi-support. The ZIF-8-rGO support is essential for regulating the size of metal NPs particles. The AuPd-MnOx composite supported onto the ZIF-8-rGO bi-support, in contrast to unsupported and mono-supported AuPd-MnOx composites, displayed a narrow size distribution of 2–3 nm [135], which could be attributed to the substantial surface area of rGO and ZIF-8 as well as the potent metal-support interaction. With a high initial turnover frequency (TOF) of 382.1 h^{-1} without any additives at 25°C, the resultant AuPd-MnOx/ZIF-8-rGO catalyst demonstrated the greatest catalytic activity for H2 production from FA dehydrogenation.

Similarly, ultrasmall Pd-Ag NPs were immobilized on a multi-support made of zirconia, porous carbon, and reduced graphene oxide (ZrO_2/C/rGO) by carbonization treatment at 800°C [136]. Ultrasmall bimetallic Pd-Ag NPs were created due to the abundant and homogeneous metal precursor nuclei made possible by the uniform dispersion of ZrO_2/C on rGO. The catalytic activity of the FA dehydrogenation over the produced PdAg@ZrO2/C/rGO catalysts was significantly influenced by the ratios of Ag to Pd, metal to UiO-66/GO, as well as the particle size of Pd-Ag NPs. The

$Pd_{0.6}Ag0.4@ZrO2/C/rGO$ nanocatalyst produced the greatest catalytic activity, with an initial TOF value of. 4500 h^{-1} at 60°C.

The oxygen evolution reaction (OER) is crucial in water splitting and metal-air batteries. Transition-metal hydroxides are the most effective electrocatalysts for this process. However, reducing the thickness of hydroxides to increase surface area often fails to increase the number of active sites because the true active sites are typically located at the edges. Current state-of-the-art OER electrocatalysts still have overpotentials larger than 200 mV at a current density of 10 mA cm^{-2} ($\eta 10$). Mesoporous metal-based materials have shown promise in hydrogen generation and storage. The unique pore structure of these materials provides high surface area and easy diffusion of reactants, allowing for efficient catalytic activity in hydrogen production reactions. Additionally, mesoporous metal oxides can be used for hydrogen storage due to their ability to adsorb hydrogen through weak interactions. These materials can also be functionalized with metal dopants or organic molecules to enhance their catalytic and hydrogen storage properties. Ongoing research in this field is focused on developing new mesoporous metal-based materials and optimizing their performance for practical applications in renewable energy technologies.

13.9.2 MESOPOROUS METAL-BASED MATERIALS FOR SOLAR CELLS

A potential solution to humanity's energy crisis and environmental pollution problems is converting the sun's plentiful energy into electrical or chemical energy. Over the past few decades, photovoltaic cells have become a well-researched and effective way to convert solar energy into electrical energy. However, because of challenging fabrication techniques, silicon-based devices still have high pricing and significant pollution issues even though they dominate the commercial market. Therefore, creating a new generation of solar technologies with less complicated manufacturing processes, lower costs, and higher power conversion efficiency is vitally important. Dye-sensitized solar cells (DSSCs) and metal halide perovskite solar cells (PSCs), which are anticipated to become commercially available in the upcoming years, are two notable alternatives.

Mesoporous metal-based materials have emerged as a promising class of materials for photovoltaic applications due to their unique structural and electronic properties. These materials exhibit a high surface area, tunable pore size, and excellent charge transport properties, making them suitable candidates for various components of solar cells, such as electron transport layers, photoanodes, and counter electrodes.

13.9.3 DSSCs

With a record efficiency of 7.3%, a photoelectrode composed of rutile TiO_2 mesoporous crystals (MCs) and TiO_2 nanoparticles has been developed and used in DSSCs. Comparing this to devices that rely solely on TiO_2 nanoparticle photoelectrodes, it is a major advancement [137]. The higher electron transport capacity through the film and the increased light scattering effect brought on by adding TiO_2 MCs account for better performance. On top of TiO_2 nanoparticle porous films, TiO_2 MCs can be used as a scattering layer to increase the specific surface area for dye loading and improve the light scattering effect for better light trapping. Through a solvothermal method, ellipsoidal TiO_2 MCs of various particle sizes have been created [138]. DSSC efficiency can be raised to 7.16%t when used as the scattering layer component on top of TiO2 nanoparticle porous films. The morphology of MCs and size also influence the efficacy of light harvesting. Spherical MCs with long-range ordered stacking patterns improve the photoanode's light scattering effect compared to spindle-like TiO2 MCs, resulting in a high short-circuit current density of 16.6 mA cm^{-2} and a record-breaking efficiency of 8.10% [139].

Mesoporous TiO_2 microspheres with exposed-face surfaces [101] have outperformed spherical ones regarding dye adsorption and sunlight reflection. The adsorption procedure is made easier by the well-matched distance between the Ti atoms on the surface and the binding carboxylate groups of dyes [139]. Using a green solvothermal technique, single anatase TiO2 microspheres were successfully created, and they displayed strong light reflectance and low transmission in the visible spectrum [140]. These microspheres have also been used as the scattering layer component of DSSCs, leading to a considerable increase in efficiency compared to DSSCs based only on P25 film. Stronger dye adsorption capacity and quicker electron transport dynamics are seen in high surface energy facets with more unsaturated bonds, such as the 111 facets in anatase TiO2 nanocubes and nano parallelepiped structures, which contribute to improved performance in bi-layered DSSCs [141].

TiO2 mesocrystals (MCs) serve as both a scattering layer and the anodes of dye-sensitized solar cells, giving them a dual purpose (DSSCs). Anatase TiO_2 has many different facets, but it has been discovered that the 001 facets improve dye adsorption, light scattering, carrier recombination rates, and electron injection from the dye to the electrode [142–144]. Mesocrystalline anatase TiO2 nanosheet arrays with {001} faces were created using topotactic conversion, and they have shown a maximum power conversion efficiency (PCE) of 7.51% in solar devices. MCs are easily converted to mesocrystalline TiO_2 with heat treatment (MSCs). For instance, mesoporous TiO_2 MCs with a high surface area were created using a straightforward evaporation-driven directed assembly process. After being annealed at 400 °C in air, they were changed into TiO_2 MSCs. Anatase TiO2 MSCs-based solid-state DSSCs have a 3.11% efficiency [145]. The carrier carrying rate of anatase TiO_2 MSCs increases with temperature while the charge density drops, resulting in equivalent efficiency for devices based on films annealed at various temperatures, according to research on the effects of temperature on anatase TiO_2 MSCs [146]. On the foundation of this study, TiO_2 MSCs with various morphologies, including ellipsoidal TiO_2 MSCs, core-shell microspheres, and in situ produced TiO_2 nanosheets, have been successfully generated, and all have demonstrated noticeably enhanced photovoltaic performance [147–149]. As effective photoanodes in DSSCs, mesoporous single crystal ZnO platelet films have also been created and employed, with a maximum efficiency of 5.63% and a lower size [150].

13.9.4   PSCs

Metal halide perovskites are a great candidate for light-active absorbers due to their extraordinary optoelectronic characteristics, which include a high light absorption coefficient, prolonged carrier lifetimes, and excellent carrier mobility. Only a few hundred nano meters of the PSC active layer are required to absorb sunlight completely, thanks to the high absorption coefficient of up to 10^4 cm^{-1}, which is shorter than the carrier diffusion length (exceeding 1 m) in the absorbing layer [150,151]. Since most photo carriers can arrive at interfaces related to selective contact layers, electron selective layers (ESLs) with high electron extraction and transport efficiencies must match the quick carrier separation dynamics in perovskite active layers. Solid-state PSCs using anatase TiO2 MSCs have shown an efficiency of 7.29%, even though TiO_2 MSCs have been found to have superior electrical characteristics compared to nanoparticles. SnO_2 MSCs are anticipated to function better than TiO_2 MSCs as ESLs in PSCs because they have considerably higher carrier mobility than TiO_2 [152]. SnO_2 MSCs were employed as ESLs in PSCs, but due to high charge recombination at the SnO2/perovskite interface, a poor efficiency of 3.76% was obtained. This was addressed by passivating interfacial imperfections by depositing a thin TiO_2 barrier layer on SnO_2 by TiCl4 treatment, yielding an efficiency of 8.54%. With an average efficiency of 12.96%, Han's team recently created printable hole-conductor-free mesoscopic PSCs based on anatase TiO2 MSCs. However, producing high-quality films with an appropriate thickness and few flaws for metal oxide MSCs in PSCs is still difficult [153,154].

Using m-SnO_2 ETL (electron transport layer) with RbF modification, photovoltaic solar cells performed remarkably well, achieving a record power conversion efficiency (PCE) of 22.72% and maintaining 90% of the initial PCE after 300 hours of maximum power point (MPP) tracking, according to a recent study [155]. These outcomes were obtained without the perovskite film's surface being passivated. Because of the low-temperature method of generating the mp-SnO_2 ETL and the low cost of RbF, combining these two components offers a practical and economical method of producing high-performance and stable PSCs. A study [156] found that adding a meso-Sn: TiO_2 electron extraction layer to PSCs significantly increased the devices' power conversion efficiency, fill factor, and hysteresis index. In particular, the fill factor climbed from 76.62 to 81.72%, the hysteresis index reduced from 0.16 to 0.03, and the power conversion efficiency rose from 16.86 to 20.55%. These results imply that Sn doping may be a promising strategy for improving the photovoltaic efficiency of PSCs.

Mesoporous metal-based solar cells are photovoltaic technology that uses a thin film of metal oxide nanoparticles to create a high surface area and porous structure. This structure allows for a large interface area with the electrolyte, enhancing the collection of electrons and improving the efficiency of the cell. These cells have shown promising results in efficiency and stability, and research is ongoing to optimize their design and performance. Overall, mesoporous metal-based solar cells have the potential to become a significant player in the renewable energy market due to their low cost and high efficiency.

13.10 CONCLUSION

In conclusion, mesoporous metal-based materials have shown great promise in energy production, storage, and conversion applications. Recent advances in this field have led to the development of new and improved materials with enhanced properties, such as high surface area, high conductivity, and tunable pore size. These materials can potentially revolutionize the energy industry by providing more efficient and sustainable energy storage and conversion solutions. However, further research is still required to optimize these materials' performance and develop scalable synthesis methods. With continued innovation and collaboration between researchers, industry, and policymakers, mesoporous metal-based materials can be crucial in achieving a cleaner, more sustainable future for our planet.

REFERENCES

1. Dolatyari L, Yaftian M, Rostamnia S. Removal of uranium(VI) ions from aqueous solutions using Schiff base functionalized SBA-15 mesoporous silica materials. *Journal of Environmental Management (2016) 169* 8–17.
2. Ghorbani M, Nowee SM, Ramezanian N, Raji F. A new nanostructured material amino functionalized mesoporous silica synthesized via co-condensation method for Pb(II) and Ni(II) ion sorption from aqueous solution. *Hydrometallurgy (2016) 161* 117–126.
3. Imessaoudene D, Bensacia N, Chenoufi F. Removal of cobalt(II) from aqueous solution by spent green tea leaves. *Journal of Radioanalytical and Nuclear Chemistry (2020) 324(3)* 1245–1253.
4. Bashir J, Chowdhury MB, Kathak RR, Dey S, Tasnim AT, Amin MA, Kaneti YV, Masud, Hossain, MK. Electrochemical fabrication of mesoporous metal-alloy films. *Materials Advances (2023) 4(2)* 408–431.
5. Tang R, Hong W, Srinivasakannan C, Liu X, Wang X, and Duan X. A novel mesoporous Fe-silica aerogel composite with phenomenal adsorption capacity for malachite green. *Separation and Purification Technology (2022) 281* 119950.
6. Farooq A, Ko C, Park Y. Sewage sludge steam gasification over bimetallic mesoporous Al-MCM48 catalysts for efficient hydrogen generation. *Environmental Research (2023) 224* 115553.
7. Kan X, Song F, Zhang G, Zheng Y, Zhu Q, Liu F, Jiang L. Sustainable design of co-doped ordered mesoporous carbons as efficient and long-lived catalysts for H2S reutilization. *Chemical Engineering Science (2023) 269* 118483.
8. Han X, Zhang T, Wang X, Zhang Z, Li Y, Qin Y, Wang B, Han A, Liu J. Hollow mesoporous atomically dispersed metal-nitrogen-carbon catalysts with enhanced diffusion for catalysis involving larger molecules. *Nature Communications (2022) 13(1)* 1–9.
9. Shah S, Famta P, Bagasariya D, Charankumar K, Sikder A, Kashikar R, Kotha AK, Chougule MB, Khatri DK, Asthana A, Raghuvanshi RS. Tuning mesoporous silica nanoparticles in novel avenues of cancer therapy. *Molecular Pharmaceutics (2022) 19(12)* 4428–4452.
10. Choudante PC, Nethi SK, Díaz-García D, Prashar S, Misra S, Gómez-Ruiz S, Patra CR. Tin-loaded mesoporous silica nanoparticles: Antineoplastic properties and genotoxicity assessment. *Biomaterials Advances (2022) 137* 212819.

11. Sargazi S, Laraib U, Barani M, Rahdar A, Fatima I, Bilal M, Pandey S, Sharma RK, Kyzas GZ. Recent trends in mesoporous silica nanoparticles of rode-like morphology for cancer theranostics: A review. *Journal of Molecular Structure (2022) 1261* 132922.

12. Ma J, Li Y, Li J, Yang X, Ren Y, Alghamdi AA, Song G, Yuan K, Deng Y. Rationally designed dual-mesoporous transition metal oxides/noble metal nanocomposites for fabrication of gas sensors in real-time detection of 3-hydroxy-2-butanone biomarker. *Advanced Functional Materials (2022) 32(4)* 2107439.

13. Chen Y, Li Y, Feng B, Wu Y, Zhu Y, Wei J. Self-templated synthesis of mesoporous Au-ZnO nanospheres for seafood freshness detection. *Sensors and Actuators B: Chemical (2022) 360* 131662.

14. Gao Y, Huang K, Long C, Ding Y, Chang J, Zhang D, Etgar L, Liu M, Zhang J, Yang J. Flexible perovskite solar cells: From materials and device architectures to applications. *JACS Energy Letters (2022) 7(4)* 1412–1445.

15. Lv J, Xie J, Mohamed AG, Zhang X, Wang YY. Photoelectrochemical energy storage materials: Design principles and functional devices towards direct solar to electrochemical energy storage. *Chemical Society Reviews (2022) 51(4)* 1511–1528.

16. Qin G, Wang X, Wang X, Cheng D, Li L. Mesoporous TiO2-ZrO2 composite film electrode for electrocatalytic reduction 2-pyridinaldehyde in ionic liquid. *Composite Interfaces. (2017) 24(3):*267–77.

17. Deng Y, Wei J, Sun Z, Zhao D. Large-pore ordered mesoporous materials templated from non-Pluronic amphiphilic block copolymers. *Chemical Society Reviews (2013) 42(9)* 4054–4070.

18. Deng X, Chen K, Tüysüz H. Protocol for the nanocasting method: Preparation of ordered mesoporous metal oxides. *Chemistry of Materials (2017) Jan 10 29(1):*40–52.

19. Malgras V, Ataee-Esfahani H, Wang H, Jiang B, Li C, Wu KC, Kim JH, Yamauchi Y. Nanoarchitectures for mesoporous metals. *Advanced Materials (2016) 28(6)* 993–101.

20. Mei P, Yamauchi Y, Pramanik M, Fatehmulla A, Adhafiri AM, Farooq WA, Bando Y, Shiddiky MJ, Kaneti YV, Lin J, Kim Y. Hard-templated preparation of mesoporous cobalt phosphide as an oxygen evolution electrocatalyst. *Electrochemistry Communications (2019) 104* 106476.

21. Saad A, Cheng Z, Zhang X, Liu S, Shen H, Thomas T, Wang J, Yang M. Ordered mesoporous cobalt–nickel nitride prepared by nanocasting for oxygen evolution reaction electrocatalysis. *Advanced Materials Interfaces (2019) 6(20)* 1900960.

22. Sundblom A, Oliveira CL, Pedersen JS, Palmqvist AE. On the formation mechanism of pluronic-templated mesostructured silica. *Journal of Physical Chemistry C (2010) 114(8)* 3483–3492.

23. Yang H, Wang DK, Motuzas J, Diniz da Costa JC. Hybrid vinyl silane and P123 template sol–gel derived carbon silica membrane for desalination. *Journal of Sol-Gel Science and Technology (2018) 85(2)* 280–289.

24. Sachse A, Grau-Atienza A, Jardim EO, Linares N, Thommes M, Garcia-Martinez J. Development of intracrystalline mesoporosity in zeolites through surfactant-templating. *Crystal Growth and Design (2017) 17(8)* 4289–4305.

25. Gutiérrez-Becerra A, Barcena-Soto M, Soto V, Arellano-Ceja J, Casillas N, Prévost S, Noirez L, Gradzielski M, Escalante JI. Structure of reverse microemulsion-templated metal hexacyanoferrate nanoparticles. *Nanoscale Research Letters (2012) 7(1)* 1–12.

26. Cheng JY, Ross CA, Smith HI, Thomas EL. Templated self-assembly of block copolymers: Top-down helps bottom-up. *Advanced Materials (2006) 18(19)* 2505–2521.

27. Guan C, Xiao W, Wu H, Liu X, Zang W, Zhang H, Ding J, Feng YP, Pennycook SJ, Wang J. Hollow Mo-doped CoP nanoarrays for efficient overall water splitting. *Nano Energy (2018) 48* 73–80.

28. Li X, Wu H, Elshahawy AM, Wang L, Pennycook SJ, Guan C, Wang J. Cactus-like NiCoP/NiCo-OH 3D architecture with tunable composition for high-performance electrochemical capacitors. *Advanced Functional Materials (2018) 28(20)* 1800036.

29. Elshahawy AM, Guan C, Li X, Zhang H, Hu Y, Wu H, Pennycook SJ, Wang J. Sulfur-doped cobalt phosphide nanotube arrays for highly stable hybrid supercapacitor. *Nano Energy (2017) 39* 162–17.

30. Kong M, Wang Z, Wang W, Ma M, Liu D, Hao S, Kong R, Du G, Asiri AM, Yao Y, Sun X. NiCoP nanoarray: A superior pseudocapacitor electrode with high areal capacitance. *Chemistry – A European Journal (2017) 23(18)* 4435–4441.

31. Mi L, Wei W, Huang S, Cui S, Zhang W, Hou H, Chen WA. A nest-like Ni@ Ni1.4Co1.6S2 electrode for flexible high-performance rolling supercapacitor device design. *Journal of Materials Chemistry A (2015) 3(42)* 20973–20982.

32. Rakesh Kumar T, Shilpa Chakra CH, Madhuri S, Sai Ram E, Ravi K. Microwave-irradiated novel mesoporous nickel oxide carbon nanocomposite electrodes for supercapacitor application. *Journal of Materials Science: Materials in Electronics (2021) 32(15)* 20374–20383.

33. Mala NA, Dar MA, Sivakumar S, Bhat KS, Sinha GN, Batoo KM. Electrochemical supremacy of cobalt-doped nickel oxide and its supercapacitor applications with its mesoporous morphology. *Journal of Materials Science: Materials in Electronics (2022) 33(14)* 11582–11590.

34. Song F, Ao X, Chen Q. Effect of heteroatom doping on the charge storage and operating voltage window of nickel-based sulfide composite electrodes in alkaline electrolytes. *Chemical Engineering Journal (2022) 427* 130885.

35. Huang C, Hu Y, Jiang S, Chen HC. Amorphous nickel-based hydroxides with different cation substitutions for advanced hybrid supercapacitors. *Electrochimica Acta (2019) 325* 134936.

36. Song F, Chen Q, Li Y, Li Y, Zhang L. High energy density supercapacitors based on porous mSiO2@Ni3S2/NiS2 promoted with boron nitride and carbon. *Chemical Engineering Journal (2020) 390* 124561.

37. Ubaidullah M, Ahmed J, Ahamad T, Shaikh SF, Alshehri SM, Al-Enizi AM. Hydrothermal synthesis of novel nickel oxide@nitrogenous mesoporous carbon nanocomposite using costless smoked cigarette filter for high performance supercapacitor. *Materials Letters (2020) 266* 127492.

38. Lu Y, Deng B, Liu Y, Wang J, Tu Z, Lu J, Xiao X, Xu G. Nanostructured Co3O4 for achieving high-performance supercapacitor. *Materials Letters (2021) 285* 129101.

39. Yuan C, Yang L, Hou L, Shen L, Zhang X, Lou XW. Growth of ultrathin mesoporous Co3O4 nanosheet arrays on Ni foam for high-performance electrochemical capacitors. *Energy & Environmental Science (2012) 5(7)* 7883–7887.

40. Pawar SA, Patil D, Shin J. Transition of hexagonal to square sheets of Co3O4 in a triple heterostructure of Co3O4/MnO2/GO for high performance supercapacitor electrode. *Current Applied Physics (2019) 19(7)* 794–803.

41. Kumar M, Subramania A, Balakrishnan K. Preparation of electrospun Co3O4 nanofibers as electrode material for high performance asymmetric supercapacitors. *Electrochimica Acta (2014) 149* 152–158.

42. Duan BR, Cao Q. Hierarchically porous Co3O4 film prepared by hydrothermal synthesis method based on colloidal crystal template for supercapacitor application. *Electrochimica Acta (2012) 64* 154–161.

43. Deng J, Kang L, Bai G, Li Y, Li P, Liu X, Yang Y, Gao F, Liang W. Solution combustion synthesis of cobalt oxides (Co3O4 and Co3O4/CoO) nanoparticles as supercapacitor electrode materials. *Electrochimica Acta (2014) 132* 127–135.

44. Chen M, Ge Q, Qi M, Liang X, Wang F, Chen Q. Cobalt oxides nanorods arrays as advanced electrode for high performance supercapacitor. *Surface and Coatings Technology (2019) 360* 73–77.

45. Liu Z, Zhou W, Wang S, Du W, Zhang H, Ding C, Du Y, Zhu L. Facile synthesis of homogeneous core-shell Co3O4 mesoporous nanospheres as high performance electrode materials for supercapacitor. *Journal of Alloys and Compounds (2019) 774* 137–144.

46. Xu Y, Ding Q, Li L, Xie Z, Jiang G. Facile fabrication of porous Co3O4 nanowires for high performance supercapacitors. *New Journal of Chemistry (2018) 42(24)* 20069–20073.

47. Wang X, Zhang N, Chen X, Liu J, Lu F, Chen L, Shao G. Facile precursor conversion synthesis of hollow coral-shaped Co3O4 nanostructures for high-performance supercapacitors. *Colloids and Surfaces A: Physicochemical and Engineering Aspects (2019) 570* 63–72.

48. Li D, Zhu S, Gao X, Jiang X, Liu Y, Meng F. Anchoring sea-urchin-like Co(OH)2 microspheres on nickel foam as three-dimensional free-standing electrode for high-performance supercapacitors. *Ionics (2021) 27(2)* 789–799.

49. Wang T, Chen HC, Yu F, Zhao XS, Wang H. Boosting the cycling stability of transition metal compounds-based supercapacitors. *Energy Storage Materials (2019) 16* 545–573.

50. Li Y, Han X, Yi T, He Y, Li X. Review and prospect of NiCo2O4-based composite materials for supercapacitor electrodes. *Journal of Energy Chemistry (2019) 31* 54–78.

51. Kore R, Lokhande B. A robust solvent deficient route synthesis of mesoporous Fe2O3 nanoparticles as supercapacitor electrode material with improved capacitive performance. *Journal of Alloys and Compounds (2017) 725* 129–138.

52. Maitra A, Das AK, Karan SK, Paria S, Bera R, Khatua BB. A mesoporous high-performance supercapacitor electrode based on polypyrrole wrapped iron oxide decorated nanostructured cobalt vanadium oxide hydrate with enhanced electrochemical capacitance. *Industrial and Engineering Chemistry Research (2017) 56(9)* 2444–2457.

53. Tufa LT, Oh S, Kim J, Jeong KJ, Park TJ, Kim HJ, Lee J. Electrochemical immunosensor using nanotriplex of graphene quantum dots, Fe3O4, and Ag nanoparticles for tuberculosis. *Electrochimica Acta (2018) 290* 369–377.

54. Jiang K, Sun B, Yao M, Wang N, Hu W, Komarneni S. In situ hydrothermal preparation of mesoporous Fe3O4 film for high-performance negative electrodes of supercapacitors. *Microporous and Mesoporous Materials (2018) 265* 189–194.

55. Veeramani V, Dinesh B, Chen SM, Saraswathi R. Electrochemical synthesis of Au–MnO 2 on electrophoretically prepared graphene nanocomposite for high performance supercapacitor and biosensor applications. *Journal of Materials Chemistry A (2016) 4(9)* 3304–3315.

56. Singh SB, Singh TI, Kim NH, Lee JH. A core–shell MnO2@ Au nanofiber network as a high-performance flexible transparent supercapacitor electrode. *Journal of Materials Chemistry A (2019) 7(17)* 10672–10683.

57. Sun S, Jiang G, Liu Y, Yu B, Evariste U. Preparation of α-MnO2/Ag/RGO hybrid films for asymmetric supercapacitor. *Journal of Energy Storage (2018) 18* 256–258.

58. Kumar A, Thomas A, Gupta A, Garg M, Singh J, Perumal G, Sujithkrishnan E, Elumalai P, Arora HS. Facile synthesis of MnO2-Cu composite electrode for high performance supercapacitor. *Journal of Energy Storage (2021) 42* 103100.

59. Xiao K, Li JW, Chen GF, Liu ZQ, Li N, Su YZ. Amorphous MnO2 supported on 3D-Ni nanodendrites for large areal capacitance supercapacitors. *Electrochimica Acta (2014) 149* 341–348.

60. Wang Y, Jiang L. Freestanding carbon aerogels produced from bacterial cellulose and its Ni/MnO2/Ni(OH)2 decoration for supercapacitor electrodes. *Journal of Applied Electrochemistry (2018) 48(5)* 495–507.

61. Deng MJ, Ho PJ, Song CZ, Chen SA, Lee JF, Chen JM, Lu KT. Fabrication of Mn/Mn oxide core–shell electrodes with three-dimensionally ordered macroporous structures for high-capacitance supercapacitors. *Energy & Environmental Science (2013) 6(7)* 2178–2185.

62. Cao J, Wang Y, Zhou Y, Ouyang JH, Jia D, Guo L. High voltage asymmetric supercapacitor based on MnO2 and graphene electrodes. *Journal of Electroanalytical Chemistry (2013) 689* 201–206.

63. Zhang J, Yang X, He Y, Bai Y, Kang L, Xu H, Shi F, Lei Z, Liu ZH. δ-MnO 2/holey graphene hybrid fiber for all-solid-state supercapacitor. *Journal of Materials Chemistry A (2016) 4(23)* 9088–9096.

64. Choi C, Sim HJ, Spinks GM, Lepró X, Baughman RH, Kim SJ. Elastomeric and dynamic MnO2/CNT core–shell structure coiled yarn supercapacitor. *Advanced Energy Materials (2016) 6(5)* 1502119.

65. Park T, Jang Y, Park JW, Kim H, Kim SJ. Quasi-solid-state highly stretchable circular knitted MnO 2@ CNT supercapacitor. *RSC Advances (2020) 10(24)* 14007–14012.

66. Liu M, Gan L, Xiong W, Xu Z, Zhu D, Chen L. Development of MnO 2/porous carbon microspheres with a partially graphitic structure for high performance supercapacitor electrodes. *Journal of Materials Chemistry A (2014) 2(8)* 2555–2562.

67. Li M, Yu J, Wang X, Yang Z. 3D porous MnO2@ carbon nanosheet synthesized from rambutan peel for high-performing supercapacitor electrodes materials. *Applied Surface Science (2020) 530* 147230.

68. Adaikalam K, Ramesh S, Santhoshkumar P, Kim HS, Park HC, Kim HS. MnO2/Co3O4 with N and S co-doped graphene oxide bimetallic nanocomposite for hybrid supercapacitor and photosensor applications. *International Journal of Energy Research (2022) 46(4)* 4494–4505.

69. Gao H, Cao S, Cao Y. Hierarchical core-shell nanosheet arrays with MnO2 grown on mesoporous CoFe2O4 support for high-performance asymmetric supercapacitors. *Electrochimica Acta (2017) 240* 31–42.

70. Wu X, Liu B. Formation of ZnCo2O4@ MnO2 core–shell electrode materials for hybrid supercapacitor. *Dalton Transactions (2018) 47(43)* 15506–15511.

71. Winter M, Brodd R. What are batteries, fuel cells, and supercapacitors?. *Chemical Reviews (2004) 104(10)* 4245–4269.

72. Wang Y, Liu B, Li Q, Cartmell S, Ferrara S, Deng ZD, Xiao J. Lithium and lithium ion batteries for applications in microelectronic devices: A review. *Journal of Power Sources (2015) 286* 330–345.

73. Armand M, Tarascon J. Building better batteries. *Nature (2008) 451(7179)* 652–657.

74. Obrovac M, Christensen L. Structural changes in silicon anodes during lithium insertion/extraction. *Electrochemical and Solid-State Letters (2004) 7(5)* A93.

75. Derrien G, Hassoun J, Panero S, Scrosati B. Nanostructured Sn–C composite as an advanced anode material in high-performance Lithium-ion batteries. *Advanced Materials (2007) 19(17)* 2336–2340.

76. Hassoun J, Derrien G, Panero S, Scrosati B. A nanostructured Sn–C composite lithium battery electrode with unique stability and high electrochemical performance. *Advanced Materials (2008) 20(16)* 3169–3175.

77. Chen J, Yang L, Fang S, Hirano SI. Ordered mesoporous Sn–C composite as an anode material for lithium ion batteries. *Electrochemistry Communications (2011) 13(8)* 848–851.

78. Chen J, Yang L, Fang S, Hirano SI. Synthesis of mesoporous Sn–Cu composite for lithium ion batteries. *Journal of Power Sources (2012) 209* 204–208.

79. Park GO, Yoon J, Shon JK, Choi YS, Won JG, Park SB, Kim KH, Kim H, Yoon WS, Kim JM. Discovering a dual-buffer effect for lithium storage: Durable nanostructured ordered mesoporous Co–Sn intermetallic electrodes. *Advanced Functional Materials (2016) 26(17)* 2800–2808.

80. Boukamp BA, Lesh GC, Huggins RA. All-solid lithium electrodes with mixed-conductor matrix. *Proceedings of the Electrochemical Society (1981) 81–4(4)* 467–476.

81. Huggins R. Lithium alloy negative electrodes. Journal of Power Sources (1999) 81–82 13–19.

82. Jia H, Gao P, Yang J, Wang J, Nuli Y, Yang Z. Novel three-dimensional mesoporous silicon for high power lithium-ion battery anode material. *Advanced Energy Materials (2011) 1(6)* 1036–1039.

83. Xing A, Zhang J, Bao Z, Mei Y, Gordin AS, Sandhage KH. A magnesiothermic reaction process for the scalable production of mesoporous silicon for rechargeable lithium batteries. *Chemical Communications (2013) 49(60)* 6743–6745.

84. Hong I, Scrosati B, Croce F. Mesoporous, Si/C composite anode for Li battery obtained by 'magnesium-thermal' reduction process. *Solid State Ionics (2013) 232* 24–28.

85. Wang X, Li Z, Li Q, Wang C, Chen A, Zhang Z, Fan R, Yin L. Ordered mesoporous SnO2 with a highly crystalline state as an anode material for lithium ion batteries with enhanced electrochemical performance. *CrystEngComm* (2013) *15(18)* 3696–3704.

86. Etacheri V, Seisenbaeva GA, Caruthers J, Daniel G, Nedelec JM, Kessler VG, Pol VG. Ordered network of interconnected SnO2 nanoparticles for excellent lithium-ion storage. *Advanced Energy Materials (2015) 5(5)* 1401289.

87. Shon JK, Kim H, Kong SS, Hwang SH, Han TH, Kim JM, Pak C, Doo S, Chang H. Nano-propping effect of residual silicas on reversible lithium storage over highly ordered mesoporous SnO2 materials. *Journal of Materials Chemistry (2009) 19(37)* 6727–6732.

88. Liu H, Wang G, Liu J, Qiao S, Ahn H. Highly ordered mesoporous NiO anode material for lithium ion batteries with an excellent electrochemical performance. *Journal of Materials Chemistry (2011) 21(9)* 3046–3052.

89. Zheng X, Wang H, Wang C, Deng Z, Chen L, Li Y, Hasan T, Su BL. 3D interconnected macro-mesoporous electrode with self-assembled NiO nanodots for high-performance supercapacitor-like Li-ion battery. *Nano Energy (2016) 22* 269–277.

90. Dupont L, Laruelle S, Grugeon S, Dickinson C, Zhou W, Tarascon JM. Mesoporous Cr2O3 as negative electrode in lithium batteries: TEM study of the texture effect on the polymeric layer formation. *Journal of Power Sources (2008) 175(1)* 502–509.

91. Liu H, Du X, Xing X, Wang G, Qiao SZ. Highly ordered mesoporous Cr2O3 materials with enhanced performance for gas sensors and lithium ion batteries. *Chemical Communications (2011) 48(6)* 865–867.

92. Poizot PL, Laruelle S, Grugeon S, Dupont L, Tarascon JM. Nano-sized transition-metal oxides as negative-electrode materials for lithium-ion batteries. *Nature (2000) 407(6803)* 496–499.

93. Shaju KM, Jiao F, Débart A, Bruce PG. Mesoporous and nanowire Co3O4 as negative electrodes for rechargeable lithium batteries. *Physical Chemistry Chemical Physics (2007) 9(15)* 1837–1842.

94. Lin Z, Yue W, Huang D, Hu J, Zhang X, Yuan ZY, Yang X. Pore length control of mesoporous Co3O4 and its influence on the capacity of porous electrodes for lithium-ion batteries. *RSC Advances (2012) 2(5)* 1794–1797.

95. Yoon S, Jo C, Noh SY, Lee CW, Song JH, Lee J. Development of a high-performance anode for lithium ion batteries using novel ordered mesoporous tungsten oxide materials with high electrical conductivity. *Physical Chemistry Chemical Physics* (2011) *13(23)* 11060–11066.

96. Chen F, Wang J, Huang L, Bao H, Shi Y. Ordered mesoporous crystalline Mo-doped WO2 materials with high tap density as anode material for lithium ion batteries. *Chemistry of Materials* (2016) *28(2)* 608–617.

97. Thackeray M. Manganese oxides for lithium batteries. *Progress in Solid State Chemistry* (1997) *25(1-2)* 1–71.

98. Thackeray MM, Johnson CS, Vaughey JT, Li N, Hackney SA. Advances in manganese-oxide 'composite' electrodes for lithium-ion batteries. *Journal of Materials Chemistry* (2005) *15(23)* 2257–2267.

99. Luo J, Zhang J, Xia Y. Highly electrochemical reaction of lithium in the ordered mesoporosus β-MnO2. *Chemistry of Materials* (2006) *18(23)* 5618–5623.

100. Jiao F, Bruce P. Mesoporous crystalline β-MnO2—a reversible positive electrode for rechargeable lithium batteries. *Advanced Materials* (2007) *19(5)* 657–660.

101. Li L, Hua P, Tian X, Yang C, Pi Z. Synthesis and electrochemical properties of two types of highly ordered mesoporous MnO2. *Electrochimica Acta* (2010) *55(5)* 1682–1686.

102. Luo JY, Wang YG, Xiong HM, Xia YY. Ordered mesoporous spinel LiMn2O4 by a soft-chemical process as a cathode material for lithium-ion batteries. *Chemistry of Materials* (2007) *19(19)* 4791–4795.

103. Jiao F, Shaju K, Bruce P. Synthesis of nanowire and mesoporous low-temperature LiCoO2 by a post-templating reaction. *Angewandte Chemie International Edition* (2005) *44(40)* 6550–6553.

104. Putnam RL, Nakagawa N, McGrath KM, Yao N, Aksay IA, Gruner SM, Navrotsky A. Titanium dioxide–surfactant mesophases and Ti-TMS1. *Chemistry of Materials* (1997) *9(12)* 2690–2693.

105. Takahashi R, Takenaka S, Sato S, Sodesawa T, Ogura K, Nakanishi K. Structural study of mesoporous titania and titanium–stearic acid complex prepared from titanium alkoxide. *Journal of the Chemical Society, Faraday Transactions* (1998) *94(20)* 3161–3168.

106. Yang P, Zhao D, Margolese DI, Chmelka BF, Stucky GD. Generalized syntheses of large-pore mesoporous metal oxides with semicrystalline frameworks. *Nature* (1998) *396(6707)* 152–155.

107. Kavan L, Rathouský J, Grätzel M, Shklover V, Zukal A. Surfactant-templated TiO2 (Anatase): Characteristic features of lithium insertion electrochemistry in organized nanostructures. *Journal of Physical Chemistry B* (2000) *104(50)* 12012–12020.

108. Crepaldi EL, Soler-Illia GJ, Grosso D, Cagnol F, Ribot F, Sanchez C. Controlled formation of highly organized mesoporous titania thin films: From mesostructured hybrids to mesoporous nanoanatase TiO2. *Journal of the American Chemical Society* (2003) *125(32)* 9770–9786.

109. Alberius PC, Frindell KL, Hayward RC, Kramer EJ, Stucky GD, Chmelka BF. General predictive syntheses of cubic, hexagonal, and lamellar silica and titania mesostructured thin films. *Chemistry of Materials* (2002) *14(8)* 3284–3294.

110. Ren Y, Hardwick L, Bruce P. Lithium intercalation into mesoporous anatase with an ordered 3D pore structure. *Angewandte Chemie International Edition* (2010) *49(14)* 2570–2574.

111. Yue W, Xu X, Irvine JT, Attidekou PS, Liu C, He H, Zhao D, Zhou W. Mesoporous monocrystalline TiO2 and its solid-state electrochemical properties. *Chemistry of Materials (2009) 21(12)* 2540–2546.

112. Attia A, Zukalová M, Rathouský J, Zukal A, Kavan L. Mesoporous electrode material from alumina-stabilized anatase TiO2 for lithium ion batteries. *Journal of Solid State Electrochemistry (2005) 9(3)* 138–145.

113. Elder SH, Gao Y, Li X, Liu J, McCready DE, Windisch CF. Zirconia-stabilized 25-Å TiO2 anatase crystallites in a mesoporous structure. *Chemistry of Materials (1998) 10(10)* 3140–3145.

114. Chae W, Lee S, Kim Y. Zirconia-stabilized 25-Å TiO2 anatase crystallites in a mesoporous structure. *Chemistry of Materials (2005) 17(12)* 3072–3074.

115. Eftekhari A. LiFePO4/C nanocomposites for lithium-ion batteries. *Journal of Power Sources (2017) 343* 395–411.

116. Santos-Pena J, Soudan P, Arean CO, Palomino GT, Franger S.S. Electrochemical properties of mesoporous iron phosphate in lithium batteries. *Journal of Solid State Electrochemistry (2006) 10(1)* 1–9.

117. Shi Z.C, Attia A, Ye W.L, Wang Q, Li Y.X, Yang Y. Synthesis, characterization and electrochemical performance of mesoporous FePO4 as cathode material for rechargeable lithium batteries. *Electrochimica Acta (2008) 53(6)* 2665–2673.

118. Yang D, Lu Z, Rui X, Huang X, Li H, Zhu J, Zhang W, Lam YM, Hng HH, Zhang H, Yan Q. Synthesis of two-dimensional transition-metal phosphates with highly ordered mesoporous structures for lithium-ion battery applications. *Angewandte Chemie International Edition (2014) 53(35)* 9352–9355.

119. Mori K, Futamura Y, Masuda S, Kobayashi H, Yamashita H. Controlled release of hydrogen isotope compounds and tunneling effect in the heterogeneously-catalyzed formic acid dehydrogenation. *Nature Communications (2019) 10(1)* 1–10.

120. Koh K, Seo JE, Lee JH, Goswami A, Yoon CW, Asefa T. Ultrasmall palladium nanoparticles supported on amine-functionalized SBA-15 efficiently catalyze hydrogen evolution from formic acid. *Journal of Materials Chemistry A (2014) 2(48)* 20444–20449.

121. Zhang M, Xiao X, Wu Y, An Y, Xu L, Wan C. Hydrogen production from ammonia borane over PtNi alloy nanoparticles immobilized on graphite carbon nitride. *Catalysts (2019) 9(12)* 32.

122. Xu L, Jin B, Zhang J, Cheng DG, Chen F, An Y, Cui P, Wan C. Facile synthesis of amine-functionalized SBA-15-supported bimetallic Au–Pd nanoparticles as an efficient catalyst for hydrogen generation from formic acid. *RSC Advances (2016) 6(52)* 46908–46914.

123. Jin MH, Park JH, Oh D, Lee SW, Park JS, Lee KY, Lee DW. Pd/NH2-KIE-6 catalysts with exceptional catalytic activity for additive-free formic acid dehydrogenation at room temperature: Controlling Pd nanoparticle size by stirring time and types of Pd precursors. *International Journal of Hydrogen Energy (2018) 43(3)* 1451–1458.

124. Jin MH, Park JH, Oh D, Park JS, Lee KY, Lee DW. Effect of the amine group content on catalytic activity and stability of mesoporous silica supported Pd catalysts for additive-free formic acid dehydrogenation at room temperature. *International Journal of Hydrogen Energy (2019) 44(10)* 4737–4744.

125. Lee DW, Jin MH, Park JC, Lee CB, Oh D, Lee SW, Park JW, Park JS. Waste-glycerol-directed synthesis of mesoporous silica and carbon with superior performance in room-temperature hydrogen production from formic acid. *Scientific Reports (2015) 5(1)* 15931.

126. Jin MH, Oh D, Park JH, Lee CB, Lee SW, Park JS, Lee KY, Lee DW. Mesoporous silica supported Pd-MnOx catalysts with excellent catalytic activity in room-temperature formic acid decomposition. *Scientific Reports (2016)* 6 33502.

127. Zhang S, Qian Y, Ahn WS. Catalytic dehydrogenation of formic acid over palladium nanoparticles immobilized on fibrous mesoporous silica KCC-1. *Chinese Journal of Catalysis (2019) 40(11)* 1704–1712.

128. Doustkhah E, Rostamnia S, Zeynizadeh B, Kim J, Yamauchi Y, Ide Y. Efficient H2 generation using thiourea-based periodic mesoporous organosilica with Pd nanoparticles. *Chemistry Letters (2018) 47(9)* 1243–1245.

129. Zhang Y, Lyu Y, Wang Y, Li C, Jiang M, Ding Y. Highly active and stable porous polymer heterogenous catalysts for decomposition of formic acid to produce H2. *Chinese Journal of Catalysis (2019) 40(2)* 147–151.

130. Zhang S, Wang H, Tang L, Li M, Tian J, Cui Y, Han J, Zhu X, Liu X. Sub 1 nm aggregation-free AuPd nanocatalysts confined inside amino-functionalized organosilica nanotubes for visible-light-driven hydrogen evolution from formaldehyde. *Applied Catalysis B: Environmental (2018) 220* 303–313.

131. Anouar A, Katir N, El Kadib A, Primo A, García H. Palladium supported on porous chitosan–graphene oxide aerogels as highly efficient catalysts for hydrogen generation from formate. *Molecules (2019) 24(18)* 3290.

132. Zhang S, Lee YR, Jeon HJ, Ahn WS, Chung YM. Pd nanoparticles on a microporous covalent triazine polymer for H2 production via formic acid decomposition. *Materials Letters (2018) 215* 211–213.

133. Yue X, Zhang L, Sun L, Gao S, Gao W, Cheng X, Shang N, Gao Y, Wang C. Highly efficient hydrodeoxygenation of lignin-derivatives over Ni-based catalyst. *Applied Catalysis B: Environmental (2021) 293* 120243.

134. Gunasekar G, Kim H, Yoon S. Dehydrogenation of formic acid using molecular Rh and Ir catalysts immobilized on bipyridine-based covalent triazine frameworks. *Sustainable Energy & Fuels (2019) 3(4)* 1042–1047.

135. Yan JM, Wang ZL, Gu L, Li SJ, Wang HL, Zheng WT, Jiang Q. AuPd–MnOx/MOF–Graphene: An efficient catalyst for hydrogen production from formic acid at room temperature. *Advanced Energy Materials (2015) 5(10)* 1500107.

136. Song FZ, Zhu QL, Yang X, Zhan WW, Pachfule P, Tsumori N, Xu Q. Metal–organic framework templated porous carbon-metal oxide/reduced graphene oxide as superior support of bimetallic nanoparticles for efficient hydrogen generation from formic acid. *Advanced Energy Materials (2018) 8(1)* 1701416.

137. Wang H, Sun L, Wang H, Xin L, Wang Q, Liu Y, Wang L. Rutile TiO2 mesocrystallines with aggregated nanorod clusters: extremely rapid self-reaction of the single source and enhanced dye-sensitized solar cell performance. *RSC Advances (2014) 4(102)* 58615–58623.

138. Di M, Li Y, Wang H, Rui Y, Jia W, Zhang Q. Ellipsoidal TiO2 mesocrystals as bifunctional photoanode materials for dye-sensitized solar cells. *Electrochimica Acta (2018) 261* 365–374.

139. Wu D, Cao K, Wang H, Wang F, Gao Z, Xu F, Guo Y, Jiang K. Unable synthesis of single-crystalline-like TiO2 mesocrystals and their application as effective scattering layer in dye-sensitized solar cells. *Journal of Colloid and Interface Science (2015) 456* 125–131.

140. Zhou Y, Wang X, Wang H, Song Y, Fang L, Ye N, Wang L. Enhanced dye-sensitized solar cells performance using anatase TiO2 mesocrystals with the Wulff construction of nearly 100% exposed {101} facets as effective light scattering layer. *Dalton Transactions (2014) 43(12)* 4711–4719.

141. Amoli V, Bhat S, Maurya A, Banerjee B, Bhaumik A, Sinha AK. Toward a quantitative correlation between microstructure and DSSC efficiency: A case study of TiO2–xNx nanoparticles in a disordered mesoporous framework. *ACS Applied Materials & Interfaces (2015) 7(47)* 26022–26035.

142. Peng JD, Lin HH, Lee CT, Tseng CM, Suryanarayanan V, Vittal R, Ho KC. Hierarchically assembled microspheres consisting of nanosheets of highly exposed (001)-facets TiO2 for dye-sensitized solar cells. *RSC Advances (2016) 6(17)* 14178–14191.

143. Peng JD, Shih PC, Lin HH, Tseng CM, Vittal R, Suryanarayanan V, Ho KC. TiO2 nanosheets with highly exposed (001)-facets for enhanced photovoltaic performance of dye-sensitized solar cells. *Nano Energy (2014) 10* 212–221.

144. Wu X, Chen Z, Lu GQ, Wang L. Nanosized anatase TiO2 single crystals with tunable exposed (001) facets for enhanced energy conversion efficiency of dye-sensitized solar cells. *Advanced Functional Materials (2011) 21(21)* 4167–4172.

145. Zhang Y, Cai J, Ma Y, Qi L. Qi L. Mesocrystalline TiO2 nanosheet arrays with exposed {001} facets: Synthesis via topotactic transformation and applications in dye-sensitized solar cells. *Nano Research (2017) 10(8)* 2610–2625.

146. Sivaram V, Crossland EJ, Leijtens T, Noel NK, Alexander-Webber J, Docampo P, Snaith HJ. Observation of annealing-induced doping in TiO2 mesoporous single crystals for use in solid state dye sensitized solar cells. *Journal of Physical Chemistry C (2014) 118(4)* 1821–1827.

147. Lin J, Zhao L, Heo YU, Wang L, Bijarbooneh FH, Mozer AJ, Nattestad A, Yamauchi Y, Dou SX, Kim J H. Mesoporous anatase single crystals for efficient Co (2+/3+)-based dye-sensitized solar cells. *Nano Energy (2015) 11* 557–567.

148. Xu M, Ruan P, Xie H, Yu A, Zhou X. Mesoporous TiO2 single-crystal polyhedron-constructed core–shell microspheres: Anisotropic etching and photovoltaic property. *ACS Sustainable Chemistry and Engineering (2014) 2(4)* 621–628.

149. Yu H, Wang L, Dargusch M. Low-temperature templated synthesis of porous TiO2 single-crystals for solar cell applications. *Solar Energy (2016) 123* 17–22.

150. Wu T, Deng G, Zhen C. Metal oxide mesocrystals and mesoporous single crystals: synthesis, properties and applications in solar energy conversion. *Journal of Materials Science & Technology (2021) 73* 9–22.

151. Huang J, Yuan Y, [...]Yan Y. Understanding the physical properties of hybrid perovskites for photovoltaic applications. *Nature Reviews Materials (2017) 2(7)* 1–19.

152. Crossland EJ, Noel N, Sivaram V, Leijtens T, Alexander-Webber JA, Snaith HJH. Mesoporous TiO2 single crystals delivering enhanced mobility and optoelectronic device performance. *Nature (2013) 495(7440)* 215–219.

153. Zhu Z, Zheng X, Bai Y, Zhang T, Wang Z, Xiao S, Yang SS. Mesoporous SnO 2 single crystals as an effective electron collector for perovskite solar cells. *Physical Chemistry Chemical Physics (2015) 17(28)* 18265–18268.

154. Xiong Y, Liu Y, Lan K, Mei A, Sheng Y, Zhao D, Han HH. Fully printable hole-conductor-free mesoscopic perovskite solar cells based on mesoporous anatase single crystals. *New Journal of Chemistry (2018) 42(4)* 2669–2674.

155. Chen Q, Peng C, Du L, Hou T, Yu W, Chen D, Shu H, Huang D, Zhou X, Zhang J, Zhang W. Synergy of mesoporous SnO2 and RbF modification for high-efficiency and stable perovskite solar cells. *Journal of Energy Chemistry (2022) 66* 250–259.

156. Chen SH, Ho CM, Chang YH, Lee KM, Wu MC. Efficient perovskite solar cells with low JV hysteretic behavior based on mesoporous Sn-doped TiO2 electron extraction layer. *Chemical Engineering Journal (2022) 445* 136761.

14 MXene-based 2D Materials for Energy Production, Conversion and, Storage Applications

Abdul Mateen, Elsayed Tag Eldin,
Majed A. Bajaber, and Muhammad Sufyan Javed

14.1 INTRODUCTION

Two-dimensional (2D) transition metal carbides, nitrides, and carbonitrides called MXenes [1], have been explored, and their development has accelerated since the first MXene ($Ti_3C_2T_x$) was discovered by Gogotsi and co-workers [2]. The general formula for the functionalized MXenes is $M_{n+1}X_nT_x$, (n = 1-4), where M stands for transition metals of d-block (Ti, Hf, Zr, Nb, V, Cr, V, Mo, etc.), X stands for carbon and/ or nitrogen, and T stands for surface terminations (-F, -Cl, -O, or -OH) [3]. Owing to the growing need for sustainable energy and the rapid use of traditional fossil fuels, a significant research effort has been dedicated to developing energy production, conversion and storage technologies that are advanced, high-performing, and environmentally benign [4]. Catalysts and electrode materials with great stability, high efficiency, and eco-friendly nature are highly significant to the technological advancement of electrochemical energy [5]. MXenes possess extremely desirable physicochemical, electrical, optical, and mechanical characteristics, outstanding oxidation resistance, high melting point, hardness, high surface area, and hydrophilic nature. Therefore, MXenes have been used in a variety of applications, including catalysis [6], solar cells [7], supercapacitors [8] and batteries [9]. Despite the growing global demand for MXene-based materials, there is still a lack of well-organized information on the research of the surface roughness, structure stability and development in the synthesis process of MXenes [10]. In addition, because of the limitations of particular electrode materials, it is possible that electrodes or catalysts prepared by single-phase MXene will not be able to fulfil the requirements of real applications [11]. To solve these issues, synthesizing MXene heterostructures might be an excellent idea to overcome drawbacks like surface roughness and stability for more effective outcomes and practical applications [12]. The development of MXenes

DOI: 10.1201/9781003318859-14

271

and their van der Waals heterostructures has been satisfyingly advanced during the past several years, allowing more researchers to access advanced technologies and materials [13]. Herein, in this chapter, we summarize and highlight the most recent advancements in MXene-based 2D materials for energy production, conversion, as well as storage applications. We discuss briefly ORR, OER, HRR, CO_2RR, NRR, solar cells, supercapacitors and batteries. Finally, we discuss the critical scientific challenges and prospects for MXene-based 2D materials for sustainable energy production, conversion, and storage applications.

14.1.1 MXene-based 2D Materials for Energy Production

MXene-based 2D materials have many catalytically active centers because of their structural flexibility and compositional diversity. The following section will summarize the most recent advances in the development of MXene-based 2D materials as electrodes for a wide range of catalytic applications, such as ORR, OER, HER, CO2RR, and NRR.

14.2 OXYGEN EVOLUTION REACTION (OER)

In general, OER is a process that involves the transfer of four electrons, and it may be characterized as [14]; $4OH^- \rightarrow 2H_2O + O_2 + 4e^-$ (in alkaline) or $2H_2O \rightarrow 4H^+ + O_2 + 4e^-$ (in acidic). As a result, an effective OER depends on the active site's ability to form strong bonds with O [15]. Because of their superior conductivity and chemical resistance, MXenes can ensure both the fast transport of electrons and the stability of OER. In general, the intrinsic OER activity of MXenes is relatively low, possibly because there are only a few active sites. In addition, doping with nitrogen provides a straightforward method for enhancing the electrochemical performance of MXenes [16]. Que's group [17] presented that N atoms may be easily inserted into MXenes (e.g., $Ti_3C_{1.8}N_{0.2}$ and $Ti_3C_{1.6}N_{0.4}$) by etching the relevant X-site MAX solid solution. The presence of several active sites in the $Ti_3C_{1.8}N_{0.2}$ material was due to the presence of N atoms, which also contributed to an increase in the material's electrical conductivity as well as wettability. Furthermore, $Ti_3C_{1.6}N_{0.4}$'s higher hydrophilicity boosted the electrolyte's accessibility and the number of available sites (Figure 14.1a). A series of linear sweep voltammetry (LSV) was used to activate the electrode before conducting electrochemical experiments. As exhibited in LSV curves (Figure 14.1b), at a scan rate of 5 mV s^{-1}, the bare carbon fiber paper (CFP) substrate did not display any catalytic activity. On the other hand, Ti_3C_2 flakes demonstrated little catalytic activity, whereas fresh $Ti_3C_{1.8}N_{0.2}$ flakes indicated an increase in activity. $Ti_3C_{1.6}N_{0.4}$ flakes showed the most excellent OER activity because of the synergistic impact of more active sites, increased electrical conductivity, and enhanced wetting owing to the insertion of nitrogen. Using layered double hydroxides (LDH) and 2D MXene with high conductivity and an active surface, Yu et al. [18] developed novel non-precious metal electrocatalysts for OER. Structure, electrical characteristics, and interfacial junctions are all outstanding in hierarchical FeNi-LDH/Ti_3C_2-MXene nanohybrid generated by the interconnected network of FeNi-LDH nanoplates onto

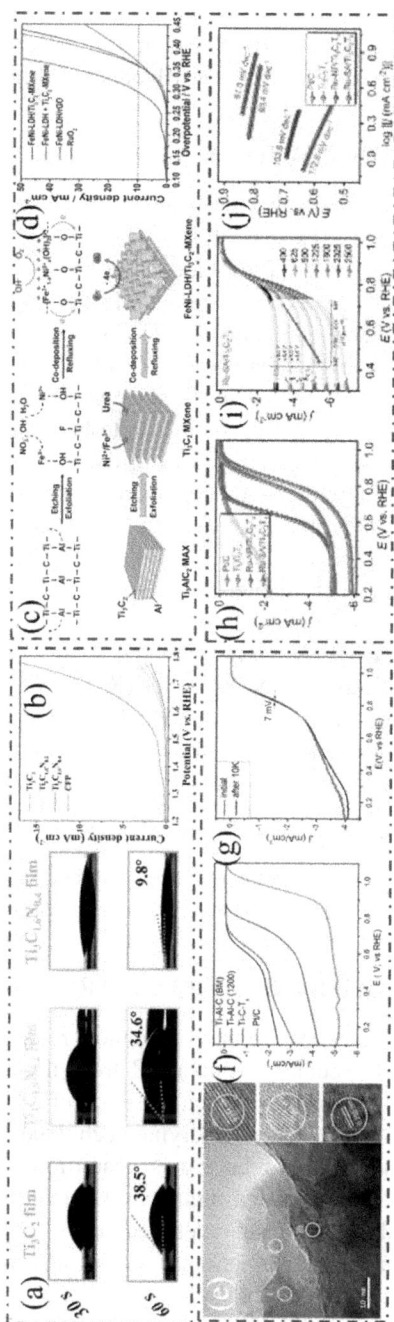

FIGURE 14.1 (a) The contact angles of films with electrolyte, (b) LSV curves. (a-b) Reproduced with permission [17], Copyright 2020, Royal Society of Chemistry. (c) Schematic diagram of 2D hierarchical FeNi-LDH/Ti$_3$C$_2$, (d) LSV curves. (c-d) Reproduced with permission [18], Copyright 2018, Elsevier. (e) HRTEM image of Ti-C-T$_x$, (f) comparisons polarization curves, (g) before/after cycles polarization curves (e-g) Reproduced with permission [19], Copyright 2018, Elsevier. (h) iR-compensated LSV curves, (i) LSV curves (inset: K-L plots), (j) Tafel plot. (h-j) Reproduced with permission [20], Copyright 2020, John Wiley & Sons.

Ti_3C_2 nanosheets (Figure 14.1c). The d-band center of the Ni/Fe atoms in FeNi-LDH/ Ti_3C_2-MXene was relocated to higher energy after hybridization with Ti_3C_2, which increased the O bonding strength. As a result, the OER's charge transfer was accelerated. While RuO_2 had a higher overpotential of 358 mV, FeNi-LDH/Ti_3C_2-MXene was shown to have a lower overpotential at 10 mA cm^{-2} (Figure 14.1d).

14.3   OXYGEN EVOLUTION REACTION (OER)

MXenes are one of the several materials that are being studied to use as developing supports and templates. For example, Parse et al. [19] explored the solid-state preparation of mixed-phase Ti-C-T$_x$. The high-resolution transmission electron microscope (HRTEM) image (Figure 14.1e) exhibits the interplanar distance of 0.253 nm (denoted as I, II, and III) associated with the (111) Ti-C's plane. The comparative polarization curves (Figure 14.1f) suggest that the Ti-C-T$_x$ electrocatalyst has a better E_{onset} and J_L value (0.87 V and 4.2 mA cm^{-2}) than the other electrocatalysts. The OER LSVs were measured both before and after the durability tests (Figure 14.1g), which showed a minor shift in half-wave potential (7 mV) in the case of Ti-C-T$_x$. Thus, the Ti-C-T$_x$ electrocatalyst is more stable in alkaline circumstances. Peng et al. [20] used single-atomic Ru (Ru-SA) sites attached to $Ti_3C_2T_x$ nanosheets, which function as electrocatalysts for catalyzing acidic OER. According to the LSV experiment findings (Figure 14.1h), Ru-SA/$Ti_3C_2T_x$ demonstrates the highest activity. It also has the most positive E_{onset} of 0.92 V vs RHE and an $E_{1/2}$ of 0.80 V vs RHE.

Figure 14.1i illustrates the ORR LSV curves recorded at various rotation speeds (400–2500 rpm). In the meantime, the Koutecky–Levich (K–L) plots derived from the LSV results exhibit good linearity at various potentials (Figure 14.1i, inset), which shows first-order reaction kinetics with respect to the O_2 concentration with a potential-independent electron transfer rate. Further, the Tafel slope of the Ru-SA/$Ti_3C_2T_x$ system is lower (60.4) than that of /$Ti_3C_2T_x$ (172.5), Ru-NP (103.6) and Pt/C (86.7) (Figure 14.1j). This indicates that the Ru-SA/$Ti_3C_2T_x$ has a fast initial electron transfer step in the ORR. These insights make it much easier to suggest the fundamentals of MXenes as a distinctive system of hybrid catalysts for specific purposes.

14.4   HYDROGEN EVOLUTION REACTION (HER)

Volmer-Heyrovsky or Volmer-Tafel are the most common routes for HER [21]. After the Volmer reaction, H_2 molecules are liberated from the electrode's interface by the Heyrovsky reaction or Tafel reaction, respectively. Que's group [22] developed a simple template-engaged aerosol drying technique for constructing MXene into aggregation-resistant 3D architectures featuring hydrophilic, conductive, as well as chemically functionalized multilevel hollow structures (mh-3D MXene) (Figure 14.2a). According to a scanning electron microscopy (SEM) image (Figure 14.2b) 3D MXene structures with spherical shapes and an average diameter of several micrometers were formed by drying MXene aerosols against polystyrene (PS) nanoparticles and exhibit honeycomb topology and many gaps on the surface. LSV curves (Figure 14.2c) show that a very low overpotential (η) of 27 mV is required to achieve a current density of 10 mA cm^{-2} (η_j=10) with rapid kinetics (Tafel slope: 41 mV dec^{-1}).

Figure 14.2d demonstrates that under a variety of pH circumstances, Pt@mh-MXene offers a significant improvement for alkaline HER. This catalyst's potential does not change over a period of 250 hours when subjected to a current density of 10 mA cm^{-2} (Figure 14.2e). Through the use of heteroatom dopants like nitrogen (N) and sulfur (S), Ramalingam et al. [23] demonstrate the coordination interaction between single Ru atoms (Ru$_{SA}$) and 2D Ti$_3$C$_2$T$_x$ supports (Figure 14.2f). It is possible that the current densities of Ru$_{SA}$-Ti$_3$C$_2$T$_x$ and Ru$_{SA}$-NS-Ti$_3$C$_2$T$_x$ catalysts are not at zero overpotential, as shown in Figure 14.2g. Therefore, the catalytic characteristics of MXenes may be modified by interactions between the metal and the support.

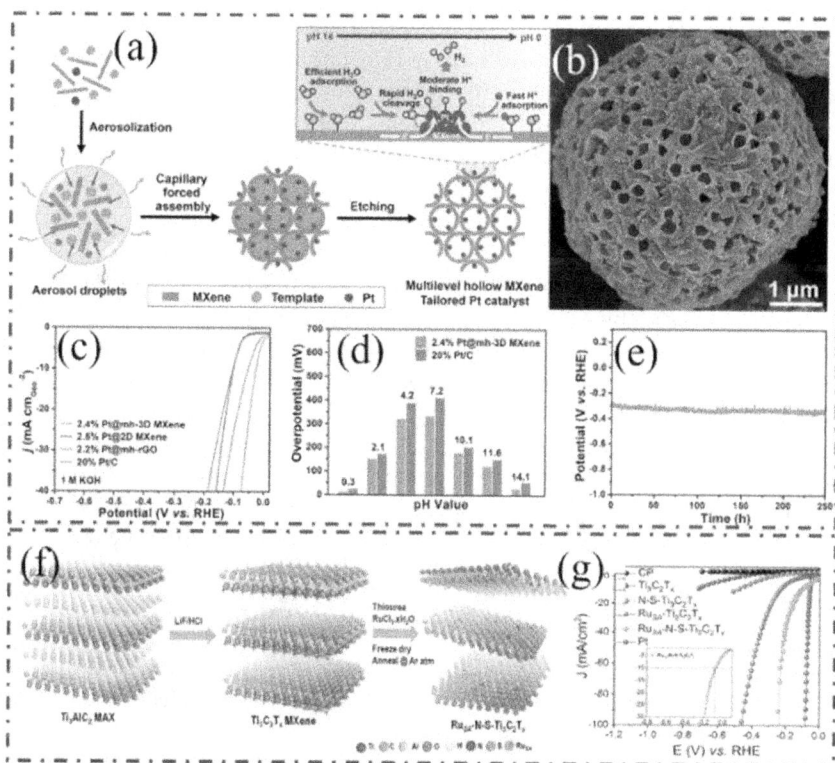

FIGURE 14.2 (a) Schematic diagram of mh-3D Mxene, (b) SEM image, (c) polarization curves, (d) comparison of 2.4% Pt@mh-3D MXene and 20% Pt/C, (e) chronopotentiometric-response. (a-e) Reproduced with permission [22], Copyright 2020, John Wiley & Sons. (f) Schematic diagram of Ru$_{SA}$-NS-Ti$_3$C$_2$T$_x$, (g) polarization curves. (f-g) Reproduced with permission [23], Copyright 2019, John Wiley & Sons.

14.5 CARBON DIOXIDE REDUCTION REACTION (CO$_2$RR)

MXene-based 2D materials are extremely important in the study of the reduction of CO$_2$. Yu's group [24] demonstrated that Ti$_3$C$_2$/Bi$_2$WO$_6$ hybrids had outstanding bulk-to-surface and interfacial charge transfer capabilities because of their short charge transport distance and wide interface contact area. Figure 14.3a visually represents the synthetic procedure for the 2D/2D heterojunction of ultrathin Ti$_3$C$_2$/Bi$_2$WO$_6$ hybrids nanosheets. Figure 14.3b shows that under the conditions of simulated solar irradiation, the photo-induced electrons of Bi$_2$WO$_6$ get excited and jump from the valence band (VB) to the conduction band (CB) of the molecule. The photo-induced electrons collected on the surface of Ti$_3$C$_2$ can then react with the CO$_2$ molecule adsorbed on the surface. Because of the one kind of 2D/2D heterostructure consisting of multiple atomic layers, photo-induced electrons may swiftly travel from the bulk of Bi$_2$WO$_6$ to the heterojunction interface and then to the surface of Ti$_3$C$_2$ (Figure 14.3c). As a direct consequence of this, the effectiveness of the photocatalytic CO$_2$ reduction may be significantly increased. Yu's group [25] presented a rice crust-like structure created by distributing TiO$_2$ nanoparticles onto a top of Ti$_3$C$_2$ (TiO$_2$/Ti$_3$C$_2$) in a homogeneous manner. The photocatalytic enhancing mechanism of TiO$_2$/Ti$_3$C$_2$ composite for CO$_2$RR is proposed and depicted in Figure 14.3d. Big specific surface area emerging from fascinating morphology assures many surface-active sites and enhances CO$_2$ adsorption capacity and the individual components contribute to the entire performance. The in situ growth on conductive 2D Ti$_3$C$_2$ forms heterogeneous surfaces can affect native materials. According to the density functional theory (DFT) simulation (Figure 14.3e), the bandgap of the TiO$_2$ was calculated to be 3.09 eV vs NHE. This result is in agreement with the light absorption spectra of the pure TiO$_2$ (TT650). The E$_{CB}$ of the TiO$_2$ was determined to be 0.25 eV vs NHE, which was more negative than the Fermi level of the Ti$_3$C$_2$ (0.04 eV vs NHE) (Figure 14.3f). The difference between the Fermi level of the Ti$_3$C$_2$ and the conduction band edge location of the TiO$_2$ was very tiny, which significantly reduced the energy barrier for electron transmission. Because of this, photogenerated electrons can easily migrate via the TiO$_2$/Ti$_3$C$_2$ interface, which results in high electron-hole separation efficiency.

14.6 NITROGEN REDUCTION REACTION (NRR)

Because of the extremely weak N$_2$ adsorption and the slow breakage of the N≡N bond, the electrochemical NRR method often suffers from low NH$_3$ yields and poor Faradaic efficiency. Recently, Tan's group [26] revealed single-atomic ruthenium-modified Mo$_2$CT$_x$ nanosheets as an effective electrocatalyst for N$_2$ fixation under ambient conditions (Figure 14.3a). Various potentials for commercial Ru/C, pristine Mo$_2$CT$_x$, and Mo$_2$CT$_x$ with varying Ru loadings were tested to determine the average NH$_3$ yield rate and Faradaic efficiency (FE) (Figure 14.3h, i), in which higher NH$_3$ yield can be seen using the SA-Ru-Mo$_2$CT$_x$ catalyst. Li et al. [27] reported fluorine-free Ti$_3$C$_2$T$_x$ nanosheets with a tiny size (50–100 nm in lateral size) for NRR (Figure 14.3j). The stability test for the Ti$_3$C$_2$T$_x$ electrode was carried out at a potential of -0.3 V (Figure 14.3k), and the Ti$_3$C$_2$T$_x$ electrode maintained its performance for a

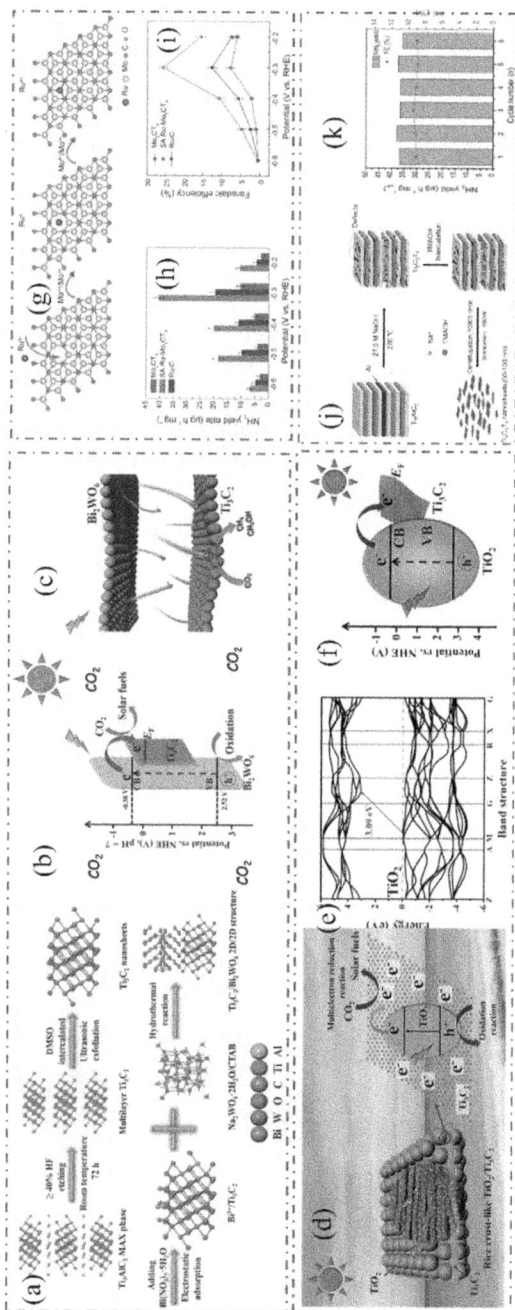

FIGURE 14.3 (a) Schematic diagram of Ti_3C_2/Bi_2WO_6, (b) energy level structure diagram, (c) photo-induced electron transport. (a-c) Reproduced with permission [24], Copyright 2018, John Wiley & Sons. (d) Schematic of photocatalytic CO_2RR, (e) TiO_2 band structure and excitation energy. (f) schematic diagram of charge carrier migration. (d-f) Reproduced with permission [25], Copyright 2018, Elsevier. (g) Schematic of SA-Ru-Mo$_2$CT$_x$ synthesis, electrochemical NRR performance (h) NH_3 yield rates, (i) NH_3 FE. (g-i) Reproduced with permission [26], Copyright 2020, John Wiley & Sons. (j) Schematic of T $i_3C_2T_x$, (k) NH_3 yield rates and FE. (j-k) Reproduced with permission [27], Copyright 2019, Royal Society of Chemistry.

period of 12 hours. Therefore, due to the size-effect as well as fluorine-free features, the performance is about two times greater than that of fluorine-based treatment.

14.7 MXENE-BASED 2D MATERIALS FOR ENERGY CONVERSIONS AND STORAGE

This section will discuss the most recent advances in the development of MXene-based 2D materials as electrodes for a wide range of energy conversion and storage applications.

14.8 SOLAR CELLS

Perovskite solar cells (PSCs) have a considerable dependence on the electron transport layer (ETL), hole transport layer (HTL), perovskite absorber, and the interfaces between these layers. Chen et al. [28] introduce the ultrathin $Ti_3C_2T_x$ quantum dots (QDs) (TQDs) with $Cu_{1.8}S$ nanocrystals (NCs) to simultaneously enhance the device performance as well as stability of PSCs. Figure 14.4a displays a schematic image of the whole device structure of the heterojunction PSCs ($cTiO_2$/$mTiO_2$-TQD/TQD-perovskite/Spiro-OMeTAD-$Cu_{1.8}S$), where, Spiro-OMeTAD stands for 2,2′,7,7′-tetrakis (N,N-di-p-methoxyphenylamine) -9,9′-spirobifluorene, $cTiO_2$ is compact

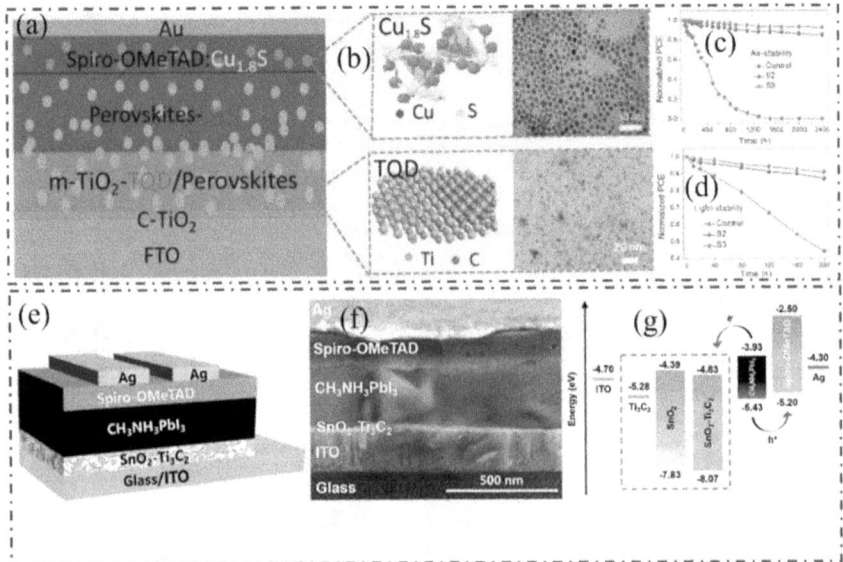

FIGURE 14.4 (a) Architecture of the PSCs, (b) TEM images, (c) stability of PCE in cells, and Light stability. (a-d) Reproduced with permission [28], Copyright 2020, John Wiley & Sons. (e) Device structure based on SnO_2-Ti_3C_2 ETL, (f) SEM image of the device, (g) schematic representation of the several layers' energy levels. (e-g) Reproduced with permission [29], Copyright 2019, Royal Society of Chemistry.

TiO_2, and m TiO_2 is mesoporous TiO_2. Figure 14.4b shows the rhombohedral-phase $Cu_{1.8}S$ NCs with an average size of 10.5 nm, and the mTiO_2 layer received the 2D single layer TQD with an average diameter of 5.1 nm and a thickness of 1.0 nm, respectively. Even after 2400 hours of exposure to the environment, the S2 and S3 devices retain roughly 85 and 93% of their initial values, respectively, compared to the Control devices (Figure 14.4c). As shown in Figure 14.4d, the S2 and S3 devices retain 87 and 90% of their initial efficiency after 200 h of illumination. Still, the Control device only retains 44% of its initial efficiency under identical test conditions. Yang et al. [29] employed Ti_3C_2 nanosheets as a conductive additive in the SnO_2 ETL for low-temperature-processed planar $CH_3NH_3PbI_3$ PSCs. The planar PSC device's cross-sectional SEM image is depicted in Figure 14.4f. An energy level diagram is illustrated in Figure 14.4g for each spectrum. As shown in Figure 14.4e, the prepared device with an architecture consisting of ITO/ETL/$CH_3NH_3PbI_3$/Spiro-OMeTAD/Ag was created to investigate Ti_3C_2,s ability to enhance the performance of PSCs, which produce a relatively high PCE of 18.34%. In addition, the SnO_2–Ti_3C_2-based PSC retained nearly 80% of the original PCE after being stored for 700 hours, which is evidence of the outstanding stability of this device.

14.9 SUPERCAPACITORS

MXenes have the potential to be used as electrode materials in supercapacitors due to their extremely high volumetric and gravimetric capacitance (up to 1500 F cm^{-3} and 400 F g^{-1}, respectively) [30–32]. Using $Ti_3C_2T_x$, Gogotsi's group [33] reported the design of two types of electrodes; macroporous $Ti_3C_2T_x$ film and $Ti_3C_2T_x$ hydrogel (Figure 14.5a). Figure 14.5b, and c show SEM of a $Ti_3C_2T_x$ MXene hydrogel and a macroporous templated $Ti_3C_2T_x$ electrode, respectively. A surface like metal oxides offers redox-active sites and pre-embedded water molecules, and 2D morphology enables ions to be highly accessible to the redox-active sites. At 10 mV s^{-1}, the capacitance of 310 F g^{-1} was seen in the cyclic voltammogram (CV) profiles (Figure 14.5d). The ion transport resistance of thick macroporous electrodes with mass loadings of 3.7 mg cm^{-2} is significantly reduced (0.04 Ωcm^2) compared to hydrogel samples with equivalent mass loadings for all potentials (Figure 14.5e). By using cetyltrimethylammonium bromide (CTAB)-grafted single-walled carbon nanotubes (SWCNTs) as inter-layer spacers, Zhang's group [34] was able to effectively construct $Ti_3C_2T_x$/SCNTs self-assembled composite. Figure 14.5f shows the schematic diagram of the synthesis process of $Ti_3C_2T_x$/SCNTs, in which the surface functional groups on the MXene sheets make them negatively charged and hydrophilic (Figure 14.5g). The CV curves (Figure 14.5h) at a scan rate of 2 mV s^{-1} show that at around 0.8 to –0.95 V, there are a pair of peaks, which could be associated with the valence change of Ti in the $Ti_3C_2T_x$ electrode. Figure 14.5i shows the computed areal capacitances at different scan rates. The $Ti_3C_2T_x$/SCNTs self-assembled composite electrode has a higher areal capacitance of 220 mF cm^{-2} (314 F cm^{-3}) at 2 mV s^{-1}.

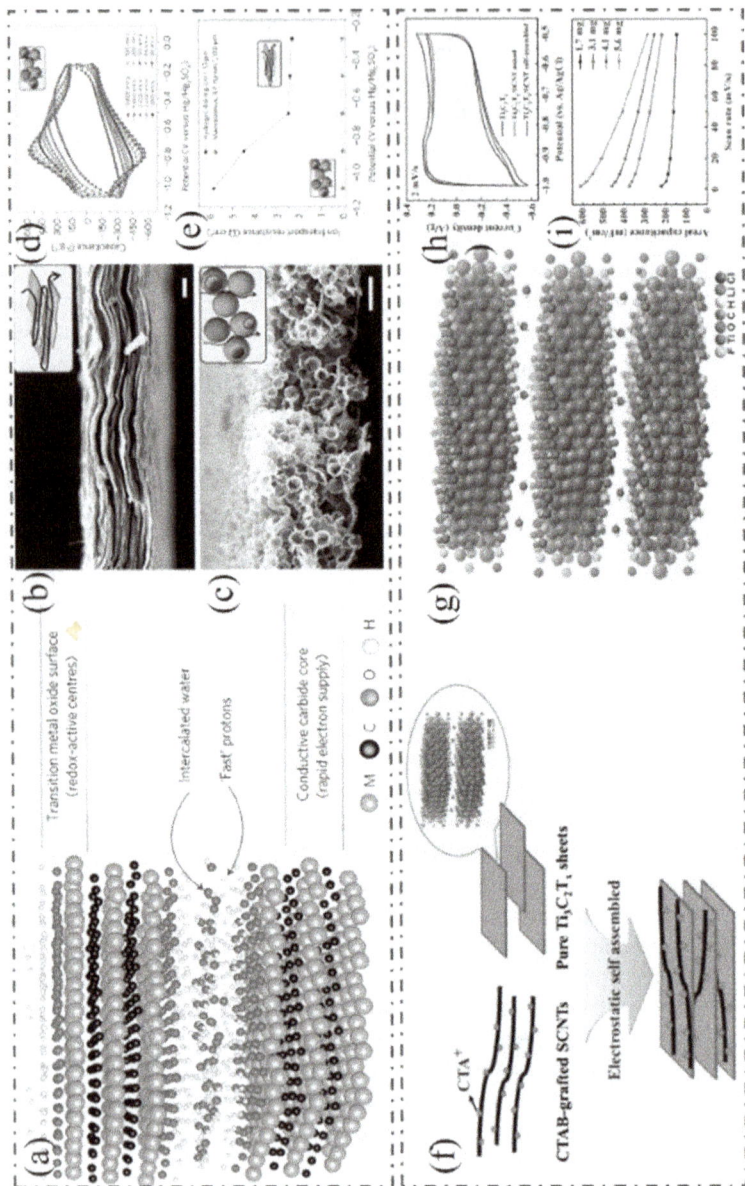

FIGURE 14.5 (a) Schematic diagram of MXene structure, SEM image of, (b) $Ti_3C_2T_x$, (d) CV profiles, (e) ion transport resistance comparison. (a-e) Reproduced with permission [33], Copyright 2017 Springer Nature. Schematics diagram of (f) the $Ti_3C_2T_x$/SCNT self-assembled, (g) the atomic architecture of the $Ti_3C_2T_x$, (h) CV profiles, (i) areal capacitance. (f-i) Reproduced with permission [34], Copyright 2018, Elsevier.

14.10 BATTERIES

2D MXenes are promising in lithium-ion batteries (LIBs) because of their preferential ion transport between interlayers (~1.3 nm) and large surface areas accessible for better ion adsorption, quick surface redox reaction, as well as excellent electrical conductivity. By using a scalable wet-chemical technique, Niu's group [35] was able to synthesize a hybrid composite material that consists of Ti_3C_2 MXene, coated by an ultrathin amorphous germanium oxide layer (GeO_x (x=1.57) @MXene) (Figure 14.6a). The thin GeO_x coating covering the MXene host can be seen in the TEM image (Figure 14.6b).

The initial capacities of 711.4 and 298.3 mAh g^{-1} were achieved when the temperature was -20 and -40 °C°, whereas 1003.8, 918.0, and 925.4 mAh g^{-1} were obtained when the temperature was 40, 60, and 80 °C, respectively (Figure 14.6c). Even after being charged and discharged after 100 cycles at a current rate of 0.2 C, the battery maintained a capacity of 333.9, 631.6, and 841.7 mAh g^{-1} while maintaining a Coulombic efficiency of 99.5, 99.2, and 99.7% at temperatures of -40, -20, and 60 °C, respectively (Figure 14.6d). Using van der Waals interactions, a generic method for

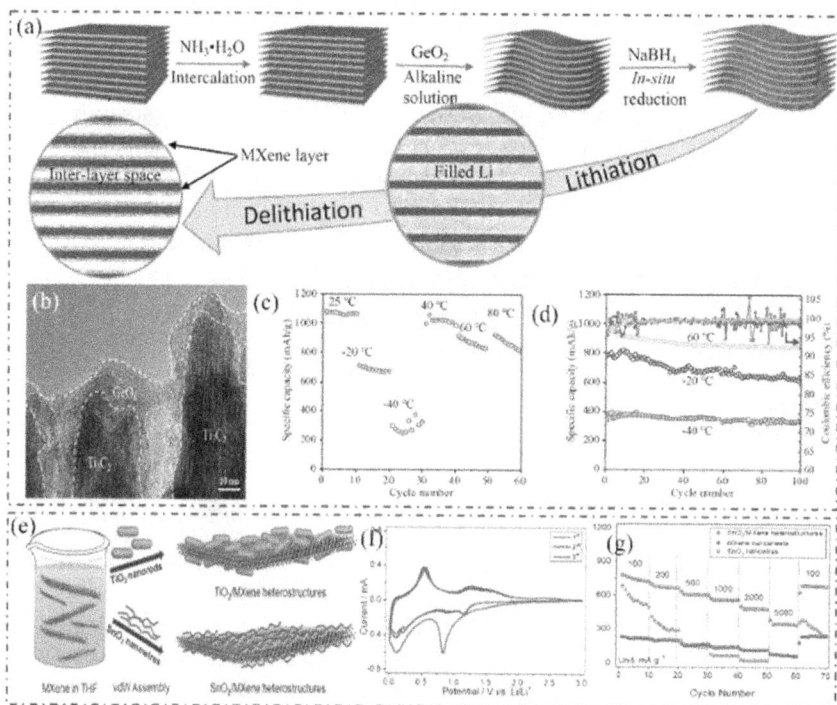

FIGURE 14.6 (a) Schematic of GeOx@MXene, (b) TEM image, (c) specific capacity, (d) cycle performance. (a-d) Reproduced with permission [35], Copyright 2020, American Chemical Society. (e) Schematic representation of SnO_2/Ti_3C_2, (f) CV profiles, (g) rate capabilities. (e-g) Reproduced with permission [13], Copyright 2018, John Wiley & Sons.

the straightforward self-assembly of SnO_2 nanowires on Ti_3C_2 nanosheets (SnO_2/Ti_3C_2) was reported by Gogotsi's group [13] (Figure 14.6e). Figure 14.6f shows two sharp peaks, and a fuzzy peak can be seen in the initial cathodic scan of the CV profile. SnO_2 is reduced to metallic Sn, and lithium is trapped between the MXene nanosheets in the solid electrolyte interphase layer, which disappears in successive scans due to these processes. The SnO_2/Ti_3C_2 heterostructures provide the largest capacities (720 and 310 mA h g^{-1} at 0.1 and 5 A g^{-1}, respectively). As soon as the current density returns to 0.1 A g^{-1}, the capacity will recover to 680 mA h g^{-1} (Figure 14.6g).

14.11 CONCLUSIONS AND SCIENTIFIC CHALLENGES

During the last decade, clean and renewable energy development and application have become a global aim and consensus. This chapter presents a detailed overview of current breakthroughs in using MXene-based 2D materials as electrodes for several applications, including energy generation, conversion, and storage. These applications include the OER, ORR, HER, CO_2RR, NRR, solar cells, supercapacitors, and batteries. Although MXenes are seen to provide excellent prospects for use in applications involving electrochemical technology; there are still several challenges; first, the kinetics of the etching processes that occur during the generation of MXenes from MAX phases is a significant factor to consider in generating MXenes of high quality and contain very few drawbacks. However, this has been addressed via the strong bonding of Al-F and hydrated Li$^+$ intercalation, which are responsible for the etching of Ti_3AlC_2 in HF/LiCl. However, Li$^+$ intercalation appears to be useless often in another MXene synthesis. Second, the ionic dynamics and charge storage processes between MXenes nanosheets are yet unknown to obtain high performance. More work is needed to use $Ti_3C_2T_x$-based hybrid structures, including layer-by-layer, cross-linking, insertion, and anchoring. Third, more in-depth in situ research of MXene-based electrochemical systems is required. An advanced synchrotron approach may be used to get atomistic-level insight into processes, construct entire 3D models of operating cells, or investigate how the electrical characteristics of surface groups or M elements change in operation. It is anticipated that the growing body of research will overcome these problems to boost further the application of MXene-based 2D materials in electrochemical technology.

ACKNOWLEDGMENTS

Authors are grateful to the scientific and technical input and support from the School of Physical Science and Technology through the Supporting Fund for Young Researchers from Lanzhou University and M.A.B. express appreciation to the Deanship of Scientific Research at King Khalid University Saudi Arabia through the research groups program under grant number R.G.P. 2/196/43.

REFERENCES

1. Tysoe, W.T., Easy alloying on flat carbides. *Nature Catalysis*, 2018. **1**(5): p. 316–317.
2. Gogotsi, Y. and P.J.S. Simon, True performance metrics in electrochemical energy storage. *Science*, 2011. **334**(6058): p. 917–918.

3. Shahzad, F., et al., Electromagnetic interference shielding with 2D transition metal carbides (MXenes). *Science*, 2016. **353**(6304): p. 1137–1140.

4. Jun, B.-M., et al., Review of MXenes as new nanomaterials for energy storage/delivery and selected environmental applications. *Nano Research*, 2019. **12**(3): p. 471–487.

5. Kim, H., Z. Wang, and H.N. Alshareef, MXetronics: Electronic and photonic applications of MXenes. *Nano Energy*, 2019. **60**: p. 179–197.

6. Seh, Z.W., et al., Two-Dimensional Molybdenum Carbide (MXene) as an efficient electrocatalyst for hydrogen evolution. *ACS Energy Letters*, 2016. **1**(3): p. 589–594.

7. Li, Z., et al., Single-Layered MXene nanosheets doping TiO2 for efficient and stable double perovskite solar cells. *Journal of the American Chemical Society*, 2021. **143**(6): p. 2593–2600.

8. Sajjad, M., et al., Recent trends in transition metal diselenides (XSe2: X = Ni, Mn, Co) and their composites for high energy faradic supercapacitors. *Journal of Energy Storage*, 2021. **43**: p. 103176.

9. Javed, M.S., et al., *The Emergence of 2D MXenes Based Zn-Ion Batteries: Recent Development and Prospects.* Small, 2022: p. 2201989.

10. Aïssa, B., et al., Transport properties of a highly conductive 2D Ti3C2Tx MXene/graphene composite. *Applied Physics Letters*, 2016. **109**(4): p. 043109.

11. Xiao, Y., J.-Y. Hwang, and Y.-K. Sun, Transition metal carbide-based materials: synthesis and applications in electrochemical energy storage. *Journal of Materials Chemistry A*, 2016. **4**(27): p. 10379–10393.

12. Pang, J., et al., Potential of MXene-Based heterostructures for energy conversion and storage. *ACS Energy Letters*, 2022. **7**(1): p. 78–96.

13. Liu, Y.T., et al., Self-assembly of transition metal oxide nanostructures on MXene nanosheets for fast and stable lithium storage. *Advanced Materials*, 2018. **30**(23): p. 1707334.

14. Suen, N.-T., et al., Electrocatalysis for the oxygen evolution reaction: Recent development and future perspectives. *Chemical Society Reviews*, 2017. **46**(2): p. 337–365.

15. Suntivich, J., et al., A perovskite oxide optimized for oxygen evolution catalysis from molecular orbital principles. *Science*, 2011. **334**(6061): p. 1383–1385.

16. Yang, C., et al., Flexible nitrogen-doped 2D titanium carbides (MXene) films constructed by an ex situ solvothermal method with extraordinary volumetric capacitance. *Advanced Energy Materials*, 2018. **8**(31): p. 1802087.

17. Tang, Y., et al., The effect of in situ nitrogen doping on the oxygen evolution reaction of MXenes. *Nanoscale Advances*, 2020. **2**(3): p. 1187–1194.

18. Yu, M., et al., Boosting electrocatalytic oxygen evolution by synergistically coupling layered double hydroxide with MXene. *Nano Energy*, 2018. **44**: p. 181–190.

19. Parse, H.B., et al., Mixed phase titanium carbide (Ti-C-Tx): A strategy to design a significant electrocatalyst for oxygen electroreduction and storage application. *Applied Surface Science*, 2018. **458**: p. 819–826.

20. Peng, X., et al., Trifunctional single-atomic Ru sites enable efficient overall water splitting and oxygen reduction in acidic media. *Small*, 2020. **16**(33): p. 2002888.

21. Li, Y., et al., Carbon doped molybdenum disulfide nanosheets stabilized on graphene for the hydrogen evolution reaction with high electrocatalytic ability. *Nanoscale*, 2016. **8**(3): p. 1676–1683.

22. Xiu, L., et al., Multilevel hollow MXene tailored low-Pt catalyst for efficient hydrogen evolution in full-pH range and seawater. *Advanced Functional Materials*, 2020. **30**(47): p. 1910028.

23. Ramalingam, V., et al., Heteroatom-mediated interactions between ruthenium single atoms and an MXene support for efficient hydrogen evolution. *Advanced Materials*, 2019. **31**(48): p. 1903841.

24. Cao, S., et al., 2D/2D heterojunction of ultrathin MXene/Bi2WO6 nanosheets for improved photocatalytic CO2 reduction. *Advanced Functional Materials*, 2018. **28**(21): p. 1800136.

25. Low, J., et al., TiO2/MXene Ti3C2 composite with excellent photocatalytic CO2 reduction activity. *Journal of Catalysis*, 2018. **361**: p. 255–266.

26. Peng, W., et al., Spontaneous atomic ruthenium doping in Mo2CTX MXene defects enhances electrocatalytic activity for the nitrogen reduction reaction. *Advanced Energy Materials*, 2020. **10**(25): p. 2001364.

27. Li, T., et al., Fluorine-free Ti3C2Tx (T = O, OH) nanosheets (~50–100 nm) for nitrogen fixation under ambient conditions. *Journal of Materials Chemistry A*, 2019. **7**(24): p. 14462–14465.

28. Chen, X., et al., Dual interfacial modification engineering with 2D MXene quantum dots and copper sulphide nanocrystals enabled high-performance perovskite solar cells. *Advanced Functional Materials*, 2020. **30**(30): p. 2003295.

29. Yang, L., et al., SnO2–Ti3C2 MXene electron transport layers for perovskite solar cells. *Journal of Materials Chemistry A*, 2019. **7**(10): p. 5635–5642.

30. Xia, Y., et al., Thickness-independent capacitance of vertically aligned liquid-crystalline MXenes. *Nature*, 2018. **557**(7705): p. 409–412.

31. Najam, T., et al., Synthesis and nano-engineering of MXenes for energy conversion and storage applications: Recent advances and perspectives. *Coordination Chemistry Reviews*, 2022. **454**: p. 214339.

32. Javed, M.S., et al., Achieving high rate and high energy density in an all-solid-state flexible asymmetric pseudocapacitor through the synergistic design of binder-free 3D ZnCo 2 O 4 nano polyhedra and 2D layered Ti 3 C 2 T x-MXenes. *Journal of Material Chemistry*, 2019. **7**(42): p. 24543–24556.

33. Lukatskaya, M.R., et al., Ultra-high-rate pseudocapacitive energy storage in two-dimensional transition metal carbides. *Nature Energy*, 2017. **2**(8): p. 17105.

34. Fu, Q., et al., Self-assembled Ti3C2Tx/SCNT composite electrode with improved electrochemical performance for supercapacitor. *Journal of Colloid and Interface Science*, 2018. **511**: p. 128–134.

35. Shang, M., et al., A fast charge/discharge and wide-temperature battery with a Germanium Oxide layer on a Ti3C2 MXene Matrix as Anode. *ACS Nano*, 2020. **14**(3): p. 3678–3686.

15 Aerogels-based Nanostructured Materials for Energy Generation, Conversion, and Storage Applications

S Alwin and X Sahaya Shajan

15.1 INTRODUCTION

The social, industrial, economic and technological developments around the world have increased global energy consumption exponentially [1, 2]. On account of facing the energy crisis, it is very crucial to consider the non-renewable nature of fossil fuels and their environmental problems. In the present scenario, the energy policies of every country are focusing on energy efficiency, rapid implementation of de-carbonization schemes and sustainable energy production through renewable resources to meet the energy demand for continuing the developments [2, 3]. Therefore, a contemporary global objective is to develop a sustainable energy technology that uses alternative energy generation and conversion processes as well as storage methods. The development of cleaner energy technologies, i.e. energy conversion devices like fuel cells, solar cells and storage devices such as batteries and supercapacitors, has a considerable impact on achieving a sustainable future [4]. However, the efficiency of these technologies mainly relies on the structure and properties of materials employed to fabricate the components of these devices [5].

Nanotechnology plays a crucial role in synthesizing nanostructured materials that offer outstanding physical and chemical properties due to confinement in size. Nanomaterials provide a high surface-to-volume ratio, favorable conducting properties and better surface properties, which are essential for the improvement of performance of the energy devices [6, 7]. Consequently, isotropic (0-D) nanostructures such as nanospheres, nanoparticles, linear (1-D) nanostructures such as nanotubes, nanorods, nanowires and planar (2-D) nanostructures such as nanosheets were extensively studied for various applications [8]. However, 1-D and 2-D nanostructures suffer from the low surface area that restricts their application in energy conversion and storage devices. The 3-D nanostructures are tailored by an anisotropic arrangement of lower-dimensional nanomaterials [9]. Aerogels are one of the best examples of

DOI: 10.1201/9781003318859-15

3-D nanostructures that offer high surface area, porosity and hierarchical network structure [10]. They are non-fluid colloidal networks derived from gels by exchanging gel liquid with air and the lightest solid materials known. Sol-gel method, emulsion method and 3-D printing are some methods adopted to obtain aerogel nanostructures [10, 11].

15.2 PROPERTIES OF AEROGEL NANOSTRUCTURES

The key property of aerogel nanostructures in accordance with energy conversion and storage devices is high specific surface area. The surface area offered by different aerogel nanostructures is between 100–1600 m^2/g [11]. The high specific surface area is favorable for both dye-sensitized solar cells and supercapacitors because the high surface area enhances the molecular adsorption/charge storage at the interface which in turn improves the performance of energy conversion and storage devices [12, 13]. Particularly in dye-sensitized solar cells, the large surface area available in aerogel facilitates the adsorption of more dye molecules which is responsible for photon absorption. When more and more dye molecules are adsorbed on the semiconductor surface, the photon absorption also increases, which improves the dye-sensitized solar cells' (DSSCs) efficiency [14]. Likewise, the performance of supercapacitors largely depends on the electrochemical properties of the electrode materials [15].

Next, the porosity of the aerogel nanostructure is very high, in the range of 80–99.8% with a mean pore diameter range of 20–150 nm [11]. The open porous structure is favorable for both DSSCs and batteries since the porous structure facilitates electrolyte diffusion and ionic mobility. For an example, in DSSCs gel polymer electrolytes are employed to avoid the leakage and evaporation of solvent. Therefore, the larger pores of aerogels help gel polymer electrolytes to penetrate inside the pores and enhance the interfacial contact [16]. In Li-ion batteries, lithium ions are intercalated/deintercalated during the charge/discharge process and hence, the high porosity with the large surface area of aerogels helps ionic mobility and improves rate capability in Li-ion batteries [13,17].

15.3 APPLICATIONS OF AEROGEL NANOSTRUCTURES IN FUEL CELLS

Fuel cells are electrochemical energy conversion devices that produce electrical energy from chemical energy without combustion. Fuel cells consist of an anode, a cathode and a polymer electrolyte. The anode is made up of a Pt/C catalyst coated on a gas diffusion layer (GDL) supported by carbon paper as a baking layer [18]. The cathode is also made up of Pt/C with different loadings. In a simple H_2-O_2 fuel cell, hydrogen is used as fuel and oxidized at anode. Oxygen is reduced at the cathode and the protons move from anode to cathode through the ionic sites of the polymer electrolyte membrane [19, 20].

As far as fuel cells are concerned, aerogels can be employed as catalyst support or filler materials in proton exchange membranes. For example, platinum (Pt) nanoparticles are deposited on the surface of SiO_2 aerogel and dispersed in Nafion solution to prepare recast membranes. The water uptake and the proton conductivity were improved for the silica-supported Pt dispersed Nafion recast membranes [21]. The presence of Pt nanoparticles on the membrane facilitates the reaction of penetrated fuel and oxidant on the surface of the membrane thereby diminishing the crossover current and improving the self-humidification attained through water molecules formed that assists the proton transfer resulting in better performance [21]. The mesopores of silica aerogel help in holding water clusters providing a shorter path length for ionic transport is found to be beneficial for proton transport through the membranes [22].

Carbon aerogels have been employed as alternative catalyst supports in fuel cells because they offer fascinating properties such as superior electronic conductivity, high surface area and pore volume due to its 3-D morphology [23]. The fuel cell performance significantly increases when the pore size of carbon aerogel increases because the larger pores of the support material provide better penetration of Nafion and increase the electrochemical surface area [24]. Hetero atom doping in carbon has been found to be an effective strategy for improving oxygen reduction activity since these nitrogen sites facilitate electron transfer [25]. Nitrogen-doped carbon/graphene aerogel was prepared and employed as a non-platinum catalyst in microbial fuel cells [26, 27]. The hierarchical porous structure of aerogel plays a vital role in microbial fuel cells that allows efficient diffusion of bacterial cells into the network structure and increases the surface area available for bacterial colonization.

Urea fuel cell (UFC) generates electricity from the electro oxidation of urea using urine or urea-containing wastes as fuel [28, 29]. The electro oxidation of urea releases 6 electrons that reduce oxygen at the cathode and the OH⁻ions formed at the cathode are transported to the anode through anion exchange membrane. The anode, cathode and overall reactions of urea fuel cell are given below [29].

Anode reaction: (15.1)

$$\underset{NH_2}{\overset{O}{\underset{\quad}{\parallel}}}\underset{NH_2}{C} + 6OH^- \longrightarrow N_2 + CO_2 + 5H_2O + 6e^-$$

Cathode reaction: (15.2)

$$O_2 + 2H_2O + 4e^- \longrightarrow 4OH^-$$

Overall reaction: (15.3)

$$2\ CO(NH_2)_2 + 3O_2 \longrightarrow 2N_2 + 2CO_2 + 4H_2O$$

Carbon nanotube (CNT) aerogel was used as catalyst support for Ni-Co bimetallic catalyst for urea oxidation. The high surface area of CNT aerogel remarkably enhanced the electro-catalytic activity of bimetallic catalyst due to the fast diffusion of urea and the uniform distribution of Ni-CO nanoparticles on the surface of aerogel facilitates electro oxidation [30].

Noble metal aerogels are a new class of promising electro catalyst materials recently developed for fuel cell applications in regard to the automotive field that requires the operation of fuel cells under stringent conditions 31, 32]. A robust electro catalyst such as metallic aerogels can circumvent the corrosion of support material and increases the durability of electrodes and the stack life. Highly porous spherical aggregates of noble metal aerogels such as Pt, Pd, Au and Ag were synthesized with a surface area ranging from 46–48 m²/ g [32, 33]. They do not contain any support materials but consist completely of catalytically active noble metal surfaces and the entire area could be used for electro catalysis. This improves the effective utilization of noble metal surfaces and decreases the metal loading and hence the cost of the device. The hierarchical porous structure of aerogel provides effective mass transport and decreases mass transport resistance. Also, the interconnected network structure offers a self-supporting character that increases the durability of the electro catalyst. The adoption of an aerogel structure reduces the cost and also exterminates the corrosion of catalyst support [34]. The bimetallic electro catalysts are also synthesized by alloying noble metals with transition metals like Fe, Co, Ni and Cu. The introduction of transition metals helps in two ways: firstly it further reduces the cost, secondly it down shifts the d-band center from the fermi level which prevents the adsorption of spectator species and makes more active sites available for oxygen adsorption to achieve a high oxygen reduction activity [35, 36].

15.4 APPLICATIONS OF AEROGEL NANOSTRUCTURES IN DYE-SENSITIZED SOLAR CELLS

Solar energy is most abundant and expected to play a crucial role in meeting the future energy demand without compromising environmental safety [37]. The photoanode comprises a wide bandgap semiconductor sensitized with dye molecules which is responsible for the absorption of photons from sunlight. After the absorption of photons, electrons are excited from the highest occupied molecular orbital (HOMO) to the lowest unoccupied molecular orbital (LUMO) followed by the injection of an electron into the conduction band of the semiconductor [38]. Then the electron reaches the conducting substrate and is transported to the counter electrode through an external circuit. The counter electrode is made up of platinized fluorinated tin oxide (FTO) substrate. The redox electrolyte contains iodide/tri-iodide redox couple that receives the electron from the counter electrode and donates the electron to regenerate excited dye molecules [39].

The surface area of the semiconductor material is a key property for DSSC applications since the amount of dye adsorption depends mainly on the surface area. Metal oxide aerogels are promising materials for DSSC applications owing to their wide bandgap, high surface area and porosity [10]. TiO_2 aerogels with a surface area ranging from 85–150 m²/g were synthesized through the sol-gel method and employed as photoanode material in DSSC [40]. The aerogel photoanode performed on par with nanocrystalline TiO_2 photoanode between 400–600 nm and performed better at wavelengths greater than 600 nm [40]. Therefore, it is proposed to use TiO_2 aerogel as a top layer for nanocrystalline TiO_2 photoanode which can increase the photon absorption at higher wavelengths 600–800 nm. A 16% improvement in power conversion efficiency was achieved with TiO_2 aerogel photoanode over P25 TiO_2 photoanode and this increment is due to the beneficial structural properties of aerogel such as high surface area, porosity and three-dimensionally interconnected pore structures [41]. Surface treatment is one of the important ways of improving DSSC performance since the physical and chemical properties of the TiO_2 undergo considerable alteration. The surface properties of the TiO_2 layer not only influence the chemisorption of dye molecules but also affect the charge separation, transport and recombination processes [42]. The chemical treatment of the TiO_2 electrode with $TiCl_4$ results in interparticle neck growth and facilitates the electron percolation through TiO_2 thereby increasing the current density [43]. The treatment of TiO_2 photoanode with aqueous nitric acid enhanced the rate of dye adsorption to 18 times faster than normal dye adsorption conditions. The HNO_3 treatment increases the positive charge density on the TiO_2 surface without affecting the TiO_2 nanostructure and improves electrostatic attraction between the TiO_2 surface and dye molecules [44].

Surface treatment of materials by physical methods such as low-pressure oxygen plasma is widely accepted for the surface modification of TiO_2 with regard to DSSC application [45]. The low-pressure oxygen plasma treatment method is extended to the TiO_2 aerogel surface also. TiO_2 aerogel was obtained through the sol-gel method followed by ambient pressure drying and subjected to 30 W oxygen plasma with different exposure time. It is found that plasma treatment has increased the specific surface of TiO_2 aerogel considerably and the pore structure becomes more uniform and rigid. X-ray photoelectron spectroscopic analysis reveals that there is an increase in the percentage of oxygen in TiO_2 for 10 minutes exposure time [46]. This suggests that additional hydroxyl groups are inserted on the TiO_2 aerogel surface during plasma treatment and this is also confirmed by dye adsorption studies. The microstructural analysis before and after plasma treatment indicates the porous network structure is preserved. The DSSC performance of plasma-treated TiO_2 aerogel as a photoanode material was investigated. The DSSC fabricated using TiO_2 aerogel with 10 minutes of plasma treatment exhibited an increased short circuit current density of 9.45 mA/cm^2 and an open circuit voltage of 0.694. The power conversion efficiency is enhanced to 3.94% compared to the efficiency of 3.08% for the DSSC fabricated using untreated TiO_2 aerogel (Figure 15.1 (a)) [47]. This improvement in the power conversion efficiency of DSSC is due to the increased amount of dye adsorption caused by the increase in specific surface area and the introduction of additional hydrophilic groups on the surface of TiO_2 aerogel accomplished through oxygen plasma treatment. Moreover, the oxygen vacancies generated by the plasma treatment facilitate the electron transport process across the electrode through the trap-detrap mechanism and decrease the diffusion time (τ_d) of electrons [47]. Conversely, the long exposure of aerogel to plasma generates a high concentration of oxygen vacancies in the aerogel structure that acts as electron trapping sites. Thus, the electron mobility decreases and the diffusion time increases when trapping sites are high in the aerogel layer.

Zinc oxide aerogel was obtained through the sol-gel method by means of epoxide-initiated gelation. The network structure of aerogel is built by the combination of 20–50 nm-sized spherical particles and hexagonal platelets. A bilayer photoanode of

TiO$_2$:ZnO mixed aerogel layer (with the ratio of 75:25, 50:50 and 25:5) as a bottom layer and an over layer of TiO$_2$ aerogel was fabricated and the performance of the material was studied in quasi-solid dye-sensitized solar cells (QSDSSCs). However, a low power conversion efficiency of 0.11% was obtained due to the dissolution of ZnO layer in dye solution [48]. Recently, a composite photoanode material consisting of mesoporous TiO$_2$ aerogel and microporous metal organic framework (MOF) was prepared and employed as a photoanode in DSSCs. The high surface area of the composite material enhanced the photon absorption due to the large amount of dye adsorption. The enhanced porosity of composite material permits better penetration of electrolytes that facilitates charge transfer across the interface [49–51].

Light scattering is another important phenomenon employed in DSSCs to improve photon absorption [52]. The incident photon to conversion efficiency (IPCE) at higher wavelengths (600–800 nm) can be effectively improved by light scattering by making the semiconductor layer with larger particles in the range of 400 nm [53]. A scattering layer is a diffusive layer made up of larger particles on the backside of the TiO$_2$ layer which increases the optical path length of light and enhances photon absorption by making more light available for absorption by dye molecules [7]. The hierarchical nanostructures developed from isotropic, linear and planar morphologies are beneficial for light scattering without compromising the amount of dye adsorption 54]. The greater light scattering effect offered by the aerogel nanostructure significantly enhances the photocurrent density. The basis for the enhanced scattering behavior of aerogel is due to the bicontinuous arrangement of the pore-solid network structure that increases the diffused reflectance of light. The hybrid photoanode of SiO$_2$-TiO$_2$ aerogel exhibited (Figure 15.1 (b)) a high power conversion efficiency of 7.57% and

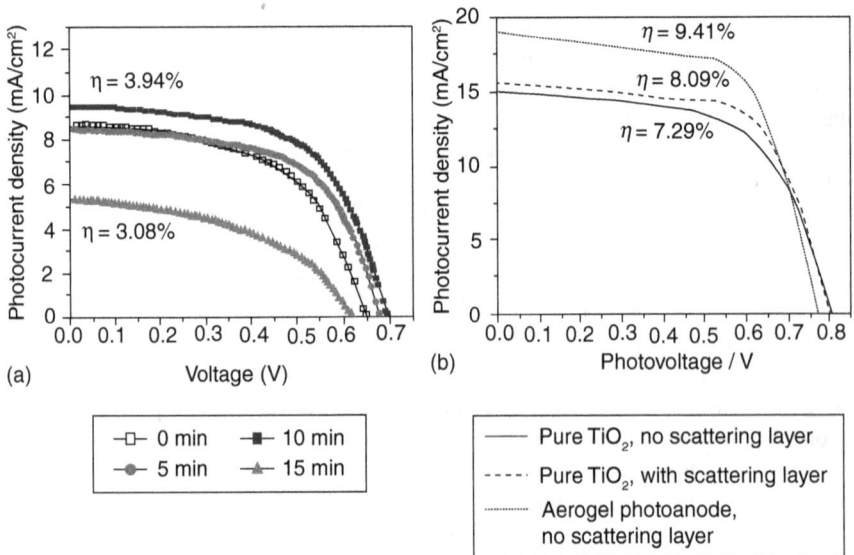

FIGURE 15.1　Effect of a) surface treatment and b) scattering effect on DSSC performance of TiO$_2$ aerogel (Reproduced from [47] and [55]).

it is increased to 9.41% after $TiCl_4$ treatment [55]. The primary scattering centers in aerogel are not simply the larger aggregates but the submicron-sized pores. Also, the scattering centers in aerogel are dispersed throughout the layer which significantly enhances the optical path length of the visible light. Besides, the scattering centers of the aerogel accommodate more dye molecules due to the high specific surface area which helps increase the photon absorption [55].

15.5 APPLICATIONS OF AEROGEL NANOSTRUCTURES IN BATTERIES

Lithium-ion batteries are one of the most advanced rechargeable batteries since the revolution of lithium-ion batteries in personal electronics and now its applications are extended in the development of electric vehicles [56]. A lithium-ion battery consists of an anode made up of graphite, a cathode made up of lithium cobalt oxide and a lithium-ion conducting electrolyte. Lithium ions are extracted from the cathode and pass through the electrolyte to the anode during charging and the anode releases Li-ions in an oxidation process and is taken up by the cathode during discharging [6]. The power density of Li-ion battery depends on the Li-ion insertion/de-insertion during charging and discharging, thus the diffusion co-efficient and the diffusion length are very important. The diffusion length can be reduced by employing nano-sized electrode materials by providing shorter transport distances [5]. At the same time, the high surface energy associated with the nanoparticles induces the particles to form agglomerates that increase the contact resistance and result in capacity fading. The high surface area of nanoparticles supports surface reactions like electro-lyte decomposition at the interface and also causes capacity fading [57]. Therefore, the hierarchical nanostructures such as the aerogel structure formed by the assembly of nanoparticles could offer shorter diffusion lengths in association with better cycle stability [6]. Vanadium pentoxide (V_2O_5) aerogel was synthesized and employed in Li and Na ion batteries and the layered structure of V_2O_5 provides intercalation host for the ions. The large interlayer spacing between the vanadium pentoxide layers assists the electrolyte diffusion and ion insertion and de-insertion during charging and discharging [58, 59]. By the application of lithium titanate aerogel as anode material in Li-ion battery and it is found that the high surface area of aerogel is beneficial for increasing the power density without sacrificing the high capacity of the electrodes. Also, the high surface area of anode material affects the insertion potential and yields a pseudo-capacitive behavior [60].

Graphene aerogels (GA) exhibit a 3-D porous network, large specific surface area, rapid ion-diffusion characteristics and multidimensional continuous electron trans-port pathways. In addition, nano-porous structures could provide extra channels for Li^+ ions and electronic transport during the insertion and de-insertion processes [61]. The porous structure of a hybrid molybdenum sulfide/graphene aerogel composite electrode promotes an electron transfer and provides a 3-D network for ion transport results increased capacity and cycle stability even in the absence of a conducting agent [62]. Similarly, the large number of pores in vanadium tetrasulfide anchored graphene aerogel anode helps the insertion-de-insertion of Li-ions [63]. The rich porosity of the aerogel structure and 3-D network maximizes the utilization of metal oxide species for catalytic reactions [64, 65]. The macropores of cobalt ferrite decorated graphene/carbon nanotube aerogel anode material acts as a buffering reservoir for Li-ion containing electrolyte suitable for rapid diffusion of Li-ions into the inner surfaces, while the mesopores offer more accessible surface area for Li-ion transport and manage the local volume changes upon Li-ion insertion/extraction [66]. Recently, silicon-embedded nanographite aerogel for Li-ion battery application. The open porous structure of aerogel governs the volume changes of silicon during lithium insertion and de-insertion process. Also, it provides channels for lithium ions and further improves the ionic conductivity. The microstructure of aerogel forms a stable solid electrolyte interface and boosts the performance [67, 68].

Metal air batteries are emerging storage devices with regard to electric vehicles that generate electrical energy from a redox reaction between metal and oxygen on a porous electrode. Therefore, electro catalysts for oxygen reduction reaction play a vital role in the performance of metal air batteries [69]. Fe and Co stabilized nitrogen-doped carbon aerogel was used as cathode material in zinc-air battery and shows high power density. Nitrogen-doped carbon shell encapsulated cobalt oxide nanoparticles anchored on nitrogen-doped graphene oxide aerogel was used as cathode in zinc-air battery and exhibits a high voltage. The interconnected porous architecture of the aerogel structure allows the permeation of reactants and the transfer of electrons and also suppresses particle agglomeration and dissolution resulting in better performance [70, 71]. Li_2S-coated boron or nitrogen-doped graphene aerogel as anode in lithium-sulfur battery enhanced the specific capacity and rate capability due to the porous structure of aerogel that facilitates the ionic and electronic transport [72].

15.6 APPLICATIONS OF AEROGEL NANOSTRUCTURES IN SUPERCAPACITORS

Supercapacitors are electrochemical storage devices that store electrical energy in the form of an electrochemical double layer between the electrode/electrolyte interfaces and exhibit high power density, rapid charging /discharging rate and long cycle life. Supercapacitors consist of two carbon-coated electrodes electrically isolated by a nonconductive material with a polymer electrolyte. When a potential is applied between the electrodes, the positive/negative ions present in the electrolyte are attracted towards the suitable electrodes of opposite charge. The capacitance of supercapacitors largely depends on the electrochemical properties of the electrolyte and the electrode [73]. Therefore, carbon materials with high surface area, tailored pore size and favorable surface chemistry can increase the energy density of supercapacitors [74, 75]. The high surface area of the aerogel has a substantial effect on the capacitance of the supercapacitor. For example, when carbon aerogel with a specific surface area of 1243 m^2/g is employed as electrode material in supercapacitor, the gravimetric capacitance is increased from 234.2 F/g [76].

The open porous structure of aerogel also helps to improve the storage capacity and rate capability of supercapacitors. Aerogel consists of a hierarchical porous

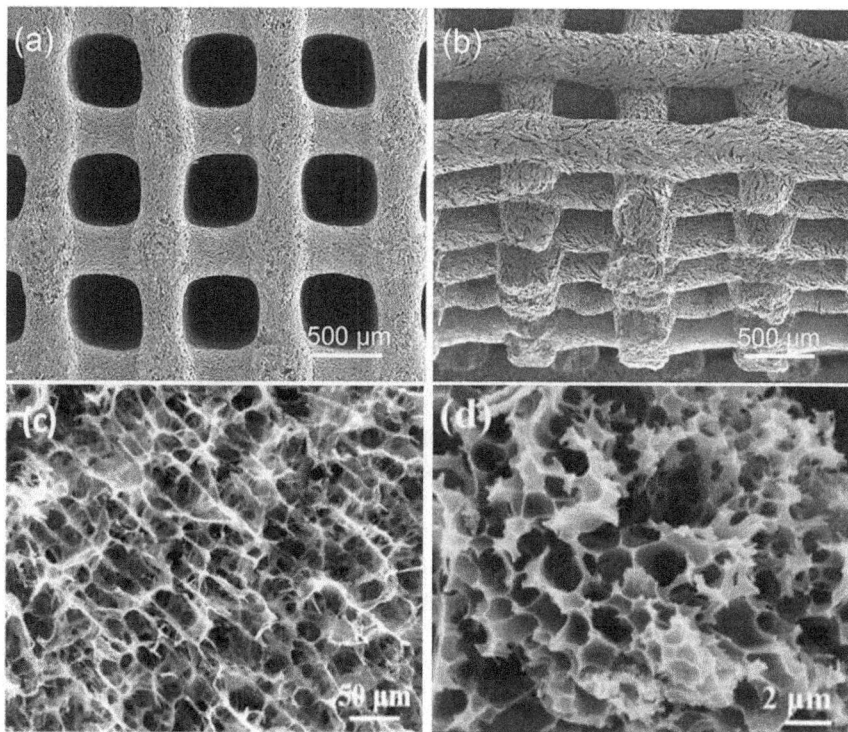

FIGURE 15.2 SEM images of carbon aerogel showing hierarchical porous structure (a), (b) and honeycomb-like structure (c), (d) (Reproduced from [78], [79] and [82]).

structure including micro, meso and macro pores that assist the ion transport, ionic diffusion and ultimately attain fast charging/discharging process [77, 78]. The hierarchical porous structure of carbon aerogels is shown in Figure 15.2 (a) and (b). The coexistence of micro and meso pores in aerogel structure facilitates the ion transport during the charging and discharging process. The correlation analysis carried out based on N_2 adsorption/desorption studies showed that the micropores and the mesopores play different roles in the electrochemical properties of the electrode materials. The correlation analysis result suggests that the specific capacitance mainly depends on the micropore volume and fairly on the mesopore volume. For instance, the electrolyte ions present in pores, the distance between micropore walls and the ions becomes short and there exists strong electrostatic adsorption resulting in high storage capacity. However, the rate capability is largely associated with mesopore volume. For example, mesopores mainly accommodate electrolyte that shortens the ion-diffusion length and delivers electrolyte ions during charging and discharging [79, 80]. The mesopores of carbon aerogel afford sufficient ion transfer pathways and the unique 3-D porous structure favors efficient contact with the electrolyte [81]. It is also found that 3-D macroporous honeycomb-like interconnected structure and well-defined channel-like pores are beneficial for decreasing the electrolyte ion-diffusion distance thus assisting the formation of more efficient electrical double layers. The honeycomb-like interconnected porous structure of carbon aerogels is shown in Figure 15.2 (c) and (d) [82]. Recently, multifunctional super-elastic, superhydrophilic carbon aerogels with ordered honeycomb-like porous structures were synthesized for supercapacitors and wearable electronic applications [83].

Carbon-based aerogels such as carbon nanotube aerogel and graphene aerogel are also promising candidates for supercapacitor applications because of their exceptional electronic conductivity, outstanding chemical stability and large specific surface area. Graphene/carbon nanotube aerogel with nanoscroll interconnected nanosheet provides fast electron and ion-diffusion pathways and also maximizes the exposure of active sites [84]. The transition metal or non-metal doping in carbon-based aerogels also favors ionic diffusion and provides a high gravimetric capacitance [85]. It is found that nitrogen doping enhances the mesoporosity of aerogel structure and generates structural defects/voids that offer excellent electrical conductivity and transport paths causing efficient electron and ion transport leading to better supercapacitor performance [86]. Transition metal oxide aerogels such as CoO_2 and MnO_2 were employed as electrode materials in supercapacitors. CoO_2 aerogel exhibits outstanding supercapacitive properties over MnO_2 aerogel. The high specific capacitance of 623 F/g was achieved for CoO_2 aerogel whereas MnO_2 electrodes provide 139 F/g at a current density of 1 A/g. This is mainly due to the variation in the specific surface area of transition metal oxide aerogels. CoO_2 aerogel offers a maximum specific surface area of 235 m^2/g whereas MnO_2 aerogel provides only 51 m^2/g. However, both aerogels provide high rate capability and excellent cycle stability because the high porosity of the aerogel structure relieves the internal strain and prevents structural changes during charging and discharging [87, 88]. TiO_2 aerogel-Co MOF composite as electrode materials showed excellent electrochemical performance in supercapacitors due to high surface area, reduced pore size, and low internal resistance of aerogel nanostructure [89].

15.7 CONCLUSIONS

This chapter demonstrates the interesting properties of aerogel nanostructure in relation to electrochemical energy conversion and storage devices. Also, the structure-property relationship of aerogel nanostructure is explored in this chapter. Particularly the effects of high surface area, open porous structure and network structure of aerogel for improving the performance of electrochemical energy conversion and storage devices are discussed. The open porous structure is more advantageous; to accommodate large dye molecules to increase photon absorption in DSSCs, better penetration of polymer electrolytes to increase ionic transport in fuel cells, provides a pathway for Li-ion transport and maintains the local volume changes during Li insertion and de-insertion and the wide range of pore size helps to achieve fast charging and discharging in supercapacitors. Hence, the attractive properties associated with the aerogel nanostructure will play a dominant role in the commercialization of electrochemical energy conversion and storage devices.

REFERENCES

[1] T. Ahmad, D. Zhang, A critical review of comparative global historical energy consumption and future demand: The story told so far, *Energy Reports*, 6 (2020) 1973–1991.

[2] S. Ghasemian, A. Faridzad, P. Abbaszadeh, A. Taklif, A. Ghasemi, R. Hafezi, An overview of global energy scenarios by 2040: Identifying the driving forces using cross-impact analysis method, *International Journal of Environmental Science and Technology*, 17 (2020) 1–24.

[3] Y. Ananthan, K. Sanghamitra K, N. Hebalkar, Silica aerogels for energy conservation and saving, *Nanotechnology for Energy Sustainability*, (2017) 937–966. doi.org/ 10.1002/9783527696109.ch38

[4] S.P. Badwal, S.S. Giddey, C. Munnings, A.I. Bhatt, A.F. Hollenkamp, Emerging electrochemical energy conversion and storage technologies, *Frontiers in Chemistry*, 2 (2014) 79.

[5] A.S. Arico, P. Bruce, B. Scrosati, J.-M. Tarascon, W. Van Schalkwijk, Nanostructured materials for advanced energy conversion and storage devices, materials for sustainable energy: A collection of peer-reviewed research and review articles from Nature Publishing Group, *Nature*, 4(5) (2011) 148–159, 366–377.

[6] Y.G. Guo, J.S. Hu, L.J. Wan, Nanostructured materials for electrochemical energy conversion and storage devices, *Advanced Materials*, 20 (2008) 2878–2887.

[7] Q. Zhang, E. Uchaker, S.L. Candelaria, G. Cao, Nanomaterials for energy conversion and storage, *Chemical Society Reviews*, 42 (2013) 3127–3171.

[8] J.N. Tiwari, R.N. Tiwari, K.S. Kim, Zero-dimensional, one-dimensional, two-dimensional and three-dimensional nanostructured materials for advanced electrochemical energy devices, *Progress in Materials Science*, 57 (2012) 724–803.

[9] J. Xu, X. Wang, X. Wang, D. Chen, X. Chen, D. Li, G. Shen, Three-dimensional structural engineering for energy-storage devices: From microscope to macroscope, *ChemElectroChem*, 1 (2014) 975–1002.

[10] S. Alwin, X. Sahaya Shajan, Aerogels: Promising nanostructured materials for energy conversion and storage applications, *Materials for Renewable and Sustainable Energy*, 9 (2020) 1–27.

[11] N. Hüsing, U. Schubert, Aerogels—airy materials: Chemistry, structure, and properties, *Angewandte Chemie International Edition*, 37 (1998) 22–45.

[12] N. Tétreault, M. Grätzel, Novel nanostructures for next generation dye-sensitized solar cells, *Energy & Environmental Science*, 5 (2012) 8506–8516.

[13] B.L. Ellis, P. Knauth, T. Djenizian, Three-dimensional self-supported metal oxides for advanced energy storage, *Advanced Materials*, 26 (2014) 3368–3397.

[14] X. Liu, J. Fang, Y. Liu, T. Lin, Progress in nanostructured photoanodes for dye-sensitized solar cells, *Frontiers of Materials Science*, 10 (2016) 225–237.

[15] S. Verma, S. Arya, V. Gupta, S. Mahajan, H. Furukawa, A. Khosla, Performance analysis, challenges and future perspectives of nickel based nanostructured electrodes for electrochemical supercapacitors, *Journal of Materials Research and Technology*, 11 (2021) 564–599.

[16] M.-S. Kang, J.H. Kim, Y.J. Kim, J. Won, N.-G. Park, Y.S. Kang, Dye-sensitized solar cells based on composite solid polymer electrolytes, *Chemical Communications*, (2005) 889–891.

[17] K. Hwang, H. Sohn, S. Yoon, Mesostructured niobium-doped titanium oxide-carbon (Nb-TiO$_2$-C) composite as an anode for high-performance lithium-ion batteries, *Journal of Power Sources*, 378 (2018) 225–234.

[18] R. O'Hayre, S-W. Cha, W. G. Colella, F B. Prinz, *Fuel Cell Fundamentals*, 3rd edition (2016) John Wiley & Sons.

[19] J. Larminie, A. Dicks, *Fuel Cell Systems Explained*, 2nd edition (2003) John Wiley & Sons.

[20] S. Alwin, S. Bhat, A. Sahu, A. Jalajakshi, P. Sridhar, S. Pitchumani, A. Shukla, Modified-pore-filled-PVDF-membrane electrolytes for direct methanol fuel cells, *Journal of The Electrochemical Society*, 158 (2010) B91.

[21] C.-H. Tsai, F.-L. Yang, C.-H. Chang, Y.W. Chen-Yang, Microwave-assisted synthesis of silica aerogel supported pt nanoparticles for self-humidifying proton exchange membrane fuel cell, *International Journal of Hydrogen Energy*, 37 (2012) 7669–7676.

[22] O.A. Pinchuk, F. Dundar, A. Ata, K.J. Wynne, Improved thermal stability, properties, and electrocatalytic activity of sol-gel silica modified carbon supported Pt catalysts, *International Journal of Hydrogen Energy*, 37 (2012) 2111–2120.

[23] J. Biener, M. Stadermann, M. Suss, M.A. Worsley, M.M. Biener, K.A. Rose, T.F. Baumann, Advanced carbon aerogels for energy applications, *Energy & Environmental Science*, 4 (2011) 656–667.

[24] A. Smirnova, X. Dong, H. Hara, A. Vasiliev, N. Sammes, Novel carbon aerogel-supported catalysts for PEM fuel cell application, *International Journal of Hydrogen Energy*, 30 (2005) 149–158.

[25] W. Yang, Y. Peng, Y. Zhang, J.E. Lu, J. Li, S. Chen, Air cathode catalysts of microbial fuel cell by nitrogen-doped carbon aerogels, *ACS Sustainable Chemistry & Engineering*, 7 (2018) 3917–3924.

[26] X. Tian, M. Zhou, C. Tan, M. Li, L. Liang, K. Li, P. Su, KOH activated N-doped novel carbon aerogel as efficient metal-free oxygen reduction catalyst for microbial fuel cells, *Chemical Engineering Journal*, 348 (2018) 775–785.

[27] Y. Yang, T. Liu, X. Zhu, F. Zhang, D. Ye, Q. Liao, Y. Li, Boosting power density of microbial fuel cells with 3D nitrogen-doped graphene aerogel electrode, *Advanced Science*, 3 (2016) 1600097.

[28] B.K. Boggs, R.L. King, G.G. Botte, Urea electrolysis: Direct hydrogen production from urine. *Chemical Communications*, 32 (2009) 4859–4861

[29] W. Xu, Z. Wu, S. Tao, Urea based fuel cells and electrocatalysts for urea oxidation. *Energy Technology*, 4 (2016) 132961337.

[30] R.M. Tesfaye, G. Das, B.J. Park, J. Kim, H.H. Yoon, Ni-Co bimetal decorated carbon nanotube aerogel as an efficient anode catalyst in urea fuel cells, *Scientific Reports*, 9 (2019) 1–9.

[31] M.K. Debe, Electrocatalyst approaches and challenges for automotive fuel cells, *Nature*, 486 (2012) 43–51.

[32] W. Liu, A.-K. Herrmann, N.C. Bigall, P. Rodriguez, D. Wen, M. Oezaslan, T.J. Schmidt, N. Gaponik, A. Eychmüller, Noble metal aerogels synthesis, characterization, and application as electrocatalysts, *Accounts of Chemical Research*, 48 (2015) 154–162.

[33] D. Wen, W. Liu, D. Haubold, C. Zhu, M. Oschatz, M. Holzschuh, A. Wolf, F. Simon, S. Kaskel, A. Eychmüller, Gold aerogels: Three-dimensional assembly of nanoparticles and their use as electrocatalytic interfaces, *ACS Nano*, 10 (2016) 2559–2567.

[34] N.C. Bigall, A.K. Herrmann, M. Vogel, M. Rose, P. Simon, W. Carrillo-Cabrera, D. Dorfs, S. Kaskel, N. Gaponik, A. Eychmüller, Hydrogels and aerogels from noble metal nanoparticles, *Angewandte Chemie International Edition*, 48 (2009) 9731–9734.

[35] S. Henning, L. Kühn, J. Herranz, J. Durst, T. Binninger, M. Nachtegaal, M. Werheid, W. Liu, M. Adam, S. Kaskel, Pt-Ni aerogels as unsupported electrocatalysts for the oxygen reduction reaction, *Journal of the Electrochemical Society*, 163 (2016) F998.

[36] C. Zhu, Q. Shi, S. Fu, J. Song, H. Xia, D. Du, Y. Lin, Efficient synthesis of MCu (M= Pd, Pt, and Au) aerogels with accelerated gelation kinetics and their high electrocatalytic activity, *Advanced Materials*, 28 (2016) 8779–8783.

[37] M. Grätzel, Recent advances in sensitized mesoscopic solar cells, *Accounts of Chemical Research*, 42 (2009) 1788–1798.

[38] A. Hagfeldt, G. Boschloo, L. Sun, L. Kloo, H. Pettersson, Dye-Sensitized solar cells, *Chemical Reviews*, 110 (2010) 6595–6663.

[39] A. Jena, S.P. Mohanty, P. Kumar, J. Naduvath, V. Gondane, P. Lekha, J. Das, H.K. Narula, S. Mallick, P. Bhargava, Dye sensitized solar cells: A review, *Transactions of the Indian Ceramic Society*, 71 (2012) 1–16.

[40] J.J. Pietron, A.M. Stux, R.S. Compton, D.R. Rolison, Dye-sensitized titania aerogels as photovoltaic electrodes for electrochemical solar cells, *Solar Energy Materials and Solar Cells*, 91 (2007) 1066–1074.

[41] Y.-C. Chiang, W.-Y. Cheng, S.-Y. Lu, Titania aerogels as a superior mesoporous structure for photoanodes of dye-sensitized solar cells, *International Journal of Electrochemical Science*, 7 (2012) 6910–6919.

[42] I. Mora-Seró, J. Bisquert, Breakthroughs in the development of semiconductor-sensitized solar cells, *Journal of Physical Chemistry Letters*, 1 (2010) 3046–3052.

[43] C.J. Barbé, F. Arendse, P. Comte, M. Jirousek, F. Lenzmann, V. Shklover, M. Grätzel, Nanocrystalline titanium oxide electrodes for photovoltaic applications, *Journal of the American Ceramic Society*, 80 (1997) 3157–3171.

[44] B. Kim, S.W. Park, J.-Y. Kim, K. Yoo, J.A. Lee, M.-W. Lee, D.-K. Lee, J.Y. Kim, B. Kim, H. Kim, Rapid dye adsorption via surface modification of TiO2 photoanodes for dye-sensitized solar cells, *ACS Applied Materials & Interfaces*, 5 (2013) 5201–5207.

[45] W.-Y. Wu, T.-W. Shih, P. Chen, J.-M. Ting, J.-M. Chen, Plasma surface treatments of TiO_2 photoelectrodes for use in dye-sensitized solar cells, *Journal of the Electrochemical Society*, 158 (2011) K101.

[46] S. Alwin, X.S. Shajan, R. Menon, P. Nabhiraj, K. Warrier, G.M. Rao, Surface modification of titania aerogel films by oxygen plasma treatment for enhanced dye adsorption, *Thin Solid Films*, 595 (2015) 164–170.

[47] S. Alwin, R. Menon, P. Nabhiraj, P. Ananthapadmanabhan, Plasma treated TiO_2 aerogel nanostructures as photoanode material and its influence on the performance of quasi-solid dye-sensitized solar cells, *Materials Research Bulletin*, 86 (2017) 201–208.

[48] S. Alwin, X.S. Shajan, Facile synthesis of 3-D nanostructured zinc oxide aerogel and its application as photoanode material for dye-sensitized solar cells, *Surfaces and Interfaces*, 7 (2017) 14–19.

[49] S. Alwin, V. Ramasubbu, X.S. Shajan, TiO_2 aerogel–metal organic framework nanocomposite: A new class of photoanode material for dye-sensitized solar cell applications, *Bulletin of Materials Science*, 41 (2018) 1–8.

[50] V. Ramasubbu, S. Alwin, E. Mothi, X.S. Shajan, TiO_2 aerogel–Cu-BTC metal-organic framework composites for enhanced photon absorption, *Materials Letters*, 197 (2017) 236–240.

[51] V. Ramasubbu, P.R. Kumar, E.M. Mothi, K. Karuppasamy, H.-S. Kim, T. Maiyalagan, X.S. Shajan, Highly interconnected porous TiO_2-Ni-MOF composite aerogel photoanodes for high power conversion efficiency in quasi-solid dye-sensitized solar cells, *Applied Surface Science*, 496 (2019) 143646.

[52] T. Deepak, G. Anjusree, S. Thomas, T. Arun, S.V. Nair, A.S. Nair, A review on materials for light scattering in dye-sensitized solar cells, *RSC Advances*, 4 (2014) 17615–17638.

[53] S. Zhang, X. Yang, Y. Numata, L. Han, Highly efficient dye-sensitized solar cells: Progress and future challenges, *Energy Environmental Science*, 6 (2013) 1443–1464.

[54] J.-K. Lee, M. Yang, Progress in light harvesting and charge injection of dye-sensitized solar cells, *Materials Science and Engineering: B*, 176 (2011) 1142–1160.

[55] X.-D. Gao, X.-M. Li, X.-Y. Gan, Y.-Q. Wu, R.-K. Zheng, C.-L. Wang, Z.-Y. Gu, P. He, Aerogel based SiO_2–TiO_2 hybrid photoanodes for enhanced light harvesting in dye-sensitized solar cells, *Journal of Materials Chemistry*, 22 (2012) 18930–18938.

[56] D. Deng, Li-ion batteries: Basics, progress, and challenges, *Energy Science & Engineering*, 3 (2015) 385–418.

[57] P. Arora, R. E. White, M. Doyle, Capacity fade mechanisms and side reactions in Lithium-Ion batteries, *Journal of the Electrochemical Society*, 145 (199) 3647.

[58] A. Moretti, F. Maroni, I. Osada, F. Nobili, S. Passerini, V_2O_5 aerogel as a versatile cathode material for lithium and sodium batteries, *ChemElectroChem*, 2 (2015) 529–537.

[59] A. Moretti, M. Secchiaroli, D. Buchholz, G. Giuli, R. Marassi, S. Passerini, Exploring the low voltage behavior of V_2O_5 aerogel as intercalation host for sodium ion battery, *Journal of the Electrochemical Society*, 162 (2015) A2723.

[60] R.P. Maloney, H.J. Kim, J.S. Sakamoto, Lithium titanate aerogel for advanced lithium-ion batteries, *ACS Applied Materials & Interfaces*, 4 (2012) 2318–2321.

[61] Z. Chen, H. Li, R. Tian, H. Duan, Y. Guo, Y. Chen, J. Zhou, C. Zhang, R. Dugnani, H. Liu, Three dimensional Graphene aerogels as binder-less, freestanding, elastic and high-performance electrodes for lithium-ion batteries, *Scientific Reports*, 6 (2016) 1–9.

[62] Y. Zhong, T. Shi, Y. Huang, S. Cheng, C. Chen, G. Liao, Z. Tang, Three-dimensional MoS_2/graphene aerogel as binder-free electrode for li-ion battery, *Nanoscale Research Letters*, 14 (2019) 1–8.

[63] L. Wu, Y. Zhang, B. Li, P. Wang, L. Fan, N. Zhang, K. Sun, Fabrication of layered structure VS4 anchor in 3D graphene aerogels as a new cathode material for lithium ion batteries, *Frontiers in Energy*, 13 (2019) 597–602.

[64] S. Chen, G. Liu, H. Yadegari, H. Wang, S.Z. Qiao, Three-dimensional MnO_2 ultrathin nanosheet aerogels for high-performance Li–O 2 batteries, *Journal of Materials Chemistry A*, 3 (2015) 2559–2563.

[65] S. Tan, Z. Yang, H. Yuan, J. Zhang, Y. Yang, H. Liu, MnO_2-decorated graphene aerogel with dual-polymer interpenetrating network as an efficient hybrid host for Li-S batteries, *Journal of Alloys and Compounds*, 791 (2019) 483–489.

[66] G. Zeng, N. Shi, M. Hess, X. Chen, W. Cheng, T. Fan, M. Niederberger, A general method of fabricating flexible spinel-type oxide/reduced graphene oxide nanocomposite aerogels as advanced anodes for lithium-ion batteries, *ACS Nano*, 9 (2015) 4227–4235.

[67] R. Patil, M. Phadatare, N. Blomquist, J. Örtegren, M. Hummelgård, J. Meshram, D. Dubal, H.K. Olin, Highly stable cycling of silicon-nanographite aerogel-based anode for lithium-ion batteries, *ACS Omega*, 6 (2021) 6600–6606.

[68] M. Phadatare, R. Patil, N. Blomquist, S. Forsberg, J. Örtegren, M. Hummelgård, J. Meshram, G. Hernández, D. Brandell, K. Leifer, Silicon-nanographite aerogel-based anodes for high performance lithium ion batteries, *Scientific Reports*, 9 (2019) 1–9.

[69] J. Zhang, Z. Xia, L. Dai, Carbon-based electrocatalysts for advanced energy conversion and storage, *Science Advances*, 1 (2015) e1500564.

[70] G. Fu, Y. Liu, Y. Chen, Y. Tang, J.B. Goodenough, J.-M. Lee, Robust N-doped carbon aerogels strongly coupled with iron–cobalt particles as efficient bifunctional catalysts for rechargeable Zn–air batteries, *Nanoscale*, 10 (2018) 19937–19944.

[71] M. Enterría, C. Botas, J.L. Gómez-Urbano, B. Acebedo, J.M.L. del Amo, D. Carriazo, T. Rojo, N. Ortiz-Vitoriano, Pathways towards high performance Na–O 2 batteries: Tailoring graphene aerogel cathode porosity & nanostructure, *Journal of Materials Chemistry A*, 6 (2018) 20778–20787.

[72] G. Zhou, E. Paek, G.S. Hwang, A. Manthiram, High-performance lithium-sulfur batteries with a self-supported, 3D Li_2S-doped graphene aerogel cathodes, *Advanced Energy Materials*, 6 (2016) 1501355.

[73] P. Forouzandeh, V. Kumaravel, S.C. Pillai, Electrode materials for supercapacitors: A review of recent advances, *Catalysts*, 10 (2020) 969.

[74] J. Castro-Gutiérrez, A. Celzard, V. Fierro, Energy storage in supercapacitors: Focus on tannin-derived carbon electrodes, *Frontiers in Materials*, 7 (2020) 217.

[75] R. Sudhakar, Aerogels utilization in electrochemical capacitors, in: *Colloids-Types, Preparation and Applications*, IntechOpen, 2020. DOI: 10.5772/intechopen.92521

[76] Z. Zapata-Benabithe, G. Diossa, C.D. Castro, G. Quintana, Activated carbon bio-xerogels as electrodes for super capacitors applications, *Procedia Engineering*, 148 (2016) 18–24.

[77] F. Zhang, T. Liu, M. Li, M. Yu, Y. Luo, Y. Tong, Y. Li, Multiscale pore network boosts capacitance of carbon electrodes for ultrafast charging, *Nano Letters*, 17 (2017) 3097–3104.

[78] B. Yao, H. Peng, H. Zhang, J. Kang, C. Zhu, G. Delgado, D. Byrne, S. Faulkner, M. Freyman, X. Lu, Printing porous carbon aerogels for low temperature supercapacitors, *Nano Letters*, 21 (2021) 3731–3737.

[79] P. Hao, Z. Zhao, J. Tian, H. Li, Y. Sang, G. Yu, H. Cai, H. Liu, C. Wong, A. Umar, Hierarchical porous carbon aerogel derived from bagasse for high performance supercapacitor electrode, *Nanoscale*, 6 (2014) 12120–12129.

[80] X. Cai, G. Tan, Z. Deng, J. Liu, D. Gui, Preparation of hierarchical porous carbon aerogels by microwave assisted sol-gel process for supercapacitors, *Polymers*, 11 (2019) 429.

[81] Y. Lv, L. Ding, X. Wu, N. Guo, J. Guo, S. Hou, F. Tong, D. Jia, H. Zhang, Coal-based 3D hierarchical porous carbon aerogels for high performance and super-long life supercapacitors, *Scientific Reports*, 10 (2020) 1–11.

[82] B. Thomas, S. Geng, M. Sain, K. Oksman, Hetero-porous, high-surface area green carbon aerogels for the next-generation energy storage applications, *Nanomaterials*, 11 (2021) 653.

[83] H. Liu, T. Xu, C. Cai, K. Liu, W. Liu, M. Zhang, H. Du, C. Si, K. Zhang, Multifunctional superelastic, superhydrophilic, and ultralight nanocellulose-based composite carbon aerogels for compressive supercapacitor and strain sensor, *Advanced Functional Materials*, 32(26) (2022) 1–12.

[84] W. Fan, Y. Shi, W. Gao, Z. Sun, T. Liu, Graphene–carbon nanotube aerogel with a scroll-interconnected-sheet structure as an advanced framework for a high-performance asymmetric supercapacitor electrode, *ACS Applied Nano Materials*, 1 (2018) 4435–4441.

[85] C. Yang, Q. Pan, Q. Jia, W. Qi, W. Jiang, H. Wei, S. Yang, R. Ling, B. Cao, Bamboo-like N/S-codoped carbon nanotube aerogels for high-power and high-energy supercapacitors, *Journal of Alloys and Compounds*, 861 (2021) 157946.

[86] M. Mirzaeian, Q. Abbas, D. Gibson, M. Mazur, Effect of nitrogen doping on the electrochemical performance of resorcinol-formaldehyde based carbon aerogels as electrode material for supercapacitor applications, *Energy*, 173 (2019) 809–819.

[87] K. Xu, X. Zhu, P. She, Y. Shang, H. Sun, Z. Liu, Macroscopic porous MnO_2 aerogels for supercapacitor electrodes, *Inorganic Chemistry Frontiers*, 3 (2016) 1043–1047.

[88] T.-Y. Wei, C.-H. Chen, K.-H. Chang, S.-Y. Lu, C.-C. Hu, Cobalt oxide aerogels of ideal supercapacitive properties prepared with an epoxide synthetic route, *Chemistry of Materials*, 21 (2009) 3228–3233.

[89] V. Ramasubbu, F.S. Omar, K. Ramesh, S. Ramesh, X.S. Shajan, Three-dimensional hierarchical nanostructured porous TiO2 aerogel/Cobalt based metal-organic framework (MOF) composite as an electrode material for supercapattery, *Journal of Energy Storage*, 32 (2020) 101750.

16 Hydrogels-based Nanostructured Materials for Energy Generation, Conversion, and Storage Applications

Zeinab Jarrahi, Soghra Ghorbanzadeh, and Ahmad Allahbakhsh

16.1 INTRODUCTION

Energy is one of the most important issues in today's world [1]. The ever-increasing population on the one hand, and the limited non-renewable energy resources, on the other hand, have turned energy into a critical issue of this century [2]. Renewable and clean energies with sustainable sources such as wind, tide, solar, etc. can be suitable alternatives for fossil fuels [3]. However, these types of energy are unstable and highly dependent on environmental conditions such as weather and geographical location [4]. Therefore, converting and storing these energies is inevitable [5]. Many initiatives have been taken to develop materials and equipment for efficient energy storage and conversion [6]. Nanostructured materials have a good potential for preparing and designing effective and optimal devices for the storage/conversion of renewable energy technologies [7]. Hydrogels and hydrogel-derived materials (Figure 16.1) offer various unique functions, including favorable ionic and electronic conductivity, electrolyte permeability, and structural flexibility to improve the long-term performance and safety of energy storage devices such as batteries and supercapacitors.

On the other hand, three-dimensional (3D) scaffolds of hydrogel-based materials have a large surface area for active sites with excellent structural stability [8, 9], which makes them attractive templates for the fabrication of advanced electrochemical and thermal energy conversion and storage systems [10], including metal-ion batteries, fuel cells, and water-splitting systems [1]. In this chapter, a comprehensive study is conducted on the synthesis of different hydrogels and their application in energy generation, conversion, and storage.

DOI: 10.1201/9781003318859-16

301

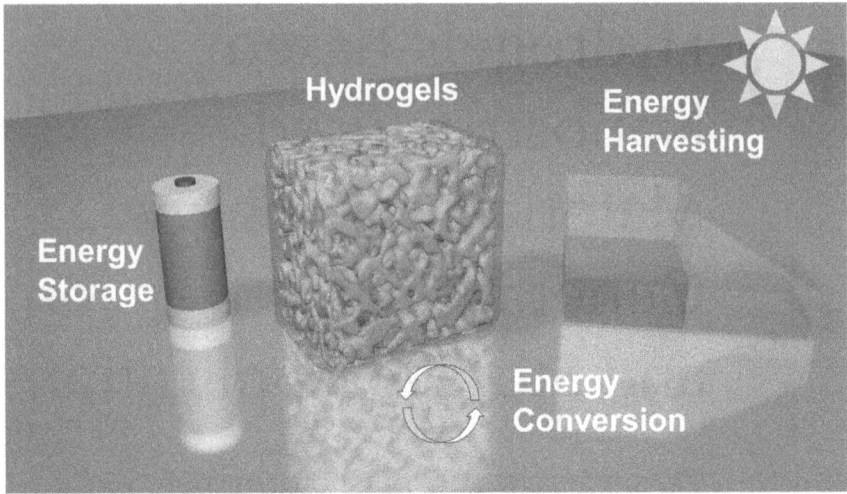

FIGURE 16.1 Applications of hydrogels for energy generation, storage, and conversion.

16.1.1　HYDROGELS

Hydrogels are cross-linked hydrophilic polymer networks saturated with water. They are mainly found as a colloidal gel in which water is the dispersing phase [11]. Hydrogels are 3D structures made of natural or synthetic polymers with high water absorption ability [12]. Hydrogels swell in water and retain a large amount of water due to the physical and/or chemical cross-linking of polymer chains [13]. Various polymeric hosts including natural products such as agar, chitosan, alginate, gelatin, and carrageenan, as well as synthetic polymers such as polyvinyl alcohol (PVA) and polyacrylic acid (PAA), have been used in hydrogels [14]. Hydrogels can have different morphological forms, including sheet, colloid, nanoparticle, and film forms. Due to this structural diversity, unique properties such as hydrophilicity, high swelling, mechanical strength, porosity, biocompatibility, and biodegradability, hydrogels are used in various research fields such as biosensors, tissue engineering, separation, water purification, pollutant removal, pharmaceutical industries, and energy [15–18]. Depending on the nature of the aqueous medium and the structure of the polymer, the gel is hydrated in different proportions. This means that the higher the number of hydrophilic groups such as alcoholic, carboxylic, and amide groups in the gel, the higher the ability of the hydrogel to uptake water. This amount varies from 10% to several times the weight of the polymeric backbone [19]. Hydrogels have remarkable flexibility and can change shape under stress and regain their original shape after the removal of the applied force [20].

Hydrogels can be classified based on their origin, network structure, preparation method, ionic charge, physical properties, and their response to the external environment/force [21]. Based on the origin of the precursor, hydrogels are divided into three categories: natural, synthetic, and hybrid (including composite hydrogels) [22–24].

Some common natural hydrogels are based on biopolymers such as collagen, agarose, alginate, chitosan, fibrin, gelatin, and polysaccharides. Moreover, some synthetic hydrogels are polyacrylamide and ethylene glycol hydrogels. Today, hybrid hydrogels are more widely used due to their water absorption, longer lifespan, and the existence of extensive raw resources [15, 25]. Moreover, composite hydrogels can present properties beyond the expected properties of their forming components [9].

Hydrogels are versatile materials that have the potential for a wide range of applications, from biosensors to drug delivery systems, and from tissue engineering scaffolds to energy storage materials [26–28]. Hydrogels with tunable structures have attracted significant research interest in the field of electrochemical energy storage, particularly as functional hydrogel electrolytes for flexible energy storage devices [29, 30]. The versatile properties of hydrogels can be utilized to impart various functions to flexible energy storage devices, resulting in the development of smart and multi-functional energy storage devices [31, 32].

Hydrogels can be classified into two categories based on the type of network structure. If the network includes covalent bonds, they are called chemical hydrogels, examples of which are polyhydroxyethyl methacrylate (PHEMA) and polymethyl methacrylate (PMMA) hydrogels [33]. If the network is formed from physical interaction, or, in other words, the network includes molecular, ionic, or hydrogen bonding, they are called physical hydrogels. Examples of these hydrogels are PVA-glycine and gelatin hydrogels. When a hydrogel contains both chemical and physical cross-linking the hydrogel is a dual network hydrogel [34].

The general classification of hydrogels, common fabrication techniques of these materials, and their main response systems are summarized in Figure 16.2.

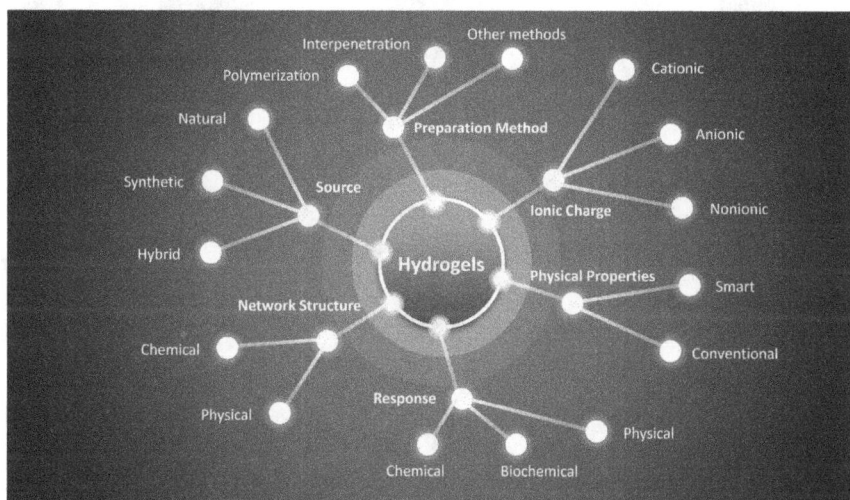

FIGURE 16.2 Classification of hydrogels.

16.1.2 Synthesis of Hydrogels

Polymeric networks form hydrogels, which are synthesized using natural or synthetic polymers (or a combination of both) and prepared using polymerization techniques by physical and/or chemical binding routes (Figure 16.3) [15].

Physically cross-linked hydrogels are 3D networks formed through non-covalent interactions between building units of the structure. Typical physical interactions include hydrophobic interactions, ionic interactions, and hydrogen bonding. Chemically cross-linked hydrogels are covalent in nature [36]. These gels show better mechanical properties because the covalent interactions are stronger than physical interactions [37]. Chemical cross-linking methods include radical polymerization, chemical reaction of complementary groups, high-energy irradiation, and the use of other functional cross-linking agents [38]. Today, with the advancement of technology, the ingredients of these gels are not limited to polymer chains. Various materials such as large organic molecules, carbon-based nanomaterials, metals, and other inorganic compounds can also be cross-linked in the form of 3D networks using strategies such as self-assembly and template-guided methods [35]. It should be noted here that among all the mentioned methods, chemical methods are the most widely used ones.

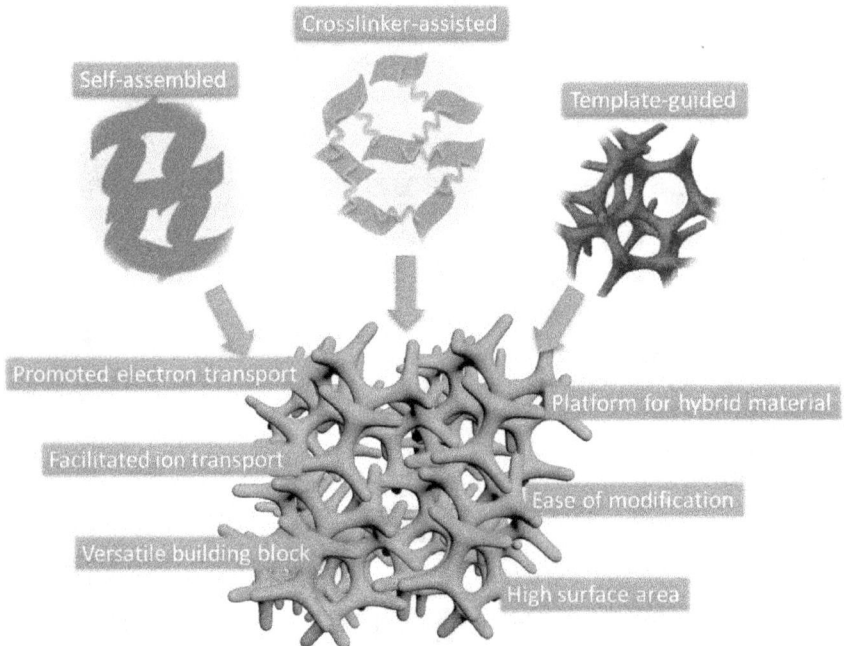

FIGURE 16.3 Conventional synthesis strategies used for gel-based nanomaterials and advantages of using these gels. Reprinted with permission from [35]. Copyright 2016 Elsevier Ltd.

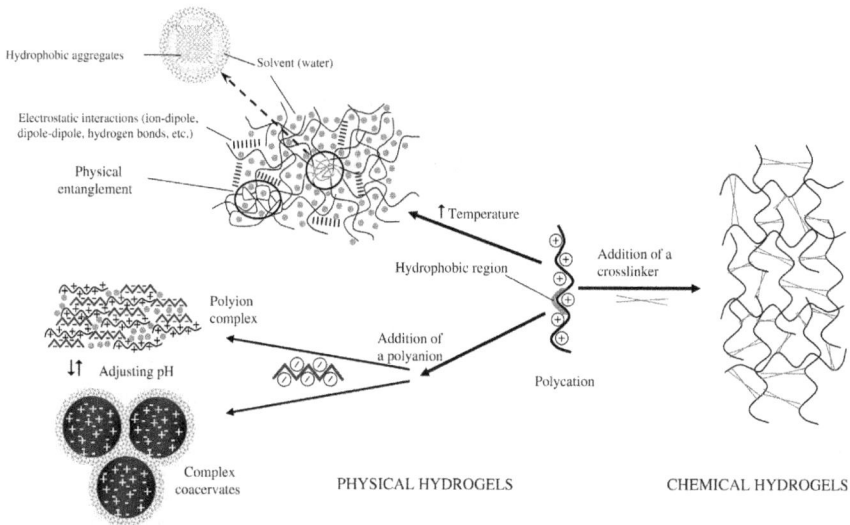

FIGURE 16.4 Mechanisms involving the formation of physical and chemical hydrogels. Reprinted with permission from [47]. Copyright 2009 Elsevier Ltd.

The most common synthetic route for the fabrication of polymer hydrogels is the free-radical cross-linking polymerization of a hydrophilic non-ionic monomer with a small amount of a cross-linker [39]. To increase their swelling capacity, an ionic comonomer can also be included in the reaction mixture [40]. Ammonium persulfate (APS), potassium permanganate ($KMnO_4$), and potassium persulfate (KPS) are some of the common crosslinkers for the fabrication of polymeric hydrogels [41, 42]. These primers help determine the path of the established cross-linking. Since the monomers for the hydrogel preparation are usually solid at the usual polymerization temperature, it is necessary to carry out the polymerization reactions in an aqueous solution.

To combine polymer chains in a chemical method, active sites can be created on the polymer backbone [43]. Various methods are used to produce the mentioned active sites, such as chemical methods, physical methods, enzymatic bonding, photochemical bonding, and plasma irradiation. Examples of hydrogels prepared by chemical methods are Gatti gum with acrylamide [42], poly(vinyl alcohol) [38], Gatti gum with acrylamide and acrylic acid [44], poly(2-hydroxyethyl methacrylate) [45], Gatti gum with acrylic acid [44], etc. Figure 16.4 shows the mechanisms of the physical and chemical formation of hydrogels [46].

Polymer hydrogels can also be prepared using the dual network method. In this method, forming physical interactions between chemically cross-linked polymeric macromolecules results in the formation of a dual network in the structure of the hydrogels. Dual network hydrogels are receiving a huge research interest for energy-related applications in recent years as these materials can present important structural features such as self-healing and smartness. Compared to the conventional polymer hydrogels, composite, nanocomposite, and hybrid hydrogels prepared using a reinforcing additive have also received much more attention [39], as the final structural properties of the hydrogel can be controlled/defined through the fabrication process by controlling the type and content of additive introduced into the structure of these hydrogels [9].

16.2 STRUCTURE, MORPHOLOGY, AND PROPERTIES OF ENERGY-RELATED HYDROGELS

16.2.1 STRUCTURE

Hydrogels can have various structural characteristics that make them great candidates for different engineering applications. Hydrogels are three-dimensional soft materials that are insoluble, cross-linked polymer networks, and can uptake or imbibe a large amount of water inside their network [48]. The key structural characteristic of hydrogels is that they have a highly hydrated polymer network that is made up of large polymer chains and is insoluble due to the cross-linking of the chains [49]. The word "network" implies that it prevents the polymer from dissolving before it is used [50]. In most cases, hydrogels are homogeneous materials with an amorphous structure [51]. Gel refers to a state that is neither completely solid nor completely liquid. This semi-solid and semi-liquid property leads to interesting properties that are not present in any pure solid or liquid [40]. Hydrogel structure and, thus, the hydrogel properties are closely related to the conditions under which the hydrogels are formed, i.e., the cross-linker concentration, the initial degree of dilution of the monomers and the chemistry of the units building the network structure [40].

16.2.2 MORPHOLOGY

Polymer hydrogels have a variety of morphologies that originate from the raw materials and their synthesis methods. All hydrogels are porous, but their pore sizes vary. In energy conversion and storage applications, to increase the rate of water absorption in the hydrogel network, the morphology of the hydrogel is presented in the form of interconnected fine holes within the 3D structure (Figure 16.5) [52]. Hydrogels are classified into microporous, mesoporous, and macroporous hydrogels in terms of pore size. In microporous hydrogels, the holes are below 2 nm, mesoporous between 2–50 nm, and macroporous above 50 nm. Therefore, micro- and mesoporous hydrogels have a high surface-to-volume ratio, which is very useful in energy applications [53].

16.2.3 PROPERTIES

The use of polymers containing hydrophilic groups for the synthesis of hydrogels for biomedical applications is very beneficial because these hydrophilic groups not only facilitate adequate water absorption but also help to interact with biological tissues. Typically, hydrogels in the fully swollen state are almost viscoelastic, soft, and rubbery, and have a low surface angle with biological fluids, which reduces the chance of a negative immune response [55]. All these factors contribute to the biocompatibility of hydrogels. Hydrogels are also usually degradable to varying degrees depending on the type of cross-linker involved.

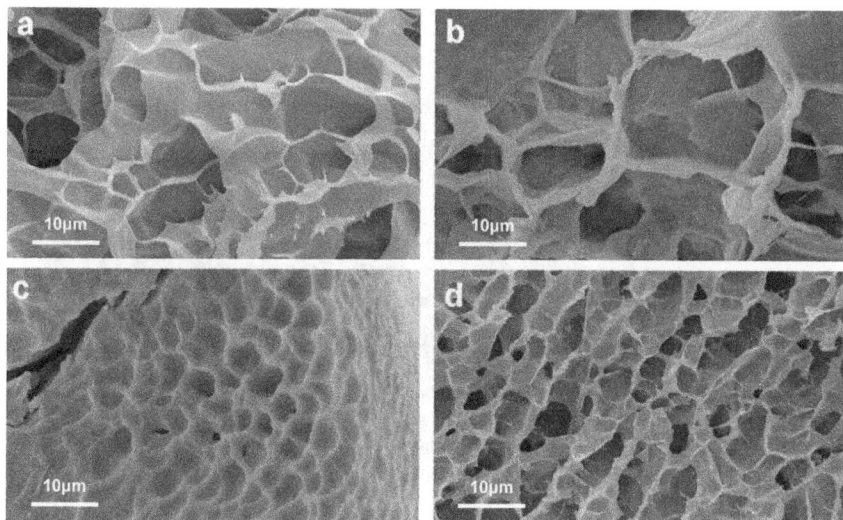

FIGURE 16.5 The morphology of the porous structure of graphene-oxide-based polyacrylamide composite hydrogels prepared without (a), and with 1 wt.% (b), 2 wt.% (c), and 3 wt.% (d) GO nanosheets against the content of acrylamide monomer. Reprinted with permission from [54]. Copyright 2012 Elsevier Ltd.

In addition, hydrogels have the property of swelling, which is the most important property of their existence. The swelling of hydrogels is done in three stages: (1) diffusion of water in the hydrogel network, (2) relaxation of polymer chains. followed by (iii) expansion of the hydrogel network. According to the Flory-Rehner theory, swelling is a function of the elastic nature of polymer chains and their compatibility with water molecules.

Hydrogels show different responses to changes in environmental stimuli, which may be broadly classified into (i) physical, (ii) chemical, and (iii) biological stimuli. Chemical and biological stimuli are internal, while physical stimuli are external, except for temperature, which may be external or internal. In addition to all these, there is a special type of smart stimuli-responsive hydrogel called shape memory hydrogel, with two properties: (1) a permanent shape and (2) a chemical or physical code that can help restore its original shape [25].

16.3 ENERGY GENERATION, CONVERSION, AND STORAGE IN HYDROGELS

Recently, hydrogels attained striking attention for energy storage and conversion applications owing to their semi-solid phase and inherent flexibility. Conventional energy storage and conversion devices have problems such as being heavy, inflexible, and unsustainability. In addition, liquid electrolyte is used in commercial devices, which is harmful. Therefore, hydrogel electrolytes can replace liquid electrolytes due

FIGURE 16.6   The 3D network of a hydrogel that can facilitate the transfer of charges along the backbone and the penetration of ions through hierarchical pores. Reprinted with permission from [35]. Copyright 2016 Elsevier Ltd.

to their properties of being semi-solid, biocompatible, biodegradable, affordable, and environmentally friendly [25].

In general, the transfer of electrons and ions is essential in energy devices. In the 3D structures of hydrogels, the transfer of electrons along the backbone and the penetration of ions through the pores are accelerated (Figure 16.6). Another point is that energy storage and conversion require physical and/or chemical active sites on surfaces and interfaces. Because hydrogels have a large surface-to-volume ratio, they can provide many active sites for interactions [35].

In the following, the applications of hydrogels in energy generation, storage, and conversion are discussed. Additionally, in Figure 16.7, the applications of hydrogels are summarized according to the properties reported in Section 16.2.3:

16.3.1   Hydrogels for Batteries

Rechargeable batteries are one of the most important energy storage devices. Batteries have three main components, including cathode, anode, and electrolyte. The usual process in the battery is the storage of charge carriers in the anode or cathode during the charging and discharging process. The function of the electrolyte is to transport charge carriers. In conventional batteries, alkali metal ions are mainly used as charge carriers. Despite the advantages such as high reactivity and high specific capacity, these ions have limitations. These limitations include low energy density, high cost of materials, and environmental problems [56]. To overcome these limitations, hydrogels were proposed. Easy fabrication, favorable ionic and electronic conductivity, suitable

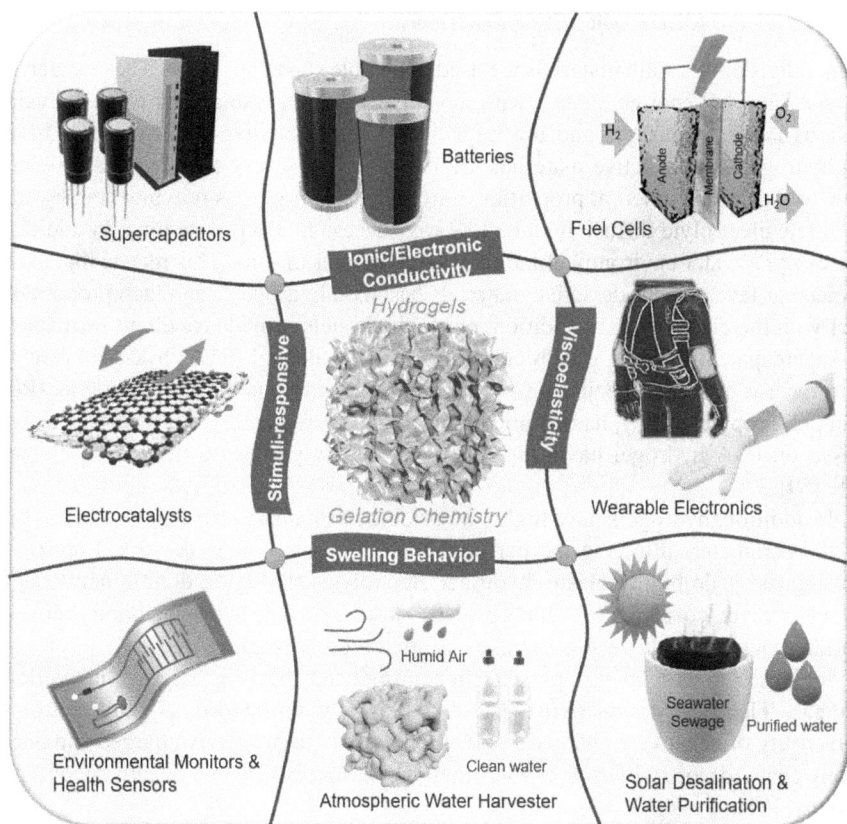

FIGURE 16.7 Gelation chemistry endows hydrogels with tunable physicochemical properties, including ionic and electronic conductivities, mechanical strength, flexibility, stretchability, stimuli responsiveness, and swelling behaviors. These properties can be integrated into hydrogel materials for a wide range of applications in energy storage, energy conversion, and sustainable clean water production. Reprinted with permission from [1]. Copyright 2020 American Chemical Society.

electrolyte-solid interphases, and improved mechanical properties are the reasons for theuperiorrity of hydrogels for use in batteries compared to conventional batteries [1]. Hydrogels can be used as electrodes, binders, and electrolytes in batteries.

Recently, electrolytes based on hydrogels have been used in batteries. For example, Lee and colleagues constructed a hierarchical polymer electrolyte. In this research, the electrolyte was made by linking polyacrylamide (PAM) on gelatin chains filled in polyacrylonitrile (PAN) matrix nanofiber. The resulting ionic conductivity is significantly higher than conventional zinc ion gel electrolytes and close to liquid electrolytes [57, 58].

16.3.2 Hydrogels for Supercapacitors

Generally, flexible solid materials are used in flexible supercapacitors. These materials cannot have full enough contact with the gel electrolyte, resulting in weak diffusion of ions in the electrolyte and low capacity. Using capacitive nanostructures based on hydrogel as new active materials are introduced as a way to solve this problem. Due to the unique inherent properties of hydrogels, soft porous polymer frameworks facilitate electrolyte diffusion and allow easy access to electrolyte ions. In addition, the existing water environment facilitates the transfer of ions. This means that at the molecular level, electrode active materials based on hydrogels can reach proper contact with the electrolyte. In addition, elastic hydrogels provide excellent formability in supercapacitors. Wang et al reported a polyvinyl alcohol (PVA) hydrogel electrolyte for use in supercapacitors. Commercial separators [such as polyethylene (PE), and polypropylene (PP)] have limited ionic conductivity. In this study, the electrolyte based on PVA hydrogel has a high ionic conductivity close to the H_2SO_4 electrolyte [59].

In addition, hydrogels have highly adjustable mechanical properties, which is one of the parameters affecting the performance of energy storage devices. Lopez and colleagues made highly elastic hydrogel electrolytes through a double network. In the electrolyte, covalent cross-links provide elasticity, while hydrogen bonds between amide groups can relieve stress caused by stretching or bending [60].

Ionizable functional groups are often introduced into the polymer network to increase electrical conductivity. These are usually introduced as polyelectrolyte (positively or negatively charged), polyampholyte (both positively charged and negatively charged), or zwitterionic (with monomers as positively and negatively charged groups) [61].

16.3.3 Hydrogels for Nanogenerators

Recently and with an increase in the electricity demand, the development of new generation portable and wearable electronic devices has attracted great research interest. Researchers have introduced different types of novel green energy sources. Among them, flexible triboelectric nanogenerators have received much attention due to their simple structure, lightweight, high output, and low cost [62].

In 2017, Xu et al. produced the first hydrogel-based triboelectric materials that used polyvinyl alcohol hydrogel as the conductor and polydimethylsiloxane as the triboelectric layer. In the same year, Pu et al. introduced the first ionic hydrogel (polyacrylamide) with high elasticity and transparency as a current collector [63].

Electrodes are also important for the integration of triboelectric materials in flexible and wearable applications. Typically, there are four types of flexible electrodes: i) Metal sheets. ii) carbon sheets. iii) conductive polymer films; Hydrogel films. Despite their high conductivity, metal sheets are not suitable for flexible and wearable applications due to their limited flexibility and stretchability. Similar problems can also be seen in carbon plates. Conductive polymer layers have better flexibility if they are produced on flexible substrates, however, in this method, the synthesis is usually complicated and the conductivity is poor. Meanwhile, hydrogels have unique properties such as high transparency, stretchability, and biocompatibility. Hydrogels increase ionic conductivity. Therefore, it is possible to precisely adjust and optimize the resistance and density of the charge carrier and select the chemical ion species used in the materials [64, 65].

16.4 HYDROGELS FOR ELECTROCATALYSIS

Electrocatalysis is an electricity-powered process that can lower the activation energy and reduce energy barriers in chemical reactions. It has been a crucial factor in addressing energy security and global warming concerns caused by the diminishing availability of fossil fuels. Hydrogel materials have garnered significant attention in electrocatalysis research due to their diverse physicochemical properties and structural advantages. They have been employed in a variety of electrocatalysis applications, such as oxygen reduction reaction, oxygen evolution reaction, hydrogen evolution reaction, and nitrogen reduction reaction.

Hydrogels and their derivatives have shown promise as active materials for electrocatalysis due to their high compositional tunability and uniform distribution of metal species. Noble-metal-free electrocatalysts have gained attention as alternatives to traditional noble metal electrocatalysts, and transition metal (M = Fe, Co, Ni) and nitrogen co-doped carbon (M-N/C) catalysts have emerged as next-generation catalysts for ORR. By carefully selecting crosslinkers and monomers, M-N-C active sites can be uniformly distributed and anchored onto carbon frameworks, creating large-scale single-atom catalysts as an alternative synthesis strategy [66].

New research has shown that electrochemical processes can be effectively catalyzed by nanostructured hydrogel electrocatalysts, such as metal and metal-free hydrogels. Inorganic hydrogels containing metals, known as metal-containing cyanogels, have been utilized for the oxidation of acidic water owing to their durable and active metal-cyano units. In their study, Fang et al. [67]. Showcased a hybrid hydrogel catalyst composed of FeCo-cyano and Ppy that could catalyze water oxidation in strongly acidic electrolytes without any apparent deterioration for up to 3000 cycles. This was achieved by polymerizing inorganic hydrogels in organic conductive hydrogel matrices, creating nanoporous polymers with an amorphous structure that facilitates rapid charge/mass transfer and increases the concentration of unsaturated metal atoms in the inorganic hydrogels, resulting in an enhanced electrochemically active surface area. Hydrogels offer a promising material platform for electrocatalysis due to their capacity for adsorption/desorption processes, porous structures that enable rapid mass transfer, conductive architectures for efficient electron transport, and abundance of structural defects and dopants that facilitate high tunability and processability. With their large surface areas, hydrogels can be used to develop cost-effective and long-lasting electrocatalysts with tunable physical and chemical properties [68].

16.5 CONCLUSIONS AND PROSPECTS

Energy has always been one of the most important topics. There have been energy storage systems from the past to today, but there is still a need for storage devices that have high performance. As alternative sources for fossil fuels, various renewable energy sources such as solar, biofuels, and fuel cells are seen, but these technologies have limitations such as charging characteristics and high production costs. Therefore, researchers seek to invent new materials whose structures are tunable and are important in energy storage materials. Recently, hydrogels have emerged as promising energy storage and conversion systems materials. Hydrogels are a unique group of 3D polymer networks with cross-linking that can hold a large number of aqueous solvents and biological fluids in their structure. Nowadays, smart hydrogels have attracted the attention of many scientists in different research fields.

REFERENCES

1. Guo, Y., et al., Hydrogels and hydrogel-derived materials for energy and water sustainability. *Chemical Reviews*, 2020. **120**(15): p. 7642–7707.
2. Chu, S., Y. Cui, and N. Liu, The path towards sustainable energy. *Nature Materials*, 2017. **16**(1): p. 16–22.
3. Allahbakhsh, A., Nitrogen-doped graphene quantum dots hydrogels for highly efficient solar steam generation. *Desalination*, 2021. **517**: p. 115264.
4. Sinsel, S.R., R.L. Riemke, and V.H. Hoffmann, Challenges and solution technologies for the integration of variable renewable energy sources—a review. *Renewable Energy*, 2020. **145**: p. 2271–2285.
5. Shaqsi, A.Z.A., K. Sopian, and A. Al-Hinai, Review of energy storage services, applications, limitations, and benefits. *Energy Reports*, 2020. **6**: p. 288–306.
6. Zhang, X., X. Cheng, and Q. Zhang, Nanostructured energy materials for electrochemical energy conversion and storage: A review. *Journal of Energy Chemistry*, 2016. **25**(6): p. 967–984.
7. Hussein, A.K., Applications of nanotechnology in renewable energies—A comprehensive overview and understanding. *Renewable and Sustainable Energy Reviews*, 2015. **42**: p. 460–476.
8. Allahbakhsh, A. and A.R. Bahramian, Self-assembled and pyrolyzed carbon aerogels: an overview of their preparation mechanisms, properties and applications. *Nanoscale*, 2015. **7**(34): p. 14139–14158.
9. Allahbakhsh, A. and M. Arjmand, Graphene-based phase change composites for energy harvesting and storage: State of the art and future prospects. *Carbon*, 2019. 148: p. 441–480.
10. Nazari, N., A.R. Bahramian, and A. Allahbakhsh, Thermal storage achievement of paraffin wax phase change material systems with regard to novolac aerogel/carbon monofilament/zinc borate form stabilization. *Journal of Energy Storage*, 2022. **50**: p. 104741.
11. Peppas, N.A. and A.S. Hoffman, Hydrogels, in *Biomaterials Science,* William R. Wagner, Shelly E. Sakiyama-Elbert, Guigen Zhang, Michael J. Yaszemski (eds). 2020, London, United Kingdom, Elsevier. P. 153–166.
12. Warren, D.S., et al., The preparation and simple analysis of a clay nanoparticle composite hydrogel. *Journal of Chemical Education*, 2017. **94**(11): p. 1772–1779.
13. Kamoun, E.A., et al., Crosslinked poly (vinyl alcohol) hydrogels for wound dressing applications: A review of remarkably blended polymers. *Arabian Journal of Chemistry*, 2015. **8**(1): p. 1–14.
14. Choudhury, N., S. Sampath, and A. Shukla, Hydrogel-polymer electrolytes for electrochemical capacitors: An overview. *Energy & Environmental Science*, 2009. **2**(1): p. 55–67.

15. Ahmed, E.M., Hydrogel: Preparation, characterization, and applications: A review. *Journal of Advanced Research*, 2015. **6**(2): p. 105–121.
16. Caló, E. and V.V. Khutoryanskiy, Biomedical applications of hydrogels: A review of patents and commercial products. *European Polymer Journal*, 2015. **65**: p. 252–267.
17. Spicer, C.D., Hydrogel scaffolds for tissue engineering: The importance of polymer choice. *Polymer Chemistry*, 2020. **11**(2): p. 184–219.
18. Chai, Q., Y. Jiao, and X. Yu, Hydrogels for biomedical applications: Their characteristics and the mechanisms behind them. *Gels*, 2017. **3**(1): p. 6.
19. Lee, J.-H. and H.-W. Kim, Emerging properties of hydrogels in tissue engineering. *Journal of Tissue Engineering*, 2018. **9**: p. 2041731418768285.
20. Batista, R.A., et al., Hydrogel as an alternative structure for food packaging systems. *Carbohydrate Polymers*, 2019. **205**: p. 106–116.
21. Guo, Y., et al., Functional hydrogels for next-generation batteries and supercapacitors. *Trends in Chemistry*, 2019. **1**(3): p. 335–348.
22. Gyles, D.A., et al., A review of the designs and prominent biomedical advances of natural and synthetic hydrogel formulations. *European Polymer Journal*, 2017. **88**: p. 373–392.
23. Catoira, M.C., et al., Overview of natural hydrogels for regenerative medicine applications. *Journal of Materials Science: Materials in Medicine*, 2019. **30**: p. 1–10.
24. Kopeček, J. and J. Yang, Smart self-assembled hybrid hydrogel biomaterials. *Angewandte Chemie International Edition*, 2012. **51**(30): p. 7396–7417.
25. Bashir, S., et al., Fundamental concepts of hydrogels: Synthesis, properties, and their applications. *Polymers*, 2020. **12**(11): p. 2702.
26. Zhang, Y.S. and A. Khademhosseini, Advances in engineering hydrogels. *Science*, 2017. **356**(6337): p. eaaf3627.
27. Zhu, T., et al., Recent advances in conductive hydrogels: Classifications, properties, and applications. *Chemical Society Reviews*, 2023. **52**(2): p. 473–509.
28. Oliveira, É.R., et al., Advances in growth factor delivery for bone tissue engineering. *International Journal of Molecular Sciences*, 2021. **22**(2): p. 903.
29. Xu, T., et al., Biopolymer-based hydrogel electrolytes for advanced energy storage/conversion devices: Properties, applications, and perspectives. *Energy Storage Materials*, 2022. **48**: p. 244–262.
30. Sardana, S., et al., Conducting polymer hydrogel based electrode materials for supercapacitor applications. *Journal of Energy Storage*, 2022. **45**: p. 103510.
31. Zhang, W., et al., Electrically conductive hydrogels for flexible energy storage systems. *Progress in Polymer Science*, 2019. **88**: p. 220–240.
32. Chan, C.Y., et al., Recent advances of hydrogel electrolytes in flexible energy storage devices. *Journal of Materials Chemistry A*, 2021. **9**(4): p. 2043–2069.
33. Ma, S., et al., A novel method for preparing poly (vinyl alcohol) hydrogels: preparation, characterization, and application. *Industrial & Engineering Chemistry Research*, 2017. **56**(28): p. 7971–7976.
34. Zeng, L.-Y., et al., Anti-freezing dual-network hydrogels with high-strength, self-adhesive and strain-sensitive for flexible sensors. *Carbohydrate Polymers*, 2023. **300**: p. 120229.
35. Shi, Y., et al., Energy gels: A bio-inspired material platform for advanced energy applications. *Nano Today*, 2016. **11**(6): p. 738–762.
36. Nazari, N., A. Allahbakhsh, and A.R. Bahramian, Analytical effective thermal conductivity model for colloidal porous composites and nanocomposites based on novolac/graphene oxide aerogels. *International Journal of Energy Research*, 2022. **46**(12): p. 16608–16628.

37. Allahbakhsh, A. and A.R. Bahramian, Novolac-derived carbon aerogels pyrolyzed at high temperatures: Experimental and theoretical studies. *RSC Advances*, 2016. **6**(76): p. 72777–72790.

38. Singhal, R. and K. Gupta, A review: Tailor-made hydrogel structures (classifications and synthesis parameters). *Polymer-Plastics Technology and Engineering*, 2016. **55**(1): p. 54–70.

39. Khalaj, M., et al., Structural, mechanical and thermal behaviors of novolac/graphene oxide nanocomposite aerogels. *Journal of Non-Crystalline Solids*, 2017. **460**: p. 19–28.

40. Okay, O., General properties of hydrogels, in *Hydrogel sensors and actuators,* Gerald Gerlach, Karl-Friedrich Arndt (eds). 2009, Berlin, Germany, Springer. P. 1–14.

41. Mondal, I.H., Graft copolymerization of nitrile monomers onto sulfonated jute-cotton blended fabric. *Journal of Applied Polymer Science*, 2003. **87**(14): p. 2262–2266.

42. Mittal, H., et al., Biosorption potential of Gum ghatti-g-poly (acrylic acid) and susceptibility to biodegradation by B. subtilis. *International Journal of Biological Macromolecules*, 2013. **62**: p. 370–378.

43. Lopez, J., et al., Designing polymers for advanced battery chemistries. *Nature Reviews Materials*, 2019. **4**(5): p. 312–330.

44. Mittal, H., et al., Preparation of poly (acrylamide-co-acrylic acid)-grafted gum and its flocculation and biodegradation studies. *Carbohydrate polymers*, 2013. **98**(1): p. 397–404.

45. Ilgin, P., et al., Adsorption of methylene blue from aqueous solution using poly(2-acrylamido-2-methyl-1-propanesulfonic acid-co-2-hydroxyethyl methacrylate) hydrogel crosslinked by activated carbon. *Journal of Macromolecular Science, Part A*, 2023. **60**(2): p. 135–149.

46. Gul, K., et al., Recent advances in the structure, synthesis, and applications of natural polymeric hydrogels. *Critical Reviews in Food Science and Nutrition*, 2022. **62**(14): p. 3817–3832.

47. Farris, S., et al., Development of polyion-complex hydrogels as an alternative approach for the production of bio-based polymers for food packaging applications: A review. *Trends in Food Science & Technology*, 2009. **20**(8): p. 316–332.

48. Gong, J.P., Friction and lubrication of hydrogels—its richness and complexity. *Soft Matter*, 2006. **2**(7): p. 544–552.

49. Lim, K.S., P. Martens, and L. Poole-Warren, Biosynthetic hydrogels for cell encapsulation, in *Functional hydrogels as biomaterials,* Jun Li, Yoshihito Osada, Justin Cooper-White (eds). 2018. Berlin, Germany, Springer. p. 1–29.

50. James, J., et al., Micro-and nano-structured interpenetrating polymer networks: State of the art, new challenges, and opportunities. Micro-and nano-structured interpenetrating polymer networks: From design to applications. *Wiley*, 2016: p. 1–27. https://onlinelibrary.wiley.com/doi/abs/10.1002/9781119138945.ch1

51. Elisseeff, J., Structure starts to gel. *Nature Materials*, 2008. **7**(4): p. 271–273.

52. Ma, S., et al., Structural hydrogels. *Polymer*, 2016. **98**: p. 516–535.

53. Alam, A., et al., Electrically conductive, mechanically robust, pH-sensitive graphene/polymer composite hydrogels. *Composites Science and Technology*, 2016. **127**: p. 119–126.

54. Shen, J., et al., Study on graphene-oxide-based polyacrylamide composite hydrogels. *Composites Part A: Applied Science and Manufacturing*, 2012. **43**(9): p. 1476–1481.

55. Fuchs, S., K. Shariati, and M. Ma, Specialty tough hydrogels and their biomedical applications. *Advanced Healthcare Materials*, 2020. **9**(2): p. 1901396.

56. Janek, J. and W.G. Zeier, A solid future for battery development. *Nature Energy*, 2016. **1**(9): p. 1–4.

57. Xu, J.J., H. Ye, and J. Huang, Novel zinc ion conducting polymer gel electrolytes based on ionic liquids. *Electrochemistry Communications*, 2005. **7**(12): p. 1309–1317.

58. Kumar, G.G. and S. Sampath, Electrochemical characterization of poly (vinylidenefluoride)-zinc triflate gel polymer electrolyte and its application in solid-state zinc batteries. *Solid State Ionics*, 2003. **160**(3–4): p. 289–300.

59. Wang, K., et al., Chemically crosslinked hydrogel film leads to integrated flexible supercapacitors with superior performance. *Advanced Materials*, 2015. **27**(45): p. 7451–7457.

60. Long, L., et al., Polymer electrolytes for lithium polymer batteries. *Journal of Materials Chemistry A*, 2016. **4**(26): p. 10038–10069.

61. Long, T., et al., Salt-mediated polyampholyte hydrogels with high mechanical strength, excellent self-healing property, and satisfactory electrical conductivity. *Advanced Functional Materials*, 2018. **28**(44): p. 1804416.

62. Rahman, M.T., et al., Biomechanical energy-driven hybridized generator as a universal portable power source for smart/wearable electronics. *Advanced Energy Materials*, 2020. **10**(12): p. 1903663.

63. Adonijah Graham, S., et al., Integrated design of highly porous cellulose-loaded polymer-based triboelectric films toward flexible, humidity-resistant, and sustainable mechanical energy harvesters. *ACS Energy Letters*, 2020. **5**(7): p. 2140–2148.

64. Chao Li, Md. Monirul Islam, Julian Moore, Joseph Sleppy, Caleb Morrison, Konstantin Konstantinov, Shi Xue Dou, Chait Renduchintala, and Jayan Thomas, Wearable energy-smart ribbons for synchronous energy harvest and storage. *Nature Communications*, 2016. **7**(1): p. 1–10. https://www.nature.com/articles/ncomms13319

65. Zohair, M., et al., Continuous energy harvesting and motion sensing from flexible electrochemical nanogenerators: Toward smart and multifunctional textiles. *ACS Nano*, 2020. **14**(2): p. 2308–2315.

66. Guo, Y., Z. Fang, and G. Yu, Multifunctional hydrogels for sustainable energy and environment. *Polymer International*, 2021. **70**(10): p. 1425–1432.

67. Fang, Z., et al., Hybrid organic–inorganic gel electrocatalyst for stable acidic water oxidation. *ACS Nano*, 2019. **13**(12): p. 14368–14376.

68. Fang, Z., et al., Inorganic cyanogels and their derivatives for electrochemical energy storage and conversion. *ACS Materials Letters*, 2019. **1**(1): p. 158–170.

Index

For Product Safety Concerns and Information please contact our EU
representative GPSR@taylorandfrancis.com
Taylor & Francis Verlag GmbH, Kaufingerstraße 24, 80331 München, Germany